Innovative Renewable Energy

Series Editor
Ali Sayigh, World Renewable Energy Congress, Brighton, UK

The primary objective of Innovative Renewable Energy book series is to highlight the best-implemented worldwide policies, projects, and research dealing with renewable energy and the environment. The books are developed and published in partnership with the World Renewable Energy Network (WREN). WREN is one of the most influential organizations in supporting and enhancing the utilization and implementation of renewable energy sources that are both environmentally safe and economically sustainable. Contributors to books in this series come from a worldwide network of agencies, laboratories, institutions, companies, and individuals, all working together towards an international diffusion of renewable energy technologies and applications. With contributions from most countries in the world, books in this series promote the communication and technical education of scientists, engineers, technicians, and managers in this field and address the energy needs of both developing and developed countries.

Each book in the series contains contributions from WREN members and covers the most up-to-date research developments, government policies, business models, best practices, and innovations from countries all over the globe. Additionally, the series publishes a collection of best papers presented during the annual and bi-annual World Renewable Energy Congress and Forum.

Ali Sayigh
Editor

Towards Net Zero Carbon Emissions in the Building Industry

Editor
Ali Sayigh
World Renewable Energy Congress
Brighton, UK

ISSN 2522-8927 ISSN 2522-8935 (electronic)
Innovative Renewable Energy
ISBN 978-3-031-15217-7 ISBN 978-3-031-15218-4 (eBook)
https://doi.org/10.1007/978-3-031-15218-4

© The Editor(s) (if applicable) and The Author(s), under exclusive license to Springer Nature Switzerland AG 2023
Chapter 6 is licensed under the terms of the Creative Commons Attribution 4.0 International License (http://creativecommons.org/licenses/by/4.0/). For further details see license information in the chapter.
This work is subject to copyright. All rights are solely and exclusively licensed by the Publisher, whether the whole or part of the material is concerned, specifically the rights of translation, reprinting, reuse of illustrations, recitation, broadcasting, reproduction on microfilms or in any other physical way, and transmission or information storage and retrieval, electronic adaptation, computer software, or by similar or dissimilar methodology now known or hereafter developed.
The use of general descriptive names, registered names, trademarks, service marks, etc. in this publication does not imply, even in the absence of a specific statement, that such names are exempt from the relevant protective laws and regulations and therefore free for general use.
The publisher, the authors, and the editors are safe to assume that the advice and information in this book are believed to be true and accurate at the date of publication. Neither the publisher nor the authors or the editors give a warranty, expressed or implied, with respect to the material contained herein or for any errors or omissions that may have been made. The publisher remains neutral with regard to jurisdictional claims in published maps and institutional affiliations.

This Springer imprint is published by the registered company Springer Nature Switzerland AG
The registered company address is: Gewerbestrasse 11, 6330 Cham, Switzerland

Preface

What is Net Zero? What does it really mean? It means that the sum of emissions by fossil fuels are balanced out by a reduction in their production and natural absorption by the oceans and land. There are multiple ways that industry, transport, and agriculture can achieve this reduction with the ultimate aim of leading to Net Zero. Of course, this is now a matter of utmost urgency, and as Helen Pirot of Fridays for Future, Berlin, has said, "it requires a drastic change from the 'we will do what we can' mentality to 'we will do what is necessary right now'." In Germany, for instance, Robert Habeck, German Minister for the Economy and Climate Protection has said that there needs to be a radical overhaul of planning and building processes in Germany and advocates a threefold increase in the speed at which CO_2 emissions are reduced. However, it is not just Germany that needs to make these changes – every country must make them.

This book shows the contribution that the building industry can and must make to achieve Net Zero. Twenty contributors from 15 countries focus on building design strategy; choice of materials and the encouragement of the use of local materials with a low carbon footprint; the use of renewable energy; energy conservation; use of greenery and appropriate aesthetics; building size and scale; building suitability for given climate; building functionality and comfort; the recycling of building materials; and adoption of appropriate green policies. No matter how small the individual reduction, each one counts towards the overall CO_2 reduction.

Brighton, UK   Ali Sayigh

Contents

The Proper Geometrical Parameters of Urban Street Profile
to Enhance Outdoor Thermal Comfort in Highland Zone of Algeria 1
Amina Naidja and Fatiha Bourbia

Thermal Behavior of Exterior Coating Texture and Its Effect
on Building Thermal Performance 23
Islam Boukhelkhal and Fatiha Bourbia

Beyond Energy Efficiency: The Emerging Era of Smart Bioenergy 43
Rachel Armstrong

Brains for Buildings to Achieve Net Zero 63
Wim Zeiler

Using Building Integrated Photovoltaic Thermal (BIPV/T) Systems
to Achieve Net Zero Goal: Current Trends and Future Perspectives..... 91
Ali Sohani, Cristina Cornaro, Mohammad Hassan Shahverdian,
Saman Samiezadeh, Siamak Hoseinzadeh, Alireza Dehghani-Sanij,
Marco Pierro, and David Moser

Simulated Versus Monitored Building Behaviours: Sample Demo
Applications of a Perfomance Gap Detection Tool in a Northern
Italian Climate... 109
Giacomo Chiesa, Francesca Fasano, and Paolo Grasso

Dynamic Simulations of High-Energy Performance Buildings:
The Role of Climatic Data and the Consideration of Climate Change ... 135
Stella Tsoka

Green Urbanism with Genuine Green Architecture:
Toward Net Zero System in New York 165
Derya Oktay and James Garrison

External Solar Shading Design for Low-Energy Buildings in Humid Temperate Climates 183
Seyedehmamak Salavatian

What It Takes to Go Net Zero: Why Aren't We There Yet? 195
Carolina Ganem-Karlen, Gustavo Javier Barea-Paci, and Soledad Elisa Andreoni-Trentacoste

The Integrated Design Studio as a Means to Achieve Zero Net Energy Buildings... 213
Khaled A. Al-Sallal, Ariel Gomez, and Ghulam Qadir

Indicators Toward Zero-Energy Houses for the Mediterranean Region.. 235
Despina Serghides, Martha Katafygiotou, Ioanna Kyprianou, and Stella Dimitriou

Toward NZEB in Public Buildings: Integrated Energy Management Systems of Thermal and Power Networks 251
Ana Beatriz Soares Mendes, Carlos Santos Silva, and Manuel Correia Guedes

The Missing Link in Architectural Pedagogy: Net Zero Energy Building (NZEB).. 283
Maryam Singery

Environmental Dimensions of Climate Change: Endurance and Change in Material Culture 293
Mona Azarbayjani and David Jacob Thaddeus

Towards Climate Neutrality: Global Perspective and Actions for Net-Zero Buildings to Achieve Climate Change Mitigation and the SDGs... 373
Mohsen Aboulnaga and Maryam Elsharkawy

Index... 435

The Proper Geometrical Parameters of Urban Street Profile to Enhance Outdoor Thermal Comfort in Highland Zone of Algeria

Amina Naidja and Fatiha Bourbia

Introduction

Urban areas are crucial to sustainable towns; outdoor spaces contribute to urban livability and vitality. They also ensure outdoor activities and pedestrian traffic. The use of urban spaces largely depends on the degree of comfort sensed in these spaces [16]. Solar and shading control have an important influence on human thermal comfort. During summer, the mutual shading between buildings is crucial for the thermal behavior of an open space and for the internal spaces next to it. Shading, either through buildings or trees, was found to be the key strategy for promoting comfort because it leads to a reduction in the absorbed radiation by a standing person [8]. Therefore, shading has an important amenity value, since it reduces the convective heat transfer from sunlit buildings and ground surfaces although the geometry of space between buildings affects the shading created in it and affects the solar radiation received by windows.Many researchers, such as Bourbia and Awbi [3], Mazouz and Zerwla [18], Naidja Amina et al. [20], Naidja Amina and Bourbia [19], Hatem Mahmoud et al. [17], and Kevin Ka-Lun Lau et al. [15], have attempted to determine the proper urban geometry that leads to safeguard shading requirement during the summer period. Nevertheless, inappropriate design of geometrical parameters of urban spaces can make shading a serious problem in winter as it can cause uncomfortable situations inside buildings and outdoors, generates cold urban public

A. Naidja (✉)
University of Oum El Bouaghi, Algeria

ABE Laboratory-University of Salah Boubnider, Constantine 3- Ali Mendjeli, Constantine, Algeria

F. Bourbia
ABE Laboratory-University of Salah Boubnider, Constantine 3- Ali Mendjeli, Constantine, Algeria

© The Author(s), under exclusive license to Springer Nature Switzerland AG 2023
A. Sayigh (ed.), *Towards Net Zero Carbon Emissions in the Building Industry*, Innovative Renewable Energy, https://doi.org/10.1007/978-3-031-15218-4_1

spaces, and increases the energy consumption for the heating of the neighboring buildings in addition to daylight lacks. According to Abraham Yezioro [24], although shading is very important in summer, the design of space between buildings should primarily consider the winter solar exposure. This is because shading in summer can be attained by dynamic solutions like deciduous trees or pergolas with a light cover that can be folding in winter. On the other hand, insolation cannot be ensuring in winter if the buildings around the open spaces block the sun completely. In this regard, [25] developed the obstruction angle rule and presented the concept of the "solar envelope." According to Knowles [13], the solar envelope can be considered as a guideline of edifice form and mass to obtain the optimum solar radiation needed in the site. This tool permits the evaluation and creation of building configurations, safeguarding solar rights of each neighboring building as well as the open spaces among them. The idea was later extended by Capeluto and Shaviv [26], who distinguished between "solar rights envelopes" and "solar collection envelopes." Okeil [21] developed a generic built form pattern named the residential solar block (RSB). He proposed the RSB as an interesting form of increasing the amount of solar rays on roofs and facades and on the ground in the cities at a latitude of 25°. More recently, the concept of solar bounding box (SBB) was introduced by Khaoula Raboudi [22] in her paper entitled "A Morphological Generator of Urban Rules of Solar Control." The SBB is the optimum volume conditioned by both the urban rules of form and solar envelope rules. Later, Isaac Guedi et al. [11] introduced the concept of parametric solar envelopes as a better tool for mediating between maximizing solar access and achieving greater built density. The latter presented an advanced flexible filtering mechanism based on specific requirements throughout the year such as weather data, site geometry, and mixed programmatic requirements. Moreover, The importance of solar insolation in winter has been studied in many research works such as Thanos N. Stasinopoulos [23], De Luca, F [6], Naidja Amina and Bourbia [19], and Koukelli et al. [14]. However, solar rights and shading requirement vary according to the latitude of the climatic zone. In Algeria, unlike vernacular urban geometries heritage, which shows a real concern and perception in planning with climate, the present urban design geometry consists of new application of urban rules, which are not usually in harmony with the climatic context desired for a given region [12]. The executive decree ($n°91-175$ of 28/05/1991) is as follows:

> In the same property, the planned buildings must be located in such a way that the openings illuminating the living dwellings are not obscured by any part of the building seen at an angle of more than 45° degrees above the horizontal plane considered the support of these openings. ([2], P.21)

Hence, the content of this decree cannot be generalized to all the national territory, because it does not specify for which climatic zone latitude will be applied, especially for the Algerian territory, which has different climatic zones with different durations of insolation from one zone to another. Moreover, there is no rigorous or even approximate climatic analysis that has covered all the national territory [2]. Under these circumstances, we attempt through this study to highlight and evaluate the shortcomings of the existing prospect urban rules in comparison with outdoor thermal comfort in the Highland zone of Algeria. In order to achieve this aim, the contemporary urban geometry has to be analyzed. The urban geometry of a town is described by a repetitive component named the urban street canyon. The urban street canyon is known as the three-dimensional spaces surrounded by a street and the constructions that border the street [9]. Therefore, an urban street canyon of an obstruction angle of 45°, representing a model of the recent prospect urban rules and building length that equals six times its height,[1] will be investigated. Subsequently, to make clear and flexible guidelines that ensure outdoor thermal comfort during summer and winter times in the Highland zone of Algeria, a simulation of fictious fabrics by varying the geometrical parameters of urban street canyon (obstruction angle and orientation) will be carried out. Then, the parametric solar envelope will be applied on the optimized street profile of the previous step. Weather data were obtained from Meteonorm 7. To guarantee that the urban geometry is the only aspect for comparison, vegetation was excluded, and materials for constructions and roads were unified. The investigation was conducted during summer and winter periods in the Highland zone of Algeria (presented by Constantine weather data). The modelling simulation was run using a parametric tool (Rhinoceros/Grasshopper/Ladybug).

The Highland Zone of Algeria

Algerian Highlands are located between the Tellian Atlas in the North and the Saharan Atlas in the south, from the border of Morocco to that of Tunisia, at more or less high altitudes of 900–1200 m. They cover an overall area of 20 million hectares. They widen from a few hundred kilometers in the Constantinois to several hundred kilometers at the Moroccan border (see Fig. 1). The climate is arid and semiarid, and rainfall is often around 300 mm but does not exceed 400 mm. Behind the shelter of the Tell Atlas, precipitation decreases quite noticeably. They become more and more irregular and few toward the south. Altitude and continentality increase temperature fluctuation between day and night. The summers are generally arid, and the winters are severe. To the east, in Constantine, the high plains are wetter, while the drought is accentuating in the western part (http://www.owlapps.net/owlapps_apps/articles?id=5753043). In order to verify and improve prospect urban rules in accordance with outdoor thermal comfort in the Highland zone, a sample of this area has to be selected. In this study, Constantine's weather data was chosen.

---

[1][1] The building length equals six times its height to meet the dimension of an urban canyon (Fazia Ali Toudert,P75-2005-)

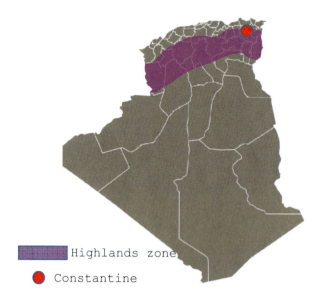

Fig. 1 Highlands zone (https://fr.wikipedia.org/wiki/Hauts_Plateaux_(Alg%C3%A9rie)

Constantine Climate

Constantine city is located in Algeria at 36.17° North and 07.23° East. The altitude is about 687 m above sea level. This town has a semiarid climate that is dry and hot in the summer with an average maximum temperature of 36 °C happening at about 3.00 pm and a humidity of about 25%, while winter is humid and cold. Moreover, the sun radiation intensities over this area are very high with clear skies and sunny periods occupying a large portion of the day. The wind direction comes relatively from the north with an average speed reaching 2.1 m/s. All of these contribute to the climatic harshness of this city (Constantine) [4].

Parametric Modelling of Urban Street Canyon

Urban design and planning are under a large impact of computation and use of digital tools. Utilization of algorithms has been recognized as an effective method to sustainable development and multi-criteria optimization difficulties. Application of genetic algorithms helps in determining the best solution, by operating certain range of parameters [27]. On this basis, a Generative Algorithm Aided Design Tool (Rhinoceros/Grasshopper/Ladybug) has been used for this research study. This latter is considered as a parametric design tool that provides a very dynamic design environment, where the designer can always discover answers by varying parameters [28].

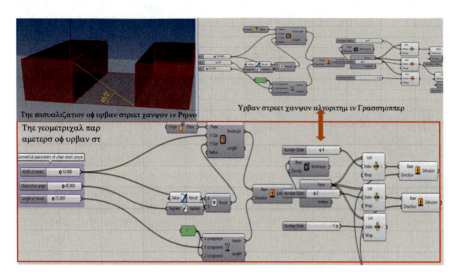

Fig. 2 The parametric definition of geometrical profile of urban street canyon

Table 1 The manageable parameters of the algorithm workflow

Parameters	Value type	Range of parameters values
Obstruction angle (a)	Float	$a \in [26.6°, 76°]$ (this range is chosen because it covered a large part of North Africa) [3]
Street width (W)	Integer	**12 m** (wide street from the view of thermal comfort is that can included streetscape elements to promote shading and good comfort conditions) P J Littlefair [16]
Building length (L)	Float	6 (Building height) **6 W (tang α)**

The investigated geometrical parameters of the urban street canyon are referencing into Grasshopper (0.9.0076) plug-in. As shown in Fig. 2, the Grasshopper plug-in is considered as a visual programming tool, which employs visual nods instead of written computer language code, thus simplifying the code generation and connection of parameters (Marko Jovanović, 2016).The scene of Rhinoceros 5 software allows the user to visualize the algorithmic definition of the geometrical parameters of the urban street canyon investigated (see Fig. 2). To acheive the purpose of this study which is the evaluation and optimization of the prospect urban rules used in relation to solar rights and outdoor thermal comfort. Table 1 summarizes the manageable parameters for the algorithm workflow.

Outdoor Thermal Comfort in Urban Street Model of Prospect Urban Rule

Outdoor thermal comfort in urban environment is a complex issue with multiple layers of concern. Outdoor thermal comfort is a composite function of atmospheric conditions and physical, physiological, psychological, and behavioral factors [1]. Therefore, the principles for creating thermally comfortable urban site are intricate and sometimes contradictory. They contain solar control in summer and solar gains in winter. The geometrical parameters of urban spaces can have a significant effect on the outdoor thermal comfort and energy performance of an urban environment [16]. Through this study, we attempt to assess the effect of prospect urban rule (obstruction angle equal 45°) on outdoor thermal comfort in the Highland zone (Constantine).To achieve this aim, the Universal Thermal Climate Index (UTCI)[2] was calculated. In this regard, a parametric design software (Grasshopper (0.9.0076)/ Ladybug[3] (0.0.62)) was used to find the algorithmic definition of outdoor thermal comfort assessment. The latter was added to the algorithmic definition of the urban street canyon model (see Fig. 16 Appendix 1). Weather data was obtained from Meteonorm 7. The developed procedure can be used for any urban environment.

As shown in Fig. 3, the average value of UTCI in high latitudes Highland zones (Constantine city) during summer period is reaching about 30.94 °C. This value indicates that the thermal sensation of walkers under those conditions is characterized by moderate heat stress (hot but not dangerous). Therefore, the application of prospect urban rule ($\alpha = 45°$) in Highlands area cannot ensure comfortable conditions of walkers during summer period. However, it provides comfortable conditions (no thermal stress) during wintertime, since the average value of UTCI is reaching about 9.27 °C (see Fig. 4). In this regard, we tried in the next step to determine the proper geometrical parameters of urban street that safeguard winter and summer outdoor thermal comfort in the Highland zone. In this way, a simulation of fictitious fabrics by varying the values of the obstruction angle and orientation of urban street model will be carried out in the next step of this study.

---

[2] The Universal Thermal Climate Index (UTCI) is a human biometeorology parameter that is used to assess the linkages between outdoor environment and human well-being. Thermal comfort indices describe how the human body experiences atmospheric conditions, specifically air temperature, humidity, wind, and radiation. There are 10 UTCI thermal stress categories that correspond to specific human physiological responses to the thermal environment. The categories related to UTCI values are as follows: above +46, extreme heat stress; +38 to +46, very strong heat stress; +32 to +38, strong heat stress; +26 to +32, moderate heat stress; +9 to +26, no thermal stress; +9 to 0, slight cold stress; 0 to −13, moderate cold stress; −13 to −27, strong cold stress; −27 to −40, very strong cold stress; below −40, extreme cold stress. https://climate-adapt.eea.europa.eu/metadata/indicators/thermal-comfort-indices-universal-thermal-climate-index-1979-2019

[3] Ladybug (0.0.62) is a plug-in of the Grasshopper (0.9.0076) (graphical algorithm editor) software for generating parametric procedures. Ladybug software allows the user to discover and survey direct relationship between elements of 3D model and environmental data over numerical and graphical data [7].

The Proper Geometrical Parameters of Urban Street Profile to Enhance Outdoor... 7

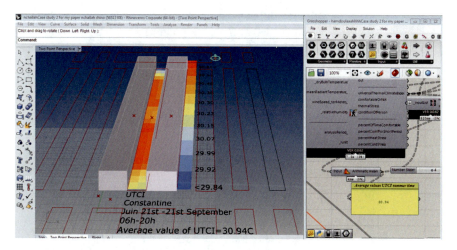

Fig. 3 The average value of UTCI during summer period in urban street of obstruction angle 45° located in Constantine

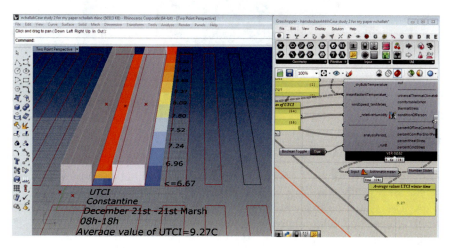

Fig. 4 The average value of UTCI during winter period in urban street of obstruction angle 45° located in Constantine

Fictious Fabric Simulation

Planning guidelines must be settled for each climatic zone towards the increase of solar access for buildings and spaces surrounding them during wintertime, also maximizing shading for buildings and urban spaces during summer. To make clear and flexible guidelines that safeguard outdoor thermal comfort during summer and winter times in Highland zones of Algeria, the obstruction angle and orientation of urban street canyon was varied. The obstruction angle (α) was varied relative to the fixed street width to create urban canyon obstruction of $\alpha = 26.6°, 45°, 56.3°, 63.4°,$

Fig. 5 Urban street canyon profile investigated

71.6°, and 76° (see Fig. 5). These values of α correspond to the values of building height/street width ratios (R = H/W) or urban canyon ratios of R = 0.5, 1, 1.5, 2, 3, and 4. These ratios cover a wide range of traditional and contemporary buildings in North Africa [3]. Correspondingly, street orientations were taken in steps of 45° from the north (S1) to east (S4) (see Fig. 6).

The Effect of Obstruction Angle and Orientation on Outdoor Thermal Comfort in Urban Street Canyon

Sun and shade have a crucial effect on outdoor thermal comfort. Urban geometry influences widely solar radiation impinged on urban street profile [19]. According to Givoni [29]:

> Narrow streets provide better shading by buildings for pedestrians on sidewalks than wide streets. (Givoni, 1997, P.369)

Fig. 6 Street orientations investigated

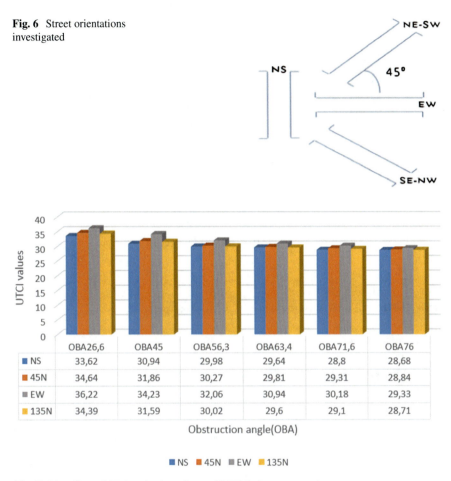

Fig. 7 The effect of OBA and orientation on UTCI during summer time

Figures 7 and 8 revealed that there is a strong positive correlation between obstruction angle (OBA) and UTCI values. Therefore, when OBA increases, UTCI values decrease. Figure 7 also displays UTCI value in urban street canyon which is oriented to the NS direction and has an obstruction angle of 76° average reaching about 28.68 °C during summer period. Hence, in comparison with the profile of current prospect urban rules (obstruction angle of 45°), the results were better for 2.26 °C, i.e., 7.30%. Although this deep profile is considered as the coolest one during summer in comparison with the other investigated profiles, it remains uncomfortable despite its high obstruction angle, since the thermal sensation in it is described as moderate heat stress (heat but not dangerous). It is worthy to note that during winter period the thermal sensation in the deep profile of obstruction angle (76°), which is oriented to NS direction, is ranged on slight cold stress (comfortable for short periods of time), since the average value of UTCI recorded during winter period in this case was 6.17 °C (see Fig. 8). This value was worse than 3.1 °C, i.e., 33.44%, in comparison with the average values of UTCI recorded in the urban street canyon of

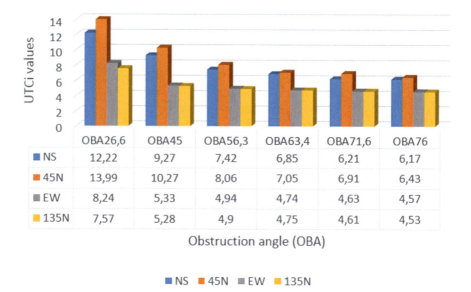

Fig. 8 The effect of OBA and orientation on UTCI values during winter period

urban prospect rule (obstruction angle 45°), in which winter outdoor thermal comfort is safeguarded (comfortable conditions). Moreover, the results also show that the orientation has a crucial effect on outdoor thermal comfort. According to Givoni, (1997):

> The direction of the streets in relation to the north determines the conditions of shade and sunshine on the face of buildings that are parallel to the street and on the sidewalks lining the streets. This affects the temperature and conditions within the buildings as well as the possibilities of protecting the pedestrians on the sidewalk from the sun in summer or of providing sunlight in the streets in winter. (Givoni, 1997, P.374)

As shown in Fig. 7, during summer time UTCI values in urban street canyons, which are mainly oriented to the east-west direction, are higher in comparison with UTCI values in urban streets, which are oriented to the north-south (NS) direction giving the lowest values. Therefore, the east-west direction has to be avoided during summer. After a comparative study on the effect of street orientation on street shade patterns, [30] concluded the following statement:

> A street grid in diagonal orientation: northeast-southwest and northwest-southeast was found to be a preferable pattern from the solar exposure aspect. It provides more shade in summer and more sun exposure in winter. (Givoni, p.373)

In another investigation, Bourbia and Awbi. (2004) concluded that diagonal street orientation NE-SW may often be a second best orientation. In another research, [31] agree that diagonal urban streets could provide better comfort conditions, since during winter period the northeast or north-west streets allow more penetration of solar access compared to the north-south urban streets.However, during summer, the NE-SW or NW-SE orientations provide more shade compared to the east-west

Table 2 Optimum obstruction angles and orientations for winter thermal comfort

Optimum Urban street profil		UTCI values		Thermal sensation	
Obstruction angle	Orientation	Winter	Summer	Winter	Summer
[26°,45°]	N 45° S	[9.27°C,....13.99°C]	[30.94°C,....34.64°C]	Comfortable conditions [+9 ; +26]	Moderate heat stress to strong heat stress [+28 ; +38]

Table 3 Optimum obstruction angles and orientations for summer thermal comfort

Optimum Urban street profil		UTCI values		Thermal sensation	
Obstruction angle	Orientation	Winter	Summer	Winter	Summer
α>76° 76°	N S	6.17°C	28.68°C	Slight cold stress [0;+9]	Moderate heat stress [+28;+32]

urban street. Consequently, Figs. 7 and 8 illustrate that summer and wintertime's urban streets, which are oriented to NE-SW directions, have better UTCI values in comparison with east-west and north-south street orientations. From the summary of this analysis, we conclude that the performance of summer and winter outdoor thermal comfort in highland areas still needs more investigation. In this regard, the parametric solar envelope will be applied in the next step of this study. To achieve this step, we summarized the best results of both winter and summer outdoor thermal comfort in the Highland zone (Constantine) (see Tables 2 and 3).

Parametric Solar Envelope Application

In order to enhance energy efficiency and outdoor thermal comfort, several famous architects and urban planners designed their buildings by drawing an envelope based on the daily sun course. R. Knowles [13] defined the solar envelope as the maximum volume, which safeguards solar rights of each buildings and spaces surrounding them during beneficial hours of winter. According to Isaac Guedi Capeluto and Boris Plotnikov [11]:

> A recent study of Niemasz, Sargent and Reinhart (2013), indicates that at least under some climate conditions and building types, **the use of traditional solar envelopes has a negative effect on total energy use including transportation as well as larger negative impact on developable density. This is partially dependent on lower densities leading to increased distance travelled which leads to increased energy use.** (Isaac Guedi Capeluto and Boris Plotnikov [11], P.397)

In this regard, Isaac Guedi Capeluto and Boris Plotnikov [11] presented the concept of parametric solar envelope as a better tool for reconciling between maximizing solar access while attaining greater built density. In another statement, Francesco De Luca and Timur Dogan [5] said:

> The solar envelope is a method used during the schematic design phase to determine the maximum volume that buildings cannot exceed to guarantee good access to direct sunlight in streets and on neighboring facades. **However, two major shortcomings exist that prohibit the use of existing solar envelope techniques in practice: They don't include the neighboring buildings in the overshadowing calculation, and they utilize a fixed start-and-end time inputs for the selection of specific hours of direct solar access.** (Francesco De Luca and Timur Dogan [5], P.817)

In this way, Francesco De Luca [6] presented the reverse solar envelope, which is considered as a new alternative manner, and computational workflow to create solar envelopes that are yielding with solar access rules but regularly compliant advanced build out volumes compared to conventional approaches that can be comprehended in consideration of several fields of the urban environment, such as roofs and building facades. Therefore, to overcome the lack of recent prospect urban rules related to outdoor thermal comfort, we attempt in this step to apply the parametric solar envelope to define the proper geometry of urban street profile based on the desired density level, latitude, and orientation. We have proved earlier in the previous step that deep profile of urban street canyons of obstruction angle (α) equal to 76° which is oriented to north south direction can enhance outdoor thermal comfort during summer period in Highland zone (Constantine); however, it drops winter outdoor thermal sensation. Therefore, the parametric solar envelope will be applied on deep profile which is oriented to north-south direction and have an obstruction angle ($\alpha = 76°$). The investigation will be conducted during summer and winter periods.

Generating Parametric Solar Envelope

According to Capeluto and Plotnikov [11], the design of the parametric solar envelope begins off with a query for which conditions do we require or want to allow direct sun access. The parametric solar envelope can be determined by underlining the situations that create a violation of solar rights, so it gives a model, which more exactly simulates real-world necessities and permits increasing of the resulting solar volume. The parametric design offers advanced building design, which is more adaptive and interactive by actively retorting to prevailing weather conditions. On

The Proper Geometrical Parameters of Urban Street Profile to Enhance Outdoor...

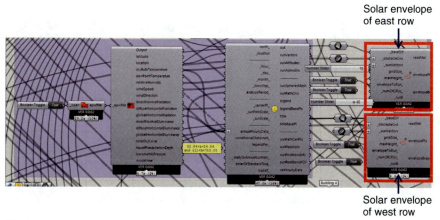

Fig. 9 The algorithm of parametric solar envelope

Table 4 Solar access conditions of generated parametric solar envelope in Highlands'area (Constantine)

Generated solar envelopes		
Condition A	**Condition A + B**	**Condition A + B + C**
The whole of the year from sunrise to sunset	Condition A+ 20 °C < Dry Bulb Temperature<24 °C	Condition A + Condition B +
Explication of the range values		
The period when solar access is required	This range is obtained by applying ASHRAE 55 (2010) standard of adaptive comfort formula on weather data of typical meteorological year (TYM 2017) (See Appendix 2)	This range is determined because a value of 568 w/m^2 is considered a minimum requirement during winter period, where it represents the maximum value of global horizontal radiation during the design day of Constantine. However, a value of 829 w/m^2 is obtained by applying [32] formula on TYM 2017 (see Appendix 3)

this regard, Sadeghipour Roudsari and Pak [22] used parametric design tool (Rhinoceros/Grasshopper/Ladybug) to outline the algorithm of the parametric solar envelope ([11], P.396) (see Fig. 9). The previous Fig. 9 shows the inventive modules in Grasshopper. Module A considers as inputs a base area that assists as the site boundary, obstacles curves, which are the neighboring edifice limits, a filtered list of sun vectors from Ladybug's Sun Path module, and a few more configuration parameters and outputs a 3D poly-surface, which is the solar envelope.

The simulated envelopes are generated to comply with the following solar access conditions (see Table 4).

After determining the solar access conditions of the generated solar envelopes in Highlands zone (Constantine), a visualization of the Sun Path with filtered suns using Ladybug is depicted in Fig. 10. The previous Fig. 10 reveals that the more we

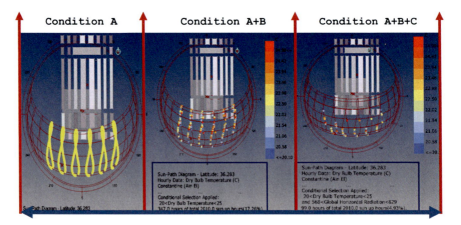

Fig. 10 Filtering process of sun vectors in Highlands zone (Constantine)

precise the conditions range, the more the number of sun hours decreases. Therefore, the more we specify the initial requirements, the more sun hours can be omitted from the calculation of the parametric solar envelope.

The Effect of Filtering Process of Sun Hours on Solar Volume Coefficient

The solar volume coefficient (average height) is defined as the solar envelope's volume to its base area ratio (SEV/A) ([23], P.9). A numerical comparison on the effect of filtering process of sun hours on solar volume coefficient (V/S) in Highlands area (Constantine) can be demonstrated by Figs. 11 and 12. The results show that the solar volume coefficient of the generated solar envelopes of the aforementioned conditions of both east and west rows increases, when the initial requirement of sun vectors is more refined. Therefore, the filtering process of sun vectors based on the conditional statement of Sun Path diagram advocated by Capeluto and Boris [11] leads to reconciliation between maximizing solar access in spaces between buildings while achieving greater built density.

The Effect of Orientation on Solar Envelope's Height

Good orientation and correct spacing between buildings improve the outdoor thermal comfort and decrease energy consumption. According to Knowles (2003):

> The designer is encouraged to differentiate building and urban form in graphic response to orientation. [33]

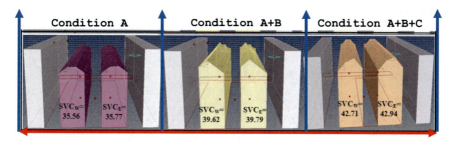

Fig. 11 The effect of filtering process of sun hours on solar envelope configuration

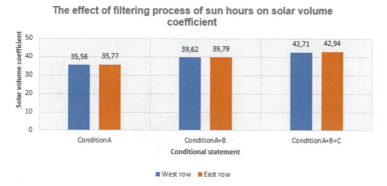

Fig. 12 The effect of filtering process of sun hours on SVC

Therefore, the direction of buildings has a decisive effect in determining the volume of solar envelope. The findings of this investigation have also revealedthat the orientation has a significant impact on the height of the solar envelope. Subsequently, the elevation of the solar envelope varies widely according to its direction, since the output of this analysis displays that in all cases investigated the north elevations have the lowest ridge in comparison with the south, east, and west facades height. This fact can be explained by the fact that the north orientation receives less solar radiation in comparison with the other ones. Therefore, the AOB depends significantly on the orientation. Table 5 summarizes the effect of orientation on the height of solar envelope (minimum and maximum).

The Effect of Sun Vectors Filtering Process on the UTCI

The present Fig. 13 reveals that solar envelope's conditional statement has a crucial effect on UTCI values, since higher SVC (solar volume coefficient) values lead to decreased UTCI values. From the summary of the simulation results shown in the previous Fig. 13, it can be concluded that the solar volume coefficient of parametric solar envelopes condition (A + B + C) is better than the solar volume coefficient of

Table 5 The effect of orientation on solar envelope's height

Conditions	Height	North East row	North West row	East East row	East West row	South East row	South West row	West East row	West West row
Cond-A	Max	36.45	36.12	37.54	37.14	37.54	37.54	37.54	37.54
	Min	30.67	30.67	32.80	32.80	32.30	32.30	31.56	31.56
Cond-A+B	Max	41.35	41.35	43.35	43.54	43.54	43.53	42.78	42.78
	Min	31.62	31.62	33.50	33.50	33.50	33.50	32.91	32.91
Cond-A+B+C	Max	47.68	47.27	47.68	47.27	47.68	47.27	47.68	47.27
	Min	35.36	35.36	37.1	37.1	37.1	37.1	37.1	37.1

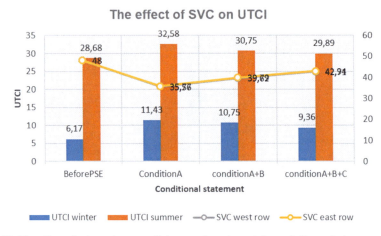

Fig. 13 The effect of solar volume coefficient on the universal thermal climate index

parametric solar envelope conditions (A; A + B) because, on one hand, it can safeguard winter outdoor comfort, and, on the other hand, the average value of UTCI during summer time reaches about 29.89 °C moderate heat stress hot but not dangerous) (see Fig. 14). These results indicate that in Highland zones summer outdoor thermal sensation, after implementing the solar volume coefficient of the parametric solar envelope condition (A + B + C), is the same thermal sensation of the walker during summertime before applying the parametric solar envelope condition (A + B + C). Therefore, the parametric solar envelope condition (A + B + C) can be considered as proper urban geometry that leads to enhanced outdoor thermal comfort in the Highland zone of Algeria (see Fig. 15).

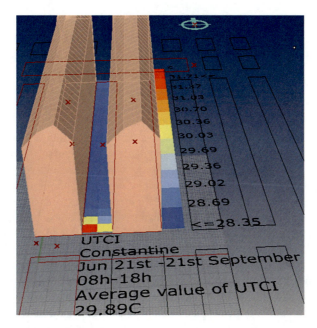

Fig. 14 UTCI values during summer time in urban street of solar envelope condition (A + B + C) in Constantine

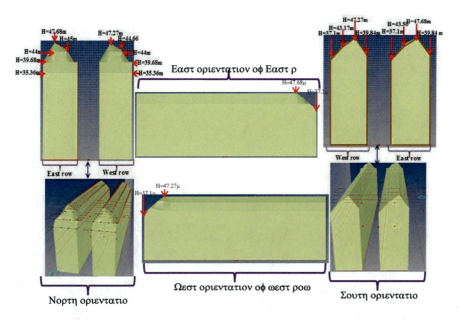

Fig. 15 The optimum urban street profile to enhance outdoor thermal comfort in the Highlands zone (Constantine)

Conclusion

Over this study, we try to assess urban geometry's effect on outdoor thermal comfort. This study will be considered as a guideline during the design process, since it aids urban designers to shape sustainable urban forms adapted to the local climate of Algerian Highlands zone (Constantine). To achieve this aim, an urban street model of obstruction angle (OBA) equal to 45° has been investigated. Afterward, the OBA and orientation have been varied relative to a fixed street width. The results of this study indicate that OBA equal to 45° safeguards winter outdoor thermal comfort, while it gives strong heat stress during summer time. The results of this study also display that deep profiles of OBA equal to 76° enhance summer outdoor thermal comfort; however, they are described by slight cold stress during winter period. Moreover, it can be comprehended that urban street direction has a crucial effect on outdoor thermal comfort, since the north-south orientation and the diagonal angle can be considered as good directions to enhance summer and winter outdoor thermal comfort. However, the east-west orientation has to be avoided, because it gives worse results of summer outdoor thermal comfort. In order to bridge the gap between summer and winter outdoor thermal comfort, the parametric solar envelope (PSE) has been applied in the next step of this research. The results indicate that the filtering process of sun vectors leads to reconciliation between shading requirement and solar rights, since the PSE of condition A + B + C gives the optimum results of UTCI values in comparison with those of the urban street profiles before applying the PSE.

Acknowledgments We would like to acknowledge the Bioclimatic Architecture and Environment laboratory (ABE) at the University of Constantine 3 for their support and advices. We would also like to acknowledge the University of Oum El Bouaghi, Algeria. The authors would like to say here their sincere thanks and deep gratitude to Doctor Farid Chaira and Doctor Karim Ayadi for checking the English grammar of this chapter.

Appendices

Appendix 1 (Fig. 16)

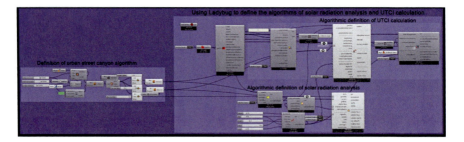

Fig. 16 Algorithmic definition of outdoor thermal comfort evaluation and solar radiation analysis

Appendix 2

According to the climate data of the city of Constantine, the annual average of air temperatures is as follows: Tm = Σ tm/12 = 16.25 °C. So: Tn = 17.8 + (0.31 × 16.25) = 22.84 + 2/−2 (comfort limit 20.84–24.84). After calculation, the comfort limit temperatures are set between 20.84 °C and 24.84 °C (Table 6).

Table 6 Dry bulb temperature (TYM-2017-)

Time	January	February	Marsh	April	May	Jun	July	August	September	October	November	December
01:00	3.70	7.40	9.49	10.38	16.23	20.30	23.58	24.23	18.05	13.56	7.94	5.47
02:00	3.11	6.81	8.73	9.56	15.32	19.28	22.47	23.19	17.23	12.79	7.34	4.97
03:00	2.52	6.23	7.98	8.91	14.84	18.71	21.85	22.46	16.40	12.02	6.74	4.45
04:00	2.25	5.90	7.58	8.49	14.29	18.17	21.26	21.91	15.93	11.62	6.42	4.18
05:00	1.95	5.60	7.19	8.10	14.01	17.77	20.84	21.47	15.48	11.22	6.12	3.91
06:00	1.72	5.39	6.92	7.88	14.00	17.83	20.82	21.31	15.19	10.95	5.89	3.70
07:00	1.52	5.17	6.83	8.36	15.31	19.45	22.17	22.08	15.39	10.97	5.65	3.55
08:00	1.49	5.28	7.85	10.06	17.05	21.42	24.14	24.01	17.30	12.31	6.08	3.60
09:00	2.70	6.96	9.63	11.73	18.97	23.46	26.26	26.20	19.27	14.21	8.15	4.99
10:00	4.42	8.83	11.52	13.46	20.83	25.48	28.40	28.36	21.48	16.28	10.18	6.68
11:00	6.21	10.60	13.31	15.09	22.53	27.29	30.33	30.29	23.38	18.15	11.92	8.30
12:00	7.68	12.11	14.89	16.48	23.95	28.76	32.02	32.04	24.87	19.76	13.41	9.68
13:00	8.85	13.23	16.14	17.59	25.06	29.92	33.35	33.41	25.95	20.97	14.59	10.69
14:00	9.74	14.09	17.03	18.37	25.85	30.74	34.29	34.39	26.67	21.65	15.34	11.34
15:00	10.17	14.50	17.48	18.78	26.30	31.19	34.81	34.95	26.97	21.86	15.52	11.64
16:00	10.05	14.46	17.57	18.80	26.37	31.25	34.93	35.08	26.88	21.58	15.01	11.39
17:00	9.26	13.85	17.07	18.37	25.99	30.91	34.58	34.68	26.17	20.71	13.90	10.46
18:00	7.99	12.71	16.06	17.46	25.09	30.03	33.67	33.68	24.98	19.43	12.44	9.28
19:00	7.37	11.56	14.68	16.20	23.84	28.72	32.33	32.23	23.48	18.36	11.79	8.70
20:00	6.73	10.87	13.77	15.07	22.40	27.13	30.71	30.66	22.46	17.51	11.07	8.11
21:00	6.12	10.18	12.88	14.07	21.18	25.70	29.21	29.29	21.42	16.69	10.43	7.59
22:00	5.49	9.50	12.00	13.08	19.91	24.25	27.69	27.93	20.41	15.86	9.75	7.01
23:00	4.86	8.80	11.08	12.03	18.65	22.79	26.18	26.56	19.40	15.00	9.08	6.48
24:00:00	4.23	8.12	10.18	11.06	17.38	21.35	24.68	25.19	18.38	14.16	8.38	5.92

Appendix 3

A value of 568 w/m^2 was chosen as a minimum requirement as it represents the radiation on December 15 (coldest month-design month) at 16.00. After calculation, the defined threshold of the amount of incident solar radiation is 829 w/m^2, obtained by the following formula: 1090.78 × 0.76 = 829 (Table 7).

Table 7 Global horizontal radiation (TYM-2017-)

Time	January	February	Marsh	April	May	Jun	July	August	September	October	November	December
01:00	0	0	0	0	0	0	0	0	0	0	0	0
02:00	0	0	0	0	0	0	0	0	0	0	0	0
03:00	0	0	0	0	0	0	0	0	0	0	0	0
04:00	0	0	0	0	0	0	0	0	0	0	0	0
05:00	0	0	0	0	0	0	0	0	0	0	0	0
06:00	0	0	0	0	3.26	7.6	2.71	0	0	0	0	0
07:00	0	0	0.84	37	127.35	153.87	117.61	58.84	10.8	1.03	0	0
08:00	0.29	6.21	87.84	205.2	310.48	343.8	298.97	240.29	177.7	99.97	14.2	1.45
09:00	75.45	136.54	270.87	364.23	500.74	536.8	491.16	439.06	356.73	258.48	150.2	88.58
10:00	202.03	289.07	447.29	518.57	671.26	711.7	665.65	619.52	549.73	422.39	278.87	200.32
11:00	314.58	420.71	601.71	650.53	793.84	838.8	809.29	758.32	681.4	537.32	373.23	294.84
12:00	380	515.75	706.55	745.2	868.1	901.6	869.87	848.58	751.63	617.23	447.17	354.97
13:00	414.55	534.68	752.06	769.57	892.71	927.83	906	884.39	755.63	633.77	482.23	373.29
14:00	429.26	555.46	740.52	757.8	867.68	904.43	881.77	864.16	725.23	595.13	463.97	362.35
15:00	367.94	495.5	663.1	673.5	788.71	822.37	797.45	789.61	636.03	485.71	371.37	311.52
16:00	259.26	369.89	535.48	553.97	661.03	697.37	674.32	664.13	515.5	346.65	241.67	218.65
17:00	129.94	237.54	365.65	399.17	492.74	545.43	517.77	491.42	336.33	191.39	106.3	82.23
18:00	6.52	80.25	173.52	227.9	306.48	359.27	340.9	295.48	159.03	37.19	2.23	0.97
19:00	0	0.5	8.77	57.87	125.84	171.17	162.74	109.81	10.1	0.1	0	0
20:00	0	0	0	0.1	3.35	12.2	11.68	0	0	0	0	0
21:00	0	0	0	0	0	0	0	0	0	0	0	0
22:00	0	0	0	0	0	0	0	0	0	0	0	0
23:00	0	0	0	0	0	0	0	0	0	0	0	0
24:00:00	0	0	0	0	0	0	0	0	0	0	0	0

References

1. Ariane, M., Nancy, S., Bjorn, H., & Nalini, C. H. (2016). Impact of shade on outdoor thermal comfort—A seasonal field study in Tempe, Arizonax. *International Journal of Biometeorology, 60*, 1849–1861. https://doi.org/10.1007/s00484-016-1172-5
2. Boucheriba, F. (2017). Effet de la Densité Urbaine Exprimée Par le "CES" et le "COS" Sur le Microclimat Urbain. Recherche d'un Indicateur Morpho-Climatique de Densité « Cas de L'habitat Individuel à la Ville d'Ain Smara . Doctoral dissertation, University of Constantine 3, Departement of Architecture.
3. Bourbia, F., & Awbi, H. B. (2004). Building cluster and shading in urban canyon for hot-dry climate. Part 2: Shading simulations. *Renewable Energy, 29*, 291–301.
4. Bourbia, F., & Boucheriba, F. (2010). Impact of street design on urban microclimate for semi-arid climate (Constantine). *Renewable Energy, 35*(2), 343–347. https://doi.org/10.1016/j.renene.2009.07.017

5. De Luca, F., & Dogan, T. (2019). A novel solar envelope method based on solar ordinances for urban planning. *Building Simulation, 12*, 817–834. https://doi.org/10.1007/s12273-019-0561-1
6. De Luca, F., Dogan, T., & Sepúlveda, A. (2021). Reverse solar envelope method. A new building form-finding method that can take regulatory frameworks into account. *Automation in Construction, 123*, 103518. https://doi.org/10.1016/j.autcon.2020.103518
7. Dragan, M., et al. (2016). Benefits of the environmental simulations for the urban planning process. In *4th International Regional eCAADe Workshop*, Novi Sad, May 2016. ISBN - 978-86-7892-807-9.
8. Elmira, J., & Priyadarsini, R. (2018). Effect of street design on pedestrian thermal comfort. *Architectural Science Review*. https://doi.org/10.1080/00038628.2018.1537236
9. Emmanuel, R. (2005). *An urban approach to climate sensitive design: Strategies for the tropics*. E & FN Spon Press. http://www.owlapps.net/owlapps_apps/articles?id=5753043
10. https://fr.wikipedia.org/wiki/Hauts_Plateaux_(Alg%C3%A9rie
11. Isaac, G. C., & Boris, P. (2017). A method for the generation of climate-based, context-dependent parametric solar envelopes. *Architectural Science Review*. https://doi.org/10.1080/00038628.2017.1331334
12. Khoukhi, M., & Fezzioui, N. (2012). Thermal comfort design of traditional houses in hot dry region of Algeria. *International Journal of Energy and Environmental Engineering, 3*(5). https://doi.org/10.1186/2251-6832-3-5
13. Knowles, R. L., & Berry, R. D. (1980). *Solar envelope concepts: Moderate density building applications*. Solar Energy Research Institute.
14. Koukelli, C., Alejandro, P., & Serdar, A. (2022). Kinetic solar envelope: Performance assessment of a shape memory alloy-based autoreactive façade system for urban Heat Island mitigation in Athens, Greece. *Applied Sciences, 12*(1), 82. https://doi.org/10.3390/app12010082
15. Lau, K. K. L., Tan, Z., Morakinyo, T. E., & Ren, C. (2022). Effects of urban geometry on mean radiant temperature. In *Outdoor thermal comfort in urban environment* (Springer briefs in architectural design and technology). Springer. https://doi.org/10.1007/978-981-16-5245-5_5
16. Littlefair, J., Santamouris, M., Alvarez, S., Dupagne, A., Hall, D., Teller, J., Coronel, J. F., & Papanikolaou, N. (2000). *Environmental site layout planning: Solar access, microclimate and passive cooling in urban areas*.
17. Mahmoud, H., Ghanem, H., & Sodoudi, S. (2021). Urban geometry as an adaptation strategy to improve the outdoor thermal performance in hot arid regions: Aswan University as a case study. *Sustainable Cities and Society, 71*, 102965. https://doi.org/10.1016/j.scs.2021.102965
18. Mazouz, S., & Zerouala, M. S. (1999). The derivation and re-use of vernacular urban space concepts. *Architectural Science Review, 42*(1), 3–13. https://doi.org/10.1080/00038628.1999.9696843
19. Naidja, A., & Bourbia, F. (2021). *Parametric study on solar control of urban spaces- spaces between buildings* (Doctoral thesis). University of Constantine 3.
20. Naidja, A., Khammar, Z., & Bourbia, F. (2017). The effect of geometrical parameters of urban street on shading requirement in hot arid climate- Contemporary urban street of Biskra- PLEA 2017 Edinburgh – 33rd International Conference on Passive and Low Energy Architecture. Cities, Buildings, People: Towards Regenerative Environments.
21. Okeil, A. (2004). *In search for energy efficient urban forms: The residential solar block. CIB World Build*. Congress Futur. Proceedings of the 5th International Conference on Indoor Air Quality, Ventilation and Energy Conservation in Buildings Proceedings.
22. Raboudi, K., & Ben Saci, A. (2013). *A morphological generator of urban rules of solar control*. Sustainable architecture for a renewable future: Proceedings of the 29th international conference on passive and low energy architecture (PLEA 2013), Munich.
23. Stasinopoulos, T. N. (2018). A survey of solar envelopes properties using solid modelling. *Journal of Green Building*. https://doi.org/10.3992/1943-4618.13.1.3
24. Yezioro, A., & Shaviv, E. (1994). A design tool for analyzing mutual shading between buildings. *Solar Energy, 52*(1), 27–37. Pergamon Press Ltd., USA.

25. Knowles, R.-L. (1974). Energy and Form. *An Ecological Approach to Urban Growth.* Cambridge, MIT Press
26. Capeluto, I. G. & Shaviv, E. (1997). *Modeling the design the urban grids and fabric with solar rights consideration.* In Proceedings of ISES 1997, Taejon, Korea.
27. Marko, J., Marko, V., Radovan, Š., Milena, S. (2016). COMPUTER AIDED CURVE AND SURFACE GENERATION IN RELATIVISTIC GEOMETRY OF HARMONIC EQUIVALENTS. *Proceedings of The 5 International Scientific Conference on Geometry and Graphics* . June 23-26, Belgrade, Serbia.2016.
28. Beirão, J., Duarte, J., Stouffs, R., Bekkering, H. (2012). Designing with urban induction patterns: a methodological approach. *Environment and Planning B: Planning and Design 39*(4), 665–682
29. Givoni, B. (1997). Performance of the "shower" cooling tower in different climates, Renewable Energy, Elsevier, vol. 10(2), 173–178.
30. R.L. Knowles, Sun Rhythm Form, MIT Press, Cambridge, MA, 1981.
31. Ali-Toudert, F., & Mayer, H. (2006). Numerical study on the effects of aspect ratio and orientation of an urban street canyon on outdoor thermal comfort in hot and dry climate. *Building and Environment, 41*, 94–108.
32. 2009 ASHRAE Handbook - Fundamentals (I-P Edition)
33. Knowles, R. L., & Kensek, K. M. (2000). *The interstitium: a zoning strategy for seasonally adaptive architecture, in: Proceedings of The PLEA 2000*, Cambridge, UK, 2000, pp. 773–774.

Thermal Behavior of Exterior Coating Texture and Its Effect on Building Thermal Performance

Islam Boukhelkhal and Fatiha Bourbia

Introduction

Climate change constitutes one of the most debated subjects of the twenty-first century. The planet experienced various phenomena generated by this change, corresponding to rising temperatures, heat waves, rising sea levels, more intense storms, and wildfires. Since 1930, further than 100,000 new chemical compounds have been developed, specifically those used in construction field with a massive lack of information or studies of their effect on health. For example, Portland cement concrete, the most frequently used material in the world (more than 10,000 million tons/year, and whose production in the next 40 years will increase by around 100%), includes chemicals and adjuvants used to modify its properties, whose effects on health and the environment are still unknown [1].

The building envelope is an outer layer that can exclude disagreeable effects while allowing those that are estimable. It plays a crucial role in improving the building's energy efficiency and the interior comfort of its occupants. The choice of the exterior envelope is an initial issue which should be considered in the sustainable design of the building. Determining the appropriate exterior cladding material and texture to optimize energy efficiency is a very important step that can be provided by multiple elements such as slate, brick, cement, plaster, and marble. These elements are used to cover, consolidate, protect, or decorate exterior walls; they can also influence energy consumption and improve indoor thermal comfort because of their texture. This chapter approaches the question of the thermal design in the building envelope through the exterior texture. The configuration of this texture can ensure solutions related to the thermal of buildings such as the minimization of the energy demand and the minimization of the hours of thermal discomfort.

I. Boukhelkhal (✉) · F. Bourbia
University of Salah Boubnider, Constantine 3 University, El Khroub, Algeria

© The Author(s), under exclusive license to Springer Nature Switzerland AG 2023
A. Sayigh (ed.), *Towards Net Zero Carbon Emissions in the Building Industry*, Innovative Renewable Energy, https://doi.org/10.1007/978-3-031-15218-4_2

How Exterior Cladding Texture Can Improve the Thermal Performance of Buildings in Hot and Arid Climate

The Algerian Great South is characterized by a rich architectural and urban heritage which merges with a way of life based on protection against climatic conditions; most of them are considered as protected architectural heritage. The strategy set in place by the ancestors was based on protecting the building against climatic conditions by covering the outer envelope of the building. This traditional decorative envelope used by these former builders in southern Algeria can play a fundamental role in improving the energy efficiency of the building. The use of these exterior coatings as a decorative element appears in various traditional architectural styles, particularly in the far south of the Algerian desert, similar to the texture shaped with bare hands in the M'Zab Valley, the texture of external coating in the form of balls projected on the wall in the region of Timimoun, the texture in the form of a cube of clay in the region of Taghit, and another type of stone crystal texture in the region of Hoggar (Fig. 1).

The question requested considers the strategy of the texture in the exterior coating, whether rough or smooth, as a solution affecting the thermal behavior and the thermal comfort of the building. In this way, it is logical to assume that covering the building with an exterior cladding, whether of a smooth or rough texture, makes it possible to avoid the exposure of the facades to intense sunlight and high temperatures and to reduce the irrational consumption of energy.

Theoretical Analysis

Several researches studied the effect of the exterior envelope on the thermal comfort of the building using many strategies as works related to the exterior coatings of the building [2–7], knowing that those researches particularly on this element are very rare. Therefore, studies made on the effect of self-shading which is considered a fundamental strategy at the level of the external envelope [8–12], other works based on the effect of the color and the albedo of the external coating in the building [13–18], or the effect of the intelligent envelope with particles of ecological materials on the thermal comfort and the energy needs in buildings [19–27]. Thus, studies were carried out on the effect of roughness and density of the exterior plant cover [28–34], in order to highlight the strategies that can be used at the level of exterior coatings.

Although all these works expose divergent results from one study to another, nevertheless, they share together reliable and useful results on the thermal behavior of the outer envelope, through the reduction of surface and air temperatures, and the reduction of the energy consumption of the building.

These theoretical syntheses can conclude that optimization in the configuration of the texture in the outer envelope can act as thermal insulation for buildings, by

Fig. 1 Traditional wall texture design in the great south of Algeria. (**a**) Ghardaia region), (**b**) Timimoun region), (**c**) Hoggar region), (**d**) Taghit region). (Source: Boukhelkhal Islam, 2015)

protecting them from solar rays and reducing the need for thermal insulation materials. The improvement solutions through textured coatings are infrequently exploited; they are suggested as a new field which solicits eminently development in the future.

Modeling Approach

In Algeria, the use of plaster texture as a decorative element appears in several times in traditional architectural styles, particularly in the southern part of the country, characterized by its hot and arid climate, and whose traditional constructions are built to face difficult climatic conditions. The objective of this study is to evaluate the thermal behavior of the exterior envelope with different textures used in exterior claddings, inspired by traditional construction methods that used self-shading walls as a cooling strategy.

In order to achieve this objective, measurements were taken to evaluate the thermal behavior of the texture in the exterior coating. The experimental system (Test Box) was set up in a private garden with open area in the city of Constantine, which is located in the northeast of Algeria (latitude: 36.9126 N, longitude: 7.0213 E) and with a semiarid climate. The figure below illustrates the methodological approach in this study (Fig. 2).

The experimental system (Test Box) is set up inside a private garden with area freed from all constructions or plantations. The measurement is taken using various instruments such as a thermal camera (FLIR), multimeter (TESTO 925), pyranometer (S-LIB-M003) with a data logger (HOBO H21 USB), weather station (Oregon Scientific) for in-situ measurements of air temperature, surface temperatures, solar radiation, wind speed and direction, relative humidity, atmospheric pressure, and precipitation rate.

The period chosen for the first phase of the study was the summer season ranging from June 23 to June 30, 2019 (08 days), which represent the longest days of the year, in where the hottest days are recorded on June 25, 26, and 27. The results are taken from 7:00 a.m. to 6:00 p.m. at an interval of 60 min. The second phase of measurements is carried out during the summer period of 2020, maintaining the same choice of period of the first phase. The results are recorded for 12 h, from 8:00 a.m. to 8:00 p.m. every 60 min.

Experimental Procedure

The first part represents an investigation in the field, comprising a series of measurements. The experiment consists in making four boxes of 1 m^3 which expose four types of external coating. The figures below give a detailed description of the textures developed under study (Figs. 3 and 4).

After an in-depth bibliographic research and investigation on the traditional texture in southern Algeria, the choice fell on four types of textures:

- Smooth texture (STB) considered as reference texture
- Rough texture (RTB)
- Crystalline texture (CTB)
- Texture of the blade texture (BTB)

Similar to the patterns shown in the figures below, the morphological configurations of the different coating textures are based on the percentage of shadow area projected over the total wall area (SO/ST). Below are the different configurations proposed in this study (Fig. 5).

The second part of this research consists in testing the effect by the incorporation of natural particles (ecological, organic, waste, and recycling components) in exterior coatings. This phase is based on a conclusion made from a state of the art on several researches in ecological materials and recycling. Finally, the choice is maintained on the rough texture in order to keep the same appearance and the

Thermal Behavior of Exterior Coating Texture and Its Effect on Building Thermal... 27

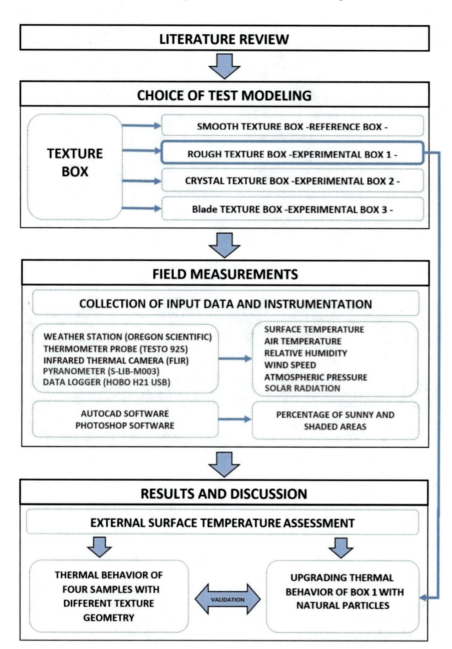

Fig. 2 Conceptual framework study. (Source: Adapted by Boukhelkhal Islam, 2021)

Fig. 3 Choice of materials for the measurement. (Source: Boukhelkhal Islam, 2019)

Fig. 4 Realization of measurement boxes. (Source: Boukhelkhal Islam, 2019)

projected shadow but with components of different mechanical and thermal characteristics.

In order to achieve this objective, a variety of particles are tested, and the selection is made according to three types with different thermal characteristics. The figure below shows a component from sand quarries, a second component produced from ecological recycling (tire waste), and another obtained from date palm waste (palm particles). These wastes are washed with distilled water to remove all impurities from their surfaces like salt and then dried in the oven. Then they were crushed and separated into different sizes, similar to sand aggregate (Fig. 6).

After various in situ measurements, the choice fell on palm particles, which showed the ability to reduce the surface temperature more than tire waste aggregates. This choice was based on the ability to retain better thermal properties in

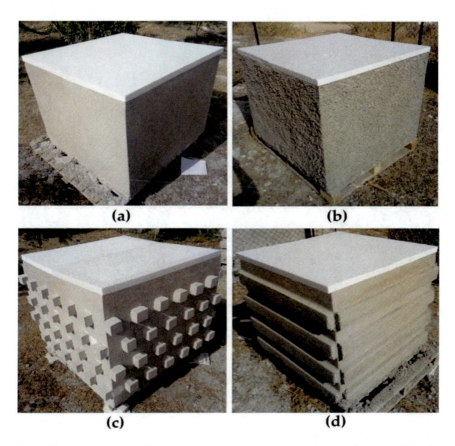

Fig. 5 The selected textures for surface temperature measurements. (**a**) smooth texture (STB), (**b**) rough texture (RTB), (**c**) crystal texture (CTB), (**d**) blade texture (BTB). (Source: Adapted by Boukhelkhal Islam, 2019)

order to be reused as mortar aggregates in the manufacture of exterior cladding textures. This material has been the subject of several researches. It has proven high heat capacity due to its ability to reduce thermal conductivity, compression, and coating weight [25]. Moreover, this material is widely available from date palms in the region of the great south of Algeria [26] (Fig. 7).

In order to evaluate the thermal behavior of the rough texture with and without palm particles, three boxes of 1 m³ each are built separately: (RTB), (RTB$_1$), and (RTB$_2$). Each box is made with different concentrations of palm particle aggregates (0%, 30%, and 70%, respectively) (Fig. 8).

Fig. 6 Evaluation of the thermal behavior of the tested aggregate. (**a**) Sand aggregates, (**b**) palm particles, (**c**) tire waste. (Source: Adapted by Boukhelkhal Islam, 2020)

Texture Effects

Effect of Texture Geometry

The first investigation of this study focused on the effect of texture geometry on the exterior surface temperature. The idea of introducing texture on the outer surface can facilitate cooling by self-shading effect. Comparing the four measured samples, the results showed variable measurements relative to all orientations. The results of the outdoor surface temperature measured by the thermal camera (FLIR) and the multimeter (TESTO 925) are shown in the figures below. Taking as an example the south facade (Figs. 9, 10, 11, and 12).

The results of the external surface temperature of the four textures studied for all orientations (north, east, south, west) are presented in the figures below (Fig. 13).

The analysis and comparison of the four textures are presented in the table below. The comparison between each type of texture with the different orientations studied (north, east, south, and west) clarifies the previous results which confirm that the texture of the coating affects the external surface temperature (Table 1).

These results show that the number of most critical hours (in red) is considerably high for the rough texture (RTB) (7 h from 10:00) in the south orientation, followed by the smooth texture (STB) with an average of 5 h, followed by the other two

Thermal Behavior of Exterior Coating Texture and Its Effect on Building Thermal... 31

Fig. 7 Surface temperature values of the three samples. (Source: Adapted by Boukhelkhal Islam, 2020)

texture types (4 h). The hottest hours are recorded between 10 a.m. and 4 p.m. The rough texture (RTB) therefore frequently indicated and exhibited the greatest number of hot hours.

Based on the results obtained, it can be concluded that there is a correlation between the shadow fraction and the external surface temperature, as mentioned in the figure below, of which the rough texture (RTB) deploys the lowest correlation compared to the others (Fig. 14).

The figure below presents a grouped histogram of the evolution of shadow fraction compared to the exterior surface temperature of the south facade in the four types of texture from 8:00 a.m. to 6:00 p.m (Fig. 15).

The figure above demonstrates that at noon, the shadow fraction rate on the south wall is identical (about 40%) for the three textures RTB, CTB, and BTB. The comparison of the external surface temperature results recorded at midday shows that these values are of the order of 52.6 °C for the blade texture (BTB), 53.9 °C for the crystalline texture (CTB), and 58.5 °C for the rough texture (RTB) although all textures have the same shadow fraction rate. This can be explained by the fact that the rough texture (RTB) has small spots of shadows not exceeding 1 cm^2 and with

Fig. 8 Rough texture with different concentration of palm particles aggregate. (Source: Adapted by Boukhelkhal Islam, 2020)

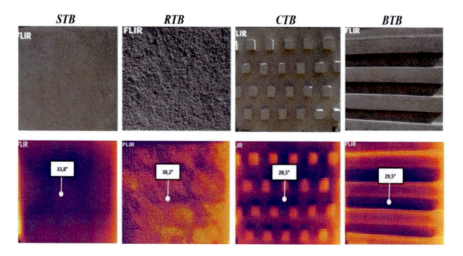

Fig. 9 External surface temperature of the south orientation at 9:00 a.m. (Source: FLIR, 2020)

random positions that favor small reflections and the absorption of solar radiation, unlike the textures in crystal (CTB) and blade (BTB), which feature an assemblage of larger, more authentic shaded surfaces that act as small canopies protecting the wall itself.

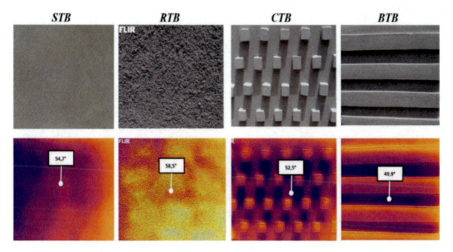

Fig. 10 External surface temperature of the south orientation at midday. (Source: FLIR, 2020)

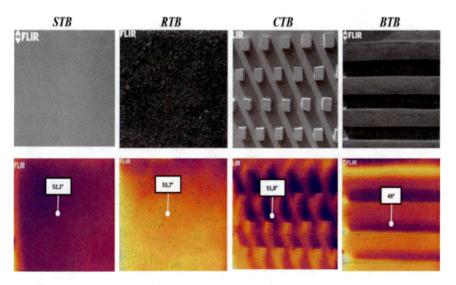

Fig. 11 External surface temperature of the south orientation at 3:00 p.m. (Source: FLIR, 2020)

Effect of Texture Particles

The results of the first phase of this study show that the rough texture registers considerably critical results for the external surface temperature. However, this texture is used for most of the exterior coatings of buildings, more particularly in Algeria. Improving the thermal behavior of this texture is essential by adding new components favorable to the harsh climate of the region. This method is carried out using natural aggregates, such as palm particles, in order to optimize thermal efficiency.

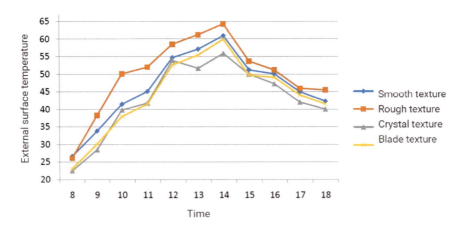

Fig. 12 External surface temperature of the south orientation. (Source: Boukhelkhal Islam, 2020)

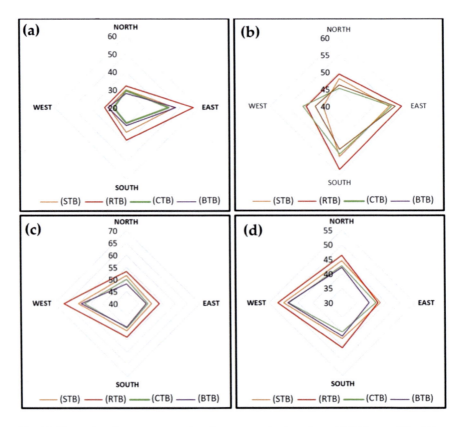

Fig. 13 External surface temperature for all textures and orientations. (**a**) 9:00 a.m., (**b**) midday, (**c**) 3:00 p.m., (**d**) 6:00 p.m. (Source: Boukhelkhal Islam, 2020)

Thermal Behavior of Exterior Coating Texture and Its Effect on Building Thermal... 35

Table 1 External surface temperature for the listed orientations (north, east, south, west)

Hours	STB	RTB	CTB	BTB
8	25,7	25,8	25,4	23,6
9	29,9	32	29,6	27,8
10	33,8	39,9	33,6	36,5
11	38,4	40,9	38,3	37,4
12	48	49,4	45,2	46,2
13	52,8	54,5	48,1	47,4
14	58,7	59,3	52,2	51
15	51,6	53,3	50	48,1
16	49,4	51,2	49,1	49,2
17	46	48	44	44,4
18	44,4	46,4	42,7	42,1

NORTH

Hours	STB	RTB	CTB	BTB
8	40	48	36,9	38,7
9	46,7	57,2	43,4	47,4
10	52,3	61,9	47,9	53,5
11	52	58,9	50,3	53,6
12	54,5	58,2	55,3	56,4
13	57,9	58,4	52,1	55,8
14	58,6	60,7	53,1	52,3
15	50,2	53,3	48,8	48
16	47,8	48,3	47	46
17	48	44	44	42,3
18	43,1	42,2	41,5	39,1

EAST

Hours	STB	RTB	CTB	BTB
8	26,6	26,1	22,5	23,1
9	33,8	38,2	28,5	30,1
10	41,5	50,1	39,8	37,9
11	45,1	52	41,8	41,5
12	54,7	58,5	53,9	52,6
13	57,1	61,2	51,7	55,4
14	60,9	64,3	55,9	59,9
15	51,2	53,7	50	49,7
16	50,1	51,2	47,3	49,1
17	45	46	42	44,1
18	42,3	45,5	40	41,5

SOUTH

Hours	STB	RTB	CTB	BTB
8	25,5	28,8	23,5	24,5
9	29,8	32,1	27,1	30,4
10	35,6	40,8	33,5	36,9
11	40,2	40,3	36,7	41
12	44,6	49,5	50,5	47
13	54,1	58,2	49,7	55,4
14	64,1	68,8	57,2	60,3
15	60	66,1	57,5	58,8
16	59,1	62	57,1	59,3
17	55,7	57	50	52
18	50	52	48,2	48,5

WEST

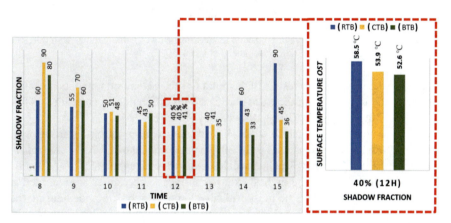

Fig. 15 External surface temperature at midday compared to SF percentage (south orientation). (Source: Adapted by Boukhelkhal Islam, 2020)

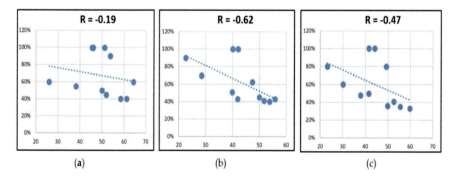

Fig. 14 Correlation results of external surface temperature (EST) and shadow fraction (SF) for (**a**) RTB, (**b**) CTB, and (**c**) BTB. (Source: Boukhelkhal Islam, 2020)

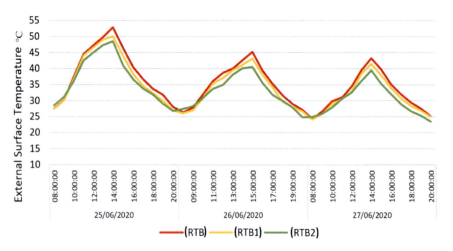

Fig. 16 External surface temperature of rough textures made up of palm particles with different concentrations of aggregates. (Source: Boukhelkhal Islam, 2020)

The figure below shows the results of the outdoor surface temperature obtained after adding to the rough texture the palm particles with different concentrations (RTB: 0%, RTB$_1$: 30%, and RTB$_2$: 70%), respectively (Fig. 16).

Palm particulate aggregate is added in different concentrations, starting with RTB of 100% sand aggregate, RTB$_1$ of 30% palm particulate and 70% sand aggregate, and finally RTB$_2$ of 70% palm particles and 30% sand aggregates, respectively. This diagram shows that the highest external surface temperatures are recorded at the reference coating RTB, the most critical value of which is recorded during the day of June 25, 2020, between 2:00 p.m. and 3:00 p.m., equal to 52.8 °C with a difference of 2.7 °C compared to RTB1 and 4.3 °C compared to RTB2.

These results prove that using palm particles in multiple densities can significantly reduce the external surface temperature at the rough texture, as the results show a difference of up to 4.3 °C.

Conclusions

Overall, the results of this work indicate significant credibility between exterior texture geometry, percent cast shadow, and exterior surface temperature. The results of this chapter show that the textured facade panels can improve the cooling, through its texture, and that there is a strong correlation between the created morphology of the texture, the shadow fraction rate, and the surface temperature. We can summarize the following:

Texture Depth

The results reveal that the textures with the least depth such as the smooth and rough texture recorded higher surface temperatures than the others. They confirm the hypothesis expressing that the deeper the surface texture, the more the surface protected by the self-shading effect increases and the greater the ratio (SF), and consequently the ability of the texture to lose heat is tall.

Organizing and Assembling the Texture

The results clearly illustrate that textures with a suitably organized layout and assembly contributed to better cooling than rough textures. The organization of texture devices seems to be more efficient and effective, creating organized breaks in a well-defined layout. This organization can provide more protection by increasing the shaded area. Moreover, this assumption explains that the arrangement and the assembly of the texture are a very effective solution in order to increase the shadow fraction rate and consequently decrease the effect of the external surface temperature.

Texture Thickness

By examining the impact of the different thicknesses of the texture on their thermal performance, this study deduced that the thicker textures such as the blade and crystal texture have a greater ratio (SF) than the thin textures such as the smooth and rough texture. These results are consistent with the study by Bergman Watt et al. (2010) [35], which reveals that as thinner the facade panel was, the faster the panel heats up, and therefore its temperature is higher.

Texture Components

The results of the second phase of this investigation clearly illustrate that the integration of natural components in the texture, such as aggregates based on palm particles, can significantly reduce the external surface temperature. The study concluded that by increasing the concentration of palm particle-based aggregates, the surface temperature decreases, and the coating cools rapidly. Therefore, the use of a concentration with more than 70% palm particles allows better surface temperature results when mixed with adjuvants to avoid deterioration caused by weather conditions.

Things to Think About

This work assists to identify the real thermal behavior in any type of texture. It creates and develops a framework for future research and designer's works to provide applicable architectural solutions for existing and new buildings and to open up new opportunities for solving overheating problems through passive design strategies. The development of this research will be to establish other similar works on the types of textures which could include numerical simulations or on other climatic regions, otherwise through other elements of buildings such as roofs, in order to assess the effect of the latter on the thermal performance of the building.

This work can also be part of the preservation and enhancement of the built heritage of southern Algeria in order to contribute to the sustainable development of these regions characterized by a hot and arid climate.

Acknowledgments We would like to acknowledge the Bioclimatic Architecture and Environment (ABE) Laboratory at the University of Constantine 3, Algeria, for the use of the equipment in this research and valuable support during the experiments. We would like also to acknowledge the Sustainable Building Design (SBD) Laboratory at the University of Liege, for the use of the equipment in this research and valuable support during the experiments and data analysis. The authors would also like to thank the University of Constantine 3, Algeria, and the Liege University, Belgium, for their assistance in administrative procedures.

Appendix

- Climate data of Constantine

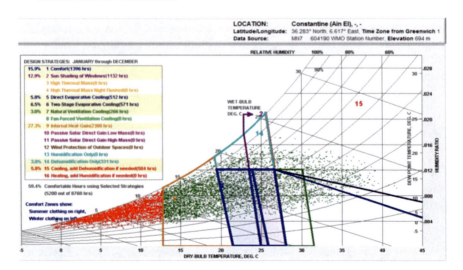

Thermal Behavior of Exterior Coating Texture and Its Effect on Building Thermal... 39

Psychrometric diagram of Constantine
(Source: Climate Consultant)

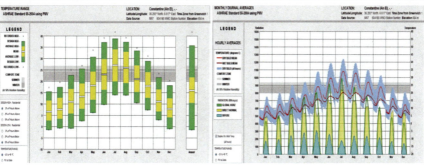

 Zone de surchauffe

Zone de confort

Isotherms of Constantine

Variation in climate data for Constantine

- Description and detail of the studied textures

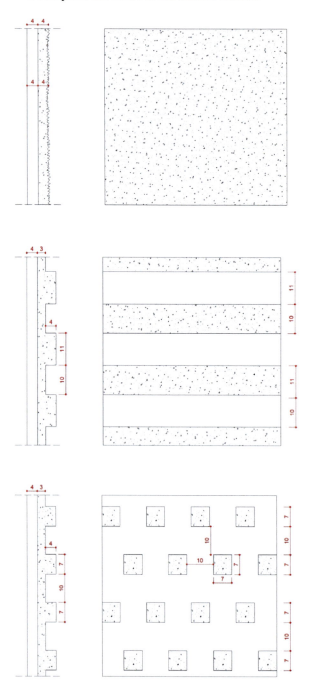

References

1. Pacheco-Torgal, F., Jonkers, H. M., Karak, N., & Ivanov, V. (2016). Introduction to biopolymers and biotech admixtures for eco-efficient construction materials. In *Biopolymers and biotech admixtures for eco-efficient construction materials*. Woodhead Publishing Edition.
2. Peeks, M., & Badarnah, L. (2021). Textured building façades: Utilizing morphological adaptations found in nature for evaporative cooling. *Biomimetics, 6*, 24. https://doi.org/10.3390/biomimetics6020024
3. Ascione, F., Bianco, N., De Masi, R. F., Mauro, G. M., & Vanoli, G. P. (2015). Design of the building envelope: A novel multi-objective approach for the optimization of energy performance and thermal comfort. *Sustainability, 7*, 809.
4. Yuxuan, Z., Yunyun, Z., Jianrong, Y., & Xiaoqiang, Z. (2020). Energy saving performance of thermochromic coatings with different colors for buildings. *Energy and Buildings, 215*, 109920.
5. Ibrahim, M., Biwole, P. H., Wurtz, E., & Achard, P. A. (2014). Study on the thermal performance of exterior walls covered with a recently patented silica-aerogel-based insulating coating. *Building and Environment, 81*, 112–122.
6. Joudi, A., Svedung, H., Cehlin, M., & Rönnelid, M. (2013). Reflective coatings for interior and exterior of buildings and improving thermal performance. *Applied Energy, 103*, 562–570.
7. Merhan, S. *Self-shading walls to improve environmental performance in desert buildings*. Available online: https://www.researchgate.net/publication/338655646. Accessed on 20 Apr 2020.
8. Capeluto, I. G. (2003). Energy performance of the self-shading building envelope. *Energy and Buildings, 35*, 327–336.
9. Alhuwayil, W. K., Mujeebu, M. A., & Algarny, A. M. M. (2019). Impact of external shading strategy on energy performance of multi-story hotel building in hot-humid climate. *Energy, 169*, 1166–1174.
10. Liu, S., Kwok, Y. T., Lau, K. K., Chan, P. W., & Ng, E. (2019). Investigating the energy saving potential of applying shading panels on opaque façades: A case study for residential buildings in Hong. *Energy and Buildings, 193*, 78–91.
11. Kandar, M. Z., Nimlyat, P. S., Abdullahi, M. G., & Dodo, Y. A. (2019). Influence of inclined wall self-shading strategy on office building heat gain and energy performance in hot humid climate of Malaysia. *Heliyon, 5*, e02077.
12. Givoni, B. (1998). *Climate considerations in building and urban design*. Wiley.
13. Mansouri, O., Bourbia, F., & Belarbi, R. (2018). Influence de la réflectivité de l'enveloppe sur la demande énergétique des bâtiments et sur le confort thermique. *Nature & Technology, A, 33*–42.
14. Shen, H., Tan, H., & Tzempelikos, A. (2011). The effect of reflective coatings on building surface temperatures, indoor environment and energy consumption — An experimental study, *43*, 573–580.
15. Cheng, V., Ng, E., & Givoni, B. (2005). Effect of envelope colour and thermal mass on indoor temperatures in hot humid climate. *Solar Energy, 78*(4 Spec. Iss), 528–534. https://doi.org/10.1016/j.solener.2004.05.005
16. Taha, H., Sailor, D., & Akbari, H. (1992). *High albedo materials for reducing cooling energy use*. Lawrence Berkeley Lab. Volume UC-530.
17. Bansal, N. K., Gargand, S. N., & Kothari, S. (1992). Effect of exterior surface colour on the thermal performance of buildings. *Building and Environment, 27*(1), 31–37.
18. Synnefa, A., Santamouris, M., & Livada, I. (2006). A study of the thermal performance of reflective coatings for the urban environment. *Solar Energy, 80*(8), 968–981.
19. Bacha, C. B., & Bourbia, F. (2016). Effect of kinetic façades on energy efficiency in office buildings Hot dry climates. In *Proceedings of the 11th Conference on Advanced Building Skins*, Bern, Switzerland, 10–11 October 2016; pp. 458–468.
20. Tarabieh, K., Abdelmohsen, S., Elghazi, Y., El-Dabaa, R., Hassan, A., & Amer, M. (2017). Parametric investigation of three types of brick bonds for thermal performance in a hot arid

climate zone. In *Design to Thrive, Plea 2017 Proceedings* (Vol. III, pp. 3699–3706). Engine Shed Tours.
21. Ercan, B., Tahira, S., & Ozkan, E. (2015). Performance-based parametric design explorations: A method for generating appropriate building components. *Design Studies, 38*, 33–53.
22. Nocera, F., Lo Faro, A., Costanzo, V., & Raciti, C. (2018). Daylight performance of classrooms in a mediterranean school heritage building. *Sustainability, 10*, 3705.
23. Rodonò, G., Sapienza, V., Recca, G., & Carbone, D. C. (2019). A novel composite material for foldable building envelopes. *Sustainability, 11*, 4684.
24. Chikhi, M., Agoudjil, B., Haddadi, M., & Boudenne, A. (2011). Numerical modelling of the effective thermal conductivity of heterogeneous materials. *Journal of Thermoplastic Composite Materials, 26*, 336–345.
25. Benmansour, N., Agoudjil, B., Gherabli, A., Kareche, A., & Boudenne, A. (2014). Thermal and mechanical performance of natural mortar reinforced with date palm fibers for use as insulating materials in building. *Energy and Buildings, 81*, 98–104.
26. Oushabi, A., Sair, S., Abboud, Y., Tanane, O., & El Bouari, A. (2017). An experimental investigation on morphological, mechanical and thermal properties of date palm particles reinforced polyurethane composites as new ecological insulating materials in building. *Case Studies in Construction Materials, 7*, 128–137.
27. Záleská, M., Pavlík, Z., Cítek, D., Jankovský, O., & Pavlíková, M. (2019). Eco-friendly concrete with scrap-tyre-rubber-based aggregate— Properties and thermal stability. *Construction and Building Materials, 225*, 709–722.
28. Benhalilou, K. (2012). L'enveloppe végétale: une alternative au rafraichissement passif cas de la façade végétale, thèse de doctorat en science, université Constantine 3.
29. Peck, S. W., Callaghan, C., Bass, B., & Kuhn, M. E. (1999). *Research report: Greenbacks from green roofs: Forging a new industry in Canada*. s.n.
30. Blanco, I., Schettini, E., & Vox, G. (2019). Predictive model of surface temperature difference between green façades and uncovered wall in mediterranean climatic area. *Applied Thermal Engineering, 163*, 114406. https://doi.org/10.1016/j.applthermaleng.2019.114406
31. Wong, N., et al. (2010). Thermal evaluation of vertical greenery systems for building walls. *Building and Environment, 45*, 663–672.
32. Zhang, L., Deng, Z., Liang, L., Zhang, Y., Meng, Q., Wang, J., & Santamouris, M. (2019). Thermal behavior of a vertical green facade and its impact on the indoor and outdoor thermal environment. *Energy and Buildings, 204*, 109502. https://doi.org/10.1016/j.enbuild.2019.109502
33. Ganji, H., Mohammad Kari, B., & Norouzian Pour, H. (2013). *Thermal performance of vegetation integrated with the building façade*. Forschungs- und StudienzentrumPinkafeld.
34. Tsoumarakis, C., et al. (2008). *Thermal performance of a vegetated wall during hot and cold weather conditions* (pp. 22–24). Dublin, s.n.
35. Bergman, W., Mitchell, S., & Salewski, V. (2010). A concept cluster. *Oikos, 119*, 89–100.

Beyond Energy Efficiency: The Emerging Era of Smart Bioenergy

Rachel Armstrong

The Power of Metabolism

> There is in biology a formula called "the equation of burning." It is one of the fundamental pair of equations by which all organic life subsists … All that is living burns. This is the fundamental fact of nature [1]

> … remember what we came for. The fire [2]

> … there is a magic deeper still which she did not know. Her knowledge goes back only to the dawn of time. But if she could have looked a little further back, into the stillness and the darkness before Time dawned, she would have read there a different incantation [3]

William Bryant Logan notices the paradox of the burning bush revealed by God to Moses and reflects on its meaning. He asks himself what it means for something to burn yet not be consumed by the process and concludes that what is being revealed is the fundamental transaction of living things—metabolism. In this process a creature processes the world around it, transforming it into energy to sustain life.

To date, the material narratives of "life" have centred on crystals (nucleotides) and molecular flows (metabolism); however, the material platform that makes them commensurate is the electron, which can combine these principles in a unifying manner that does not homogenise outcomes but enables variation. While electrons are considered classical agents in conventional physics and electronics behaving as flowing particles, they are also quantum phenomena behaving in nonclassical ways such as quantum tunnelling or demonstrating superimposition and entanglement (Einstein's spooky action at a distance).

More than the flow of "electricity" or "energy", the transfer of electrons around charged (or "valent") atomic nuclei is coupled with physical change.

R. Armstrong (✉)
Department of Architecture, Campus Sint-Lucas, Brussels/Ghent, Belgium
e-mail: Rachel.armstrong@kuleuven.be

Electron transport chains enable these unique reduction/oxidation reactions and comprise the oldest systems of "life". Having freed themselves from the solid structures of the rocks as geochemically produced organic molecules during the Hadean period, these energy-harnessing systems became portable in solution (acid/base couplings) and in the air. Embodying the enlivened reactivity of the living realm, electron transfer chains enabled lively matter to resist reaching the "brute" ground state of relative equilibrium and became enfolded within the internal environment of the earliest cells.

Chemist Ben McFarland emphasises the importance and material creativity of these forces:

> ... when molecules fit together, they only care about two things: shape and charge. Shape is familiar—all atoms are spheres that can stack together like a supermarket display of oranges—but charge is unusual. Unless you work with wires or rub your feet across shag carpeting, you don't normally see charge imbalances at our macro-level. But at the nanometre level, charge moves things around. Each atom is made of heavy protons with a positive charge and light electrons with a negative charge. When these charges are symmetric and balanced, the overall charge is neutral, but when they fall askew, a chain of domino effects can start, and chemistry can happen [4]

The changes produced by metabolism maintain the creature's own body where downstream products—often smaller, biologically reusable molecules—are released back into the world for other life forms to make use of them.

Over aeons, the first creatures that populated the world such as bacteria and archaea have made developing a robust metabolism their speciality. In the case of bacteria this has happened at the expense of structural organisation, and they have evolved remarkable abilities that keep that electron transfer chain going. The most minimal kind of metabolism is produced by a continual flow of electrons, which prevents enlivened matter from reaching an energetic ground state or functional "death". The bacterial species *Shewanella* and *Geobacter* directly harvest electrons from rocks and metals to make the universal energy storage molecule called ATP. This pared-down process is quite alien to all other life forms as they don't need a carbon source (or sugar) for respiration. This means these microbes can thrive indefinitely by eating electrons from one electrode, using them as a source of energy, and then discarding them to the other electrode [5].

Electron transfer is not a solipsistic activity and can connect bodies at a distance from each other, offering a communications medium that enables microbes—and other organisms, like ourselves—to talk at a distance using action-potential mechanisms, which is characteristic of brain tissue. Waves of potassium-driven electrical activity traveling with constant strength enable communities of microbes to propagate signals at around 3 mm/h in tissue-like formations called biofilms, while our own brains work much faster at 100 m/s. In all species, these electrical signals enable the synchronisation of activities across large expanses and are much more powerful than their chemical counterparts as communications systems—which may be likened to "the difference between shouting from a mountaintop and making an international phone call" [6]. The ability to operate at a distance from a locus of metabolism introduces notions of time, space, and anticipation—the foundations of all complex thought.

How we think about and use electricity today is encapsulated by two very different approaches.

"Life" became equated with electricity through Luigi Galvani's (1737–1798) "animal electric fluid" experiments, where he demonstrated the presence of bioelectricity in frog dissections. Regarding this force as responsible for the vitalisation of tissues, his findings were largely enfolded into the life sciences to begin the neurobiological revolution.

Responding to Galvani's experiments, Alessandro Volta (1745–1827) found other ways of producing electricity using metal electrodes and chemical sources which generated high voltages. The catalyst for the electronics revolution Volta's research provided the power and intelligence that underpins the modern industrial age, around which we have imagined and design our relationship with electricity.

Galvani's organic electricity operates through a very different quality of electron flow than Volta's. It is slower, more agile, less forceful, and can be handled by atomic "jugglers",[1] so bioelectrical transactions occur that perform metabolic work, inviting a new set of imaginaries than industrialisation, or modernism. As "life", not top-down control, or consumption, is at the heart of these principles, these electron currencies are constrained by biological principles, establishing the possibility of a new thermoeconomics as the foundation for a regenerative society [7]. When produced metabolically, the natural bioavailability of electrons establishes limits for production systems so that matter and energy are coupled (not cleaved) and are synthesised in a circular context. This means that every material ecology can be strategically metabolised using bioelectrical systems to perform all kinds of useful work, without "borrowing" unlimited resources from next generations or elsewhere.[2]

The Frankenstein Fallacy

The Frankenstein fallacy assumes that if we throw enough electrical power at an inanimate body, then it will spring to life.

When a forceful river of new electrons is launched at a troupe of jugglers, however, it puts the juggling act off balance and may stop it altogether. Electrotherapies only work when the influx of juggling objects (electrons) can boost the actions of dysfunctional jugglers that have dropped too many balls and are not performing at all well. The power to respond to the influx remains with the jugglers, as they are still motivated, and sudden bombardment with new juggling objects may persuade

---

[1] The "juggler" metaphor is used to indicate the dynamic orbits within an atom that enable the movement of electrons.

[2] For example, through their combustion fossil fuel-based systems overwhelm contemporary ecosystems with "old" sources of carbon.

them to keep going. The moment that the jugglers give up, then no matter how forceful the flow new objects into the system is, they will not start again.

As Galvani's dissections were recently prepared and the creatures newly deceased, the atomic jugglers were still willing to move objects around for a while, proving his animal electricity hypothesis. However, the force of electrons used in Volta's setup is not designed for organic bodies but brute[3] metal ones. Forming a matrix through which outer electrons can move freely instead of orbiting their respective atoms, metals are excellent conductors of electricity and heat, but their electrons are not in a context where they can perform the work of life (where atoms are spontaneously transformed by electron flow). Massive unsustainable flows of energy in conductive materials are therefore needed to animate "brute" matter, and once the circuit is turned off, this vigour is not sustained.

The Long Latency of the Bioelectrical Revolution

Galvani's organic electricity operates through a very different quality of electron flow than Volta's. It is slower, more agile, and less forceful. It can be handled by atomic jugglers, so bioelectrical transactions occur that perform metabolic work, which invite a new set of imaginaries than industrialisation or modernism. As "life," not top-down control, or consumption, is at the heart of these principles, then these electron currencies are constrained by biological principles, establishing the possibility of a new thermoeconomics. When produced metabolically, the natural bioavailability of electrons establishes limits for production systems so that matter and energy are coupled (not cleaved) within a circular context. This means that every material ecology can be strategically metabolised using bioelectrical systems to perform all kinds of useful work, without "borrowing" unlimited resources from next generations or elsewhere.[4]

Significantly, bioelectricity can cross the mechanical and organic divide, permeating both platforms, but operates at much lower power levels than generated by fossil fuels or renewables. What it lacks in quantity, however, it makes up for in the quality of its operations, inviting an era of low-power (bio)electronics.

In 1911, Michael Cressé Potter brought these organic and mechanical electrical worlds together in a "living" battery, or microbial fuel cell (MFC), that used the vital processes of *Saccharomyces* bacteria to produce several hundred millivolts of energy [9]. Acting as biocatalysts, the microbes convert the chemical energy of organic matter from waste streams into electrons for as long as they continue to be fed. Each "cell" consists of two compartments, the anode and the cathode, which are separated by a proton-exchange membrane. Bacteria anaerobically oxidise the

---

[3] In a letter to Richard Bentley, Isaac Newton uses the term "brute" to refer to an inert body [8].

[4] For example, through their combustion fossil fuel-based systems overwhelm contemporary ecosystems with "old" sources of carbon.

organic matter in the anode chamber to release electrons that flow through an external circuit to provide electrical power. Acidic protons are also produced, dissolve into solution, and pass through the membrane into the cathode, where they react with oxygen to produce fresh water. This highly mediated relationship sets up a power-sharing relationship across mechanical and natural bodies that is neither entirely biological nor exclusively mechanical. Constituting the essence of zoë, the microbial fuel cell blurs the relationship between organism and machine to form a type of cyborg "being" with microbial flesh that thrives on different types of organic fuel. Supported and directed by a technical environment, they perform a range of metabolic tasks at room temperature such as cleaning wastewater, generating bioelectricity, and detoxifying pollutants.

While bioelectrical systems like the microbial fuel cell (MFC) cannot compete with the sheer power of other electricity generating systems (renewables, fossil fuels), their (material) circularity is unsurpassed providing essential natural limits to our consumption. Metabolic transactions have helped build the soil and air, creating the biological economy that founded a carbon-based exchange between living things. Life has sustained itself thus since the dawn of biogenesis. These are not, however, the principles on which modern society and the building industry are based.

Buildings are designed to shelter us from uncertainty and danger outside. In the modern world we spend 90% of our time in these spaces. While buildings are erected as barriers that control what comes in and out of our artificial worlds, the natural world necessarily interpenetrates them. Fluid elements leak through windows and doors while shifting layers of metabolism that comprise our very acts of daily life consume resources. In deep time, urban ecosystems turn over the many bodies and buildings that form its landscapes, complicating the relationships between them and rendering impermanent the very notion of a city. Percolated by the flow of space and time, no walls can stop our lives from being changed by the world's events. Spending about 2% of the total economic value of our homes per annum, we maintain the illusion of security through distributed acts of building maintenance. Drawing comfort from our separatism and human exceptionalism, we've grown blind to those systems and resources beyond the domestic sphere that we consume daily. Measured and capitalised, a variety of natural utilities and services appear in our kitchens and bathroom, seemingly from nowhere, making themselves available for our consumption. Placing only financial demands upon us, they never ask to be replenished. So, our distance from nature grows, and our exclusively human bodies seem ever more secure in their integrity, identity, and independence from the world outside.

Tearing up the anthropocentric order of things, the climate crisis shatters these cosy architectural conventions and fully exposes us to the actual "outside". Although such an outcome has been inevitable for many decades, we remain woefully unprepared. The Anthropocene, a cultural epoch of our own design, has force fed carbon-industrial foie gras[5] on the world's ecosystems in the name of progress, extinguishing

---

[5] *Foie gras* means "fatty liver".

the very forms of life that we depend most on. The most vulnerable and smallest agents have suffered first: pollinators, reefs, fertile soils, "weeds", clean air, and oceans, setting in motion a system of collapse that characterises the Sixth Great Extinction and has damaged the planet's fundamental capacity for self-repair. With collapsing ecosystems, the comforts of modernity are no longer available in the way they used to be, and the systems by which we trade and govern no longer apply. "Freedom, justice, equality, truth—all these things are distant memories by now, as is an age where people took such things for granted. The idea of civilization as we know it has come to an end" [10]. Masked by our extreme exploitation of every resource, with few gestures of re-investment, we have crippled planetary resilience to the point where geosystems are transitioning into new kinds of order that are not of human design. More than increasing outbreaks of extreme weather events, pandemics, wildfires, quakes, heatwaves, droughts, mudslides, and tornadoes—all heightened by the climate crisis—they embody a "profound mutation" in our relationship with the world [11].

> When I was around bustling crowds of people, I saw death and destruction. When I walked on dry land, I saw floods. I imagined wild animals, especially snakes, getting out of the zoos in the aftermath of natural disasters. I worried about how we would treat each other in the face of such calamity. I doubted it would be kind [12].

Although we are consuming our planet at such a speed that it cannot renew itself, our response to this situation is shockingly feeble, and this paralysis is deeply troubling.

> I think in many ways that we autistic are the normal ones and the rest of the people are pretty strange. They keep saying that climate change is an existential threat and the most important issue of all. And yet they just carry on like before. If the emissions have to stop, then we must stop the emissions. To me that is black or white. There are no grey areas when it comes to survival. Either we go on as a civilisation or we don't [13].

There are far more questions than answers to this predicament. While we've stolen untold resources from the planet and even from our own children [14], it's not too late to start giving our descends their future back. When it once appeared that total global warming should be kept below 2 °C or as close as possible to 1.5 °C above pre-industrial revolution levels, it seemed reasonable to simply reduce our emissions as well as reusing and recycling resources at "sustainable" levels within the present economic order. Governments and businesses have even promoted Green New Deals of various descriptions, but even the most radical of these operate from within the systems that continue to contribute to climate change. Expressing sympathy for the cause is an empty gesture as their capacity for radical change is limited. With our present trajectory set to go above 3 °C or 4 °C, "tipping points" in our planetary system are visibly being reached, which means that radical breaks from the current system of production, consumption, and economic order are urgently needed. Exactly how long we have before the world becomes unliveable is unknown: for some, that point is already here. In the best-case scenario, we may only have 10 years to transform global society to reasonably limit the catastrophic impacts of climate change [15].

> Things are already bad. They are already getting worse. This report reveals—and, for many of us, confirms—that we're not doing nearly enough to stop things from getting damn apocalyptic [16].

The scale of the challenge is immense. Rather than being able to take actions ourselves, it appears the ways of dealing with the crisis are somewhere out there where "moonshot" technofixes abound. Geoengineers aim to prevent glaciers from collapsing using scaffolding made from sand; oceanic fertilisation is set to absorb vast quantities of carbon dioxide by stimulating the growth of phytoplankton; stratospheric mirrors and cloud manipulators strive to prevent sunlight from reaching the ground, and innovators have developed a whole range of products including ocean trash-eating robots, carbon-fixing materials, and artificial trees. While exciting and welcome, all these developments rely on institutions and agencies that operate beyond our own realm of influence.

> The dominant narrative around climate change tells us that it's our fault. We left the lights on too long, didn't close the refrigerator door, and didn't recycle our paper. I'm here to tell you that is bullshit. If the light switch was connected to clean energy, who the hell cares if you left it on? The problem is not so much the consumption — it's the supply. And your scrap paper did not hasten the end of the world.
>
> Don't give in to that shame. It's not yours. The oil and gas industry is gaslighting you [16].

Since our house is on fire [13], we must start the reconfiguring of our lives in the home. While we already limit our domestic energy consumption, recycle our waste, and reduce our consumption, as in the case of the Green New Deals, all these approaches are inescapably entrenched within our present economic frameworks and comfort zones. Even from the comfort of the domestic realm, we are treading the same pathway towards planetary destruction that has already been mapped out by industry—albeit more slowly, self-consciously, and considerately. To achieve a radical break from our present trajectory, a reimagining of those activities of daily life that maintain us, zoë, is needed, alongside a reappraisal of the values that uphold the good quality of life we seek, bios.

As citizens of modern societies, we are bound by those frameworks that shape how we live and work. Our non-innocence in the present situation means we must consider allies in the more than human world to meet the present challenges. While applications of fire powered by fossil fuels have brought great innovations into our homes, they have not conferred the kind of wisdom needed to bring about planetary enlivening. To step beyond these malignant processes requires a new kind of technology and fundamental metabolism that is not built on consumption but tempers the production of work with environmental enlivening. Such a technological system seems far away, corporate, or even magical, but, in fact, it is so close, obvious, and reliable that it is taken for granted. While everyday events within our living world seem commonplace—leaves seasonally sprout and fall, food is digested, wounds heal, the young become old—when considered from a cosmic perspective, each and every one of them are extraordinary.

How We Will Incorporate Nature into Our Homes and Cities

Only by working with the biological processes of metabolism can we close the loop of consumption to stay within the resource limits of a site, with the benefit of actively managing our waste. We are not the masters of this knowledge domain, but microbes are.

Microbes have a unique individual identity, as well as a larger collective one. Some are commensals, some symbiotic, and around 1400 species are pathogens. While this may seem like a large number, they account for much less than 1% of the total number of microbial species on the planet [17]. The most remarkable characteristic of a microbe is not their form but their fluid metabolism, which is modulated via the environment through redox chemistry, which is the environmental program that enables microbes to rapidly alter their metabolic (and genetic) networks that they can use to completely transform their environment.

> Life loves redox chemistry because it is mild and controllable [18]

More than the flow of "electricity" or "energy", electron transfer is coupled with physical change. The loss or gain of electrons, or change in the charge of a molecule, alters its physical properties. Comprising the oldest systems of "life", biological electron transport chains freed themselves from the solid structures of the rocks during the Hadean period. Taking the form of geochemically produced organic molecules, these energy-harnessing systems became portable in solution (acid/base couplings) and in the air. Embodying the enlivened reactivity of the living realm, electron transfer chains now enabled lively matter to resist reaching the "brute" ground state of relative equilibrium to become enfolded within the internal environment of the earliest, leaky compartment of cells, resulting in the kinds of dynamic chemical interactions that comprise the living realm.

Introducing an Emerging Platform: Microbial Technology

You may not think of nature as a technology, but radical new insights about the microbial foundations of the living world means we can understand and work with them as a creative and regenerative force.

> Although the tiniest bacterial cells are incredibly small, weighing less than 10^{-12} grams, each is in effect a veritable micro-miniaturized factory containing thousands of exquisitely designed pieces of intricate molecular machinery, made up altogether of one hundred thousand million atoms, far more complicated than any machinery built by man and absolutely without parallel in the non-living world [19]

Consequently, a new age of "living" technologies is emerging. Powered by microbes, these "living" technology platforms can process our waste, turning it into electricity and cleaned water and can detoxify our surroundings. Orchestrated microbial activities can therefore form the utilities systems of our buildings, generate new, organic materials [20], and dramatically alter the impacts of human development, so our collective daily activities are bioremediating rather than harming the environment.

Historical Relationship with Technology Frames Our Future Trajectories

Dancing with the demon of fire, the sacred being that gave us the first artificial metabolism, we could exceed the wet heat of the flesh experienced through proximity to metabolising bodies. Fire is a demanding god that consumes all it is fed to ashes and in return gives us power far beyond our biological means. It's greedy, dry, harsh combustion readily transformed organic substances such as fuel and food, from one sort into another. Gathering around this magical transformer and its unnatural light, early human societies sought to instrumentalise its multiple powers and set out to direct its actions towards specific tasks—from generating heat to finding their way in the dark and making food more palatable and digestible. While raw foods yield just 30% or 40% of their nutrients, cooked food releases everything. Suddenly, having a large brain stopped being an evolutionary liability and became an asset. By gaining the necessary nutrients more easily, people could start to imagine better ways to hunt, live, develop culture, produce art, and invent early technologies—all the things that made us who we are now [21]. Nomadic peoples quickly acquired the knowledge for making open fires, while settlers formalised these spaces within the area of communal activity that became known as the kitchen—the traditional and symbolic heart of the home.

Increasingly instrumentalised by regulating the flow of air, organising space, selecting the right materials, and designing implements to instruct fire in different ways, each culture developed their own processes and rituals. Using heat to prepare food was most the important activity for communities, being associated with rituals of gathering, preparation, cooking, dining, cleaning, socialising, ablutions, and the disposal of leftovers. While powerful, fire was also a voracious entity demanding food—predominantly wood, maintenance, and regulation so that in its voracity, it did not leap from its place in the home to burn the entire house down—which it frequently did. Early hearths were therefore made of clay or stone, their main purpose being to enclose the fire. Looking after this beast was a commitment. Demanding constant attention, it needed stoking in the morning, modulating during the day and turning safely back into glowing embers at night. Entangled with the preparation of food, the complexity of these activities was balanced by what those people responsible for them could achieve in a day. Tempered by a baseline liveability that varied from home to home, the activities around the fireplace shaped the designs of hearths and ovens, informed where kitchens were located, and spawned the development of other important household areas associated with the preparation of food such as the pantry, orchard, garden, larder, icehouse, root cellar, and medicinal herb garden. Like electrical capacitors, the kitchen fire's work could also be stored, and the earliest containers of its cooked products were animal skins, woven baskets, or gourds.

Traditionally a matriarch ran the household economics and domestic affairs, where the kitchen was central both as a workshop but also a meeting place between classes and organisational power structures. For ancient Greeks and Romans, the kitchen was designed as a separate house, with clearly defined preparation areas for turning food from city gardens into meals. By medieval times, wealthy European homes had large kitchens with dedicated rooms such as for storing utensils, pantry, cold storage, and a buttery. More commonly, kitchens were the centre for communal cooking, dining, and social activities. During the Renaissance, a range of technologies of food preparation flourished for new types of cooking that coaxed fire's participation in different ways using the spit, gridirons, ewers, salvers, and huge cauldrons. These innovations catalysed new methods for food preparation and storage during the winter months using a range of desiccating techniques like curing, drying, smoking, pickling, and salting.

Nineteenth-century kitchens embraced the Industrial Revolution, whose fire thrived on a new kind of concentrated energy obtained from fossil fuels. This freshly unleashed demon lurked in petroleum basins formed from decomposed flesh still dreaming of tarry insurrection against the rotting Sun. [22] Its capacity for intense burning intensified the demands of fire and made possible new materials such as cast iron, which could tolerate drastic temperature swings and was an ideal medium for casting into complex, prefabricated parts decorated with surface ornament. Benjamin Thompson, better known as Count Rumford, revolutionised the technology of fire and design of the kitchen by taking the open cooking fire from the hearth and constraining it within a cast iron box. Topped by a flat, perforated surface, accessed through round ports of different sizes, the fire contained below could be spatially directed to heat specially designed pots and pans that were designed for these new ranges. Requiring less surveillance and space, the machine-like fireplace, however, obliged an industrial metabolism to catalyse its intense activities. Natural gas, used by the lighting industry, followed by advanced in electrical power sources, enabled the development of lighter, smaller, and even more effective appliances that could burn beyond the full fury of an open hearth—even in the smallest of homes [23]. Even from within their boxes, the twin demons of fire and fossil fuel continue to indulge their appetites. Assumed to be securely under human control, their endless demands began to consume our world.

The rise of modern agriculture and modern manufacturing fuelled the twentieth century marketplace, rendering obsolete the need for self-sufficiency in food production. By the 1930s, the number of kitchen staff as well as the size of kitchens and pantries was reduced in most homes. A rapidly changing modern workforce spent less time in the home and more at various kinds of workplaces, where convenience foods were bought and consumed, eliminating the need for the socially productive role of the kitchen. While it remains the symbolic centre of the home, the twenty-first century kitchen's foundational technology of fire is assimilated into global power grids, and no longer draws us like moths, to its flame. If we are to curtail its consumption of our planet, then we must find new kinds of energy, or power, with which we can ally that enable us to live better with the available resources of the word and unmake our pact with demons.

MFCs directly contribute towards sanitation improvement and double up as remote energy sources, fuelled by waste, for low-power applications in rural, off-grid settings. The potential for simplifying complex processes, through digitisation and interactive decision-making apps, could result in wider interest, enhanced public acceptance, demand for, uptake of, and investment in large-scale development of the technology in the market.

This overall approach of harnessing microbial metabolisms also opens up new frontiers in low-power electronics that can be run by bioelectricity produced by the system, as well as the science of programming microbial biofilms[6] and consortia,[7] which establish the principles for designing metabolic "apps", where microbes that do not normally work together are able to collaborate for the first time, enriching the potential for metabolic design.

Living Architecture's combined effects can alter the character of our living spaces—from environments that merely consume resources—to spaces that link the webs of life and decay to provide for our needs, refresh our atmosphere, and reduce the circulation of toxic compounds, such as detergents in our waterways. Like the exchanges between the first forms of life, the way that webs of biochemical exchanges are orchestrated can increase the liveability of a space by, for example, sequestering heavy metals and breaking down nitrous oxide. Beyond their capacity to act as remedial systems, by actively transforming domestic wastes into useful outputs they generate resource, which enables them to be accordingly revalued within the domestic economy.

The ability to choose and nurture the specific metabolism of our living spaces will alter our lifestyles and the way we inhabit our homes. As "living" architecture is founded on the mutual relationships between microbes and humans, how we care for our living spaces will deeply influence their performance. With the advent of many more "apps", residents will be able to select and nurture a unique character for their homes that—in exchange for being sensitive to the system's wellness—is capable of meeting specific needs such as producing medicines, recycling organic matter into edible biomass, or producing liveable amounts of electricity and heat. While local conditions and resource availability will constrain the performance of metabolic "apps", innovative combinations may enable a small range of possibilities to achieve unexpectedly desired outcomes, such as generating bioluminescence to provide low-level, mood-elevating lighting. In whatever way "living" architectures are deployed, they will require our conscious engagement with them, so they can learn about and respond to our needs. "Living" architectures will wake up with us, go to sleep when we do, will cope with our intimate habits, and will even be

---

[6] A biofilm is a mixed cohabiting community of microorganisms.

[7] Microbial consortia are populations of one kind of microorganisms and are not mixed as in biofilms.

re-enlivened by our return after a holiday break—"as if" they are "pleased" to see us. Through the care and conscious design of their metabolisms, inhabitants of living architectures will differently understand the character of our homes, what they can do, and establish our responsibilities for them.

Domestic buildings will no longer be a projection of our own desires upon them but beings in their own right that belong as much to a community as we do. Through a choreography of material relationships, different degrees of freedom for metabolic "apps" will be increasingly able to carry out different kinds of functions within homes and community spaces. Capable of evolving through space, time, and alongside their relationships with other beings, their metabolic exchanges will be at the heart of an environmentally concerned architectural practice that gives rise to an increasingly lively world.

The existence of these metabolic approaches and programmable, "living" technologies that can be incorporated into domestic spaces, as in the Living Architecture project, has implications for the development of our homes and cities, since they change the concept of sustainability. Exceeding the conventional disciplinary limits of architecture in making a building, they expand the practice to restore the health of the biosphere through the incorporation of living technologies into our homes, buildings, and cities. No longer designing for bounded plots, architects are concerned with the impact of their designs and interventions on whole bioregions.

Citizens are empowered to customise and modulate the impacts of their lifestyle choices from a diverse range of metabolic apps, which are a form of economy and key to customising the performance of microbial colonies to meet people's needs, while enabling significant changes in energy flow, water, and waste through homes and cities. By adopting such technologies, buildings will require richer networks of elemental infrastructures that modulate the flow of water, air, and material resources, to become sites of nutrient recycling and resource processing. When networked together, these combined, programmable metabolic processors have the potential to alter the nature of resource utilisation in cities.

As the technology develops, specific building metabolisms will be available from a diverse range of microbial and bioreactor choices. The aim is not to homogenise or universalise the ways that cities and their people live but, through the programmability of microorganisms, diversify approaches in ways that enhance the quality of urban environments. We also need to look at much larger-scale applications of microbes and extend our understanding of them, so they violate binary divisions and distinctions between the built and natural environments—like artificial reefs and mangroves—that increase biodiversity, clean the water, and stabilise land erosion. With the right governance, networks of bioprocessing units will begin to benefit ecosystems, remediate our ailing atmospheres, and, ultimately, improve the health and well-being of our cities and their inhabitants.

Benefits for generations to come may include increasing the biological quality of urban environments through the production of microbial biomass, eventually even replacing all fossil fuel-based appliances with organic systems. The major "waste" outputs of such "living" cities are the antithesis to modern cities such as cleaner waterways and composts. Enabling more green plants in metropolitan locations, which are climate moderators, in turn conserves water, improves soil and air quality, produces oxygen, absorbs carbon dioxide, and traps dust particles. In effect, the incorporation of "living" metabolic technologies into our daily routines gives us a chance to make human development more compatible with nature.

The "Living" Home: Our Households Become a Circular Economy

> We ate the birds. We ate them. We wanted their songs to flow up through our throats and burst out of our mouths, and so we ate them. We wanted their feathers to bud from our flesh. We wanted their wings, we wanted to fly as they did, soar freely among the treetops and the clouds, and so we ate them [24].

Our modern lifestyles promote a culture of obligate consumption. In our homes we produce almost nothing for ourselves, and as producers of this world we are obsolete being responsible not only for widespread environmental destruction—from deforestation to industrial-scale agriculture and mining practices—but also for the associated loss of biodiversity that our insatiable appetites promote.

How might a bioelectrically centred culture of life that works with the creative constraints of metabolism appear?

The *Living Architecture project* is a "living" combined utilities infrastructure that can turn liquid household waste, like urine and grey water, into valuable resources (electricity, biomass, water, reclaiming phosphate from washing-up liquids and removing poisonous gases from the air) that can be reused in the household (Fig. 1). This movie envisages the prototype's installation into a modern building. Here, it cuts down on electricity and utilities bills, as well as the amount of untreated waste we put into the environment. Potentially replacing fossil fuels as the main source of energy in a home, this series of linked bioprocessors can charge a 12 V battery supply.

When combined with renewables, microbial technologies create value by bioremediating our waste and even produce an ecological currency for exchange, just by the activities of daily living. So by eating, going about our routines, and doing our ablutions, the wastes we produce have economic value—even when our lives are spent at home. How a home could be constructed from a posthuman household inhabited only by microbes and the digital ghosts of the past, present, and future was prefigured in the installation 999 years, 13 sqm (the future belongs to ghosts), a collaboration between Cecile B. Evans and Rachel Armstrong (2019), using a microbial fuel cell array to power the infrastructure of the space and projecting a way forward through which the transformation towards our emerging ecological era can begin (Fig. 2).

Valuing the contributions by all who carry out the work of life—the different microbial units that make up the Living Architecture system—enables those that are not usually regarded as economically productive in a capitalist economy to take part in an ecological economy. Re-centring the site of value creation within the domestic

Fig. 1 Fully inoculated Living Architecture "wall" and apparatus installed at the University of the West of England, Bristol. (Photograph courtesy of the Living Architecture project (2019))

sphere, our homes become wealth generators. Inhabitants now have choices to make about how they use this ecological resource—perhaps they can reduce their own living costs, but, maybe too, they can donate some of their well-earned resource (formerly called "waste") to help others and activate the commons.

Active Living Infrastructure: Controlled Environment (ALICE)

While Living Architecture establishes a metabolic economy for transactions between humans and microbes, the Active Living Infrastructure: Controlled Environment (ALICE) prototype (2019–2021) generates the foundations for collaboration with microbes. Using electrons produced by the anaerobic biofilm of microbial fuel cells as "data" provides a direct link between bacterial metabolism and electronic systems that can interpret and visualise this data. Possessing a very particular kind of environmental intelligence, bacterial data can reveal a great deal about the character of a place, where a technologised approach can generate a relatable communications interface (Fig. 3).

Fig. 2 The installation 999 years 13 sqm (the future belongs to ghosts) is a collaboration between Cecile B. Evans and Rachel Armstrong for the Is This Tomorrow? Exhibition at the Whitechapel Gallery, London. (Photograph courtesy Rolf Hughes (2019))

Typically, microbial activity is deciphered using the tools of biochemistry, but in human terms the interpretation process is quite slow. Tapping into the much faster electron flows within biofilms, however, provides a direct way of understanding the behaviour of a microbial population at any given moment and, depending on the sensitivity of electrodes, creates the possibility of developing a communications platform between human and microbe (Fig. 4). Electrical activity from the biofilm was a source for both power and data, which was translated by software into animations that conveyed the overall status of the biofilm in relatable terms. Audiences could, therefore, respond to the microbial behaviour—not by looking at unpleasant "slime" (the natural "face" of microbial colonies)—but by interacting with appealing forms on a familiar screen-based interface. Participants could play with resident microbes through data and performance in an exploratory exchange—as if they were a pot plant or even a pet. This world of "Mobes"—a characterful term coined for the data-based representations of microbes—offers a simple, probiotic approach to interspecies communication within the highly situated realm of microbes, in a relatable manner that could even become part of our everyday routines. Being in conversation—rather than "exploiting" microbes—means we may start to learn along with them through their ability to generate clear and direct signals and data that relate to shared concerns, like transforming waste streams into household resources based on new value systems that invite different kinds of (house)work and domestic routines for our living spaces.

Fig. 3 ALICE installation at the Digital Design Weekend, Victoria and Albert Museum, London, 26–7 September 2021, a cyborgian entity powered entirely by microbes with a concurrent online "life" which embodies the bio-digital platform through the integration of microbial and artificial intelligences with biological and technical bodies. (Courtesy of the ALICE consortium: Ioannis Ieropoulos, Julie Freeman, & Rachel Armstrong)

The "Power Plant": Activating the Commons

While *Living Architecture* shows that a circular economy of the household is implementable, we have not yet figured out the economies of scale. It is likely more efficient for many households to contribute to a shared resource through their waste and what better site to process this than a garden—turning them, literally, into *power plants*? By scaling microbial operations in ways to serve a whole community, we can provide access to resources that can be allocated according to need. Working with Hungarian company Organica Inc. that designs urban wastewater gardens for municipal use, starting with human sludge, which is passed through a series of vats that break down the organic matter using the microbes on the roots of plants, our group proposes to introduce the electricity-generating microbial technology called the microbial fuel cell—a "living" organic battery—into this system, which can also be powered by the organic waste. This means such an installation could not only treat the human residues for between 5 and 30,000 residents with no access to a formal sewage system but that sludge could also generate enough bioelectricity to power mobile phones, provide LED street lighting, power Wi-Fi transmitters, and activate screen displays that enable citizens to access—for example—websites to online council services, while also enjoying the benefits of a public garden and

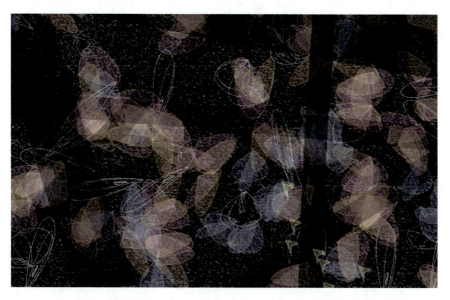

Fig. 4 "Mobes", from the ALICE website (http://alice-interface.eu) showing dynamic, interactive, graphical representations of microbes. (Courtesy of the ALICE consortium (2021))

harvesting the organic produce it provides as you would in an allotment. By creating an opportunity for citizens to share space and find things in common with each other, this platform is a big stepping point towards activating the commons, where the organic matter we call "waste" becomes a shared and flexible resource. Through harnessing those natural processes, which are taking place continually in the soil, this sludge can be turned into useable end products (vegetables, cleaned water, electricity) as well as provide a pleasant public space.

Microbial Sentinels: Keeping Us Safe

One of the intriguing aspects of working with our waste is that microbial technologies can not only make it useful but safe. During the coronavirus pandemic, one way of testing the prevalence of the virus in the broader population was to sample material from the sewers. As you have just seen, microbial technologies are powered by this very waste and, as it were, can keep an "eye on" its composition. Even more importantly, these stable communities of microbes can intervene when needed. This graph shows that when organic waste is recirculated through a microbial fuel cell, it not only produces bioelectricity but also removes pathogens, becoming a first-line, external immune system comprising a breakthrough for the sanitation industry, with very interesting possibilities for how we manage viral circulation during times of pandemic, and in the longer term, these same systems can play an active role in monitoring our health—either at both the individual level and also in the community [25].

Natural Bioremediation in Urban Spaces: Revitalising Brownfield Sites

If we take these microbial technologies out into other parts of the city, we can also strategically use them to remove toxic substances like heavy metals and noxious organic compounds from post-industrial sites. By laying down a membrane and washing the soil in situ, microbes can do the cleaning for us while contributing to the natural environmental healing process. As contamination is just another form of "food" to microbes, we harvest their bioelectricity to power sensors while they metabolise toxins, and using artificial intelligence to observe the cleaning process, we can gather data, link it to other "smart" systems, and so enable the entire bioremediation process to be computationally regulated.

Towards the Off-Grid City: Empowered and Mobile Communities

Ultimately, microbial technologies free us from the static infrastructures of modern cities and the utilities they provide. At a time of climate crisis and the displacement of peoples everywhere—such as during times of severe flooding—having access to clean water, shelter, power, and sanitation can literally save lives. Bill Gates' vision for microbial technologies is to build mobile settlements that are entirely off grid and can house up to 30,000 people—and the underpinning technology Pee Power has demonstrated this potential at Glastonbury for the last 5 years. So, while Gates' vision is specifically for refugees in developing countries, these principles are also transferrable to our Western homes, communities, and commons. Perhaps the most radical step suggested by these microbial technologies is to challenge our basic assumptions about the baselines of what we need to live comfortably and healthily. Imagine the reduced impact of human development if every home that is now connected to a 230 V grid could operate comfortably on a 12 V battery supply. While this would require innovation in some of the things that we do every day that we solve by consuming a lot of energy, like washing machines and fridges, these same tasks could be done differently, such as using advanced new materials to help with refrigeration and finding alternatives to mechanical agitation like ultrasound to carry out this housework. The hardest part to altering our impacts is changing our thinking, our habits, and our concepts of what a "good life" actually entails. Whether we like it, or not, the rules for living on this planet have changed. The good news is that microbial technologies can help us make the necessary adjustments to work within the carrying capacity of our lands, draw on our collective creativity, and help us find much, much ways of better working alongside nature. The bad news is that as yet, nobody has demonstrated at scale that this is possible in Western communities, and so this requires the participation of adventurous and bold pioneers.

Conclusion

Whether we accept it or not, we're all in mourning. For the futures we wanted to leave our children, for the children we opted not to have, for the devastation of countless species, for lost savings, loved ones or homes, which may have already been taken away from us. In such troubling times, liveable ways forwards are needed that are within our power to enact.

By considering the nature of life through a dynamic system that is capable of uniting material and ephemeral realms, electron flow enables the design and implementation of new approaches for addressing our ecological stressed world by making matter livelier. Beyond the conservationist notions of reducing consumption, such necessary activities of survival are transformed into regenerative acts where the materiality of metabolism itself becomes the arbiter of what activities we can perform—previously called our carrying capacity. Setting such natural limits to our daily routines is not about reducing our quality of engagement with the world but establishes new rituals of care where we do not just consume our surroundings but, in every living act, can give something priceless back to our incredible, vibrant world.

Acknowledgements Living Architecture is Funded by the EU Horizon 2020 Future Emerging Technologies Open programme (2016-2019) Grant Agreement 686585 a consortium of six collaborating institutions—Newcastle University, University of Trento, University of the West of England, Spanish National Research Council, Explora Biotech, and Liquifer Systems Group. The Active Living Infrastructure: Controlled Environment (ALICE) project is funded by an EU Innovation Award for the development of a bio-digital "brick" prototype, a collaboration between Newcastle University, Translating Nature, and the University of the West of England (2019–2021) under EU Grant Agreement no. 851246. 13sqm, 999 years (the future belongs to ghosts), installation by Cecile B. Evans and Rachel Armstrong at the Whitechapel Gallery at the group exhibition "Is this Tomorrow?", London (2019), was possible with contributions by bioengineering team: Ioannis Ieropoulos (lead; University of the West of England), Simone Ferracina (University of Edinburgh), Rolf Hughes (Newcastle University), Pierangelo Scravaglieri (Newcastle University), Jiseon You (University of the West of England), Arjuna Mendis (University of the West of England), Tom Hall (University of the West of England), and Patrick Brinson (University of the West of England); microbial fuel cell bioreactor brick installation design: Pierangelo Scravaglieri and Jiseon You, under the guidance of Ioannis Ieropoulos; structure designer: Dominik Arni; structure fabricator: Weber Industries; contributing writer: Amal Khalaf; animator: Tom Kemp; composer: Mati Gavriel; research and production assistance: Anna Clifford; installation team: Richard Hards, Hady Kamar; sponsorship from Personal Improvement Ltd. and Living Architecture (EU Grant Agreement no. 686585); in-kind support provided by Andrew Hesketh; Audioviz (UK FogScreen), the Bristol BioEnergy Centre at the Bristol Robotics Laboratory, and their research into alternative, sustainable sources of power for the home and infrastructure.

References

1. Logan, W. B. (2007). *Dirt: The ecstatic skin of the earth* (p. 3). W. W. Norton & Company.
2. Golding, W. (1954). *Lord of the flies* (p. 161). Faber and Faber.
3. Lewis, C. S. (1961). *The lion the witch and the wardrobe* (p. 148). Geoffrey Bles.
4. McFarland, B. (2016). *A world from dust: How the periodic table shaped life* (p. 3). Oxford University Press.
5. Brahic, C. Meet the electric life forms that live on pure energy. *New Scientist*. July 16 2014 [online]. Available at: https://institutions.newscientist.com/article/dn25894-meet-the-electric-life-forms-that-live-on-pure-energy/
6. Popkin, G. Bacteria use brainlike bursts of electricity to communicate. *Quanta Magazine*, September 5 2017. [Online]. Available at: https://www.quantamagazine.org/bacteria-use-brainlike-bursts-of-electricity-to-communicate-20170905/
7. Garrett, T. J., Grasselli, M., & Keen, S. (2020). Past world economic production constrains current energy demands: Persistent scaling with implications for economic growth and climate change mitigation. *PLoS ONE, 15*(8), e0237672.
8. Newton, I. (2007). Original letter from Isaac Newton to Richard Bentley. *The Newton Project*. [online] http://www.newtonproject.ox.ac.uk/view/texts/normalized/THEM00258
9. Potter, M. C. (1911). Electrical effects accompanying the decomposition of organic compounds. *Proceedings of the Royal Society B, 571*(84), 260–276.
10. Haque, U. The age of shock: Why the coronavirus should be a wake-up call about how our civilisation ends. *Medium, Eudaimonia*, 2 April 2020. [online]. Available at: https://eand.co/the-age-of-shock-81e56c7a7699.
11. Latour, B. (2018). *Down to earth: Politics in the new climactic regime* (p. 8). Polity Press.
12. Heglar, M. A. The big lie we're told about climate change is that it's our own fault. *Vox*, Volume 27 November 2018 [online]. Available at: https://www.vox.com/first-person/2018/10/11/17963772/climate-change-global-warming-natural-disasters
13. Thunberg, G. *Our house is on fire. Greta Thunberg, 16, urges leaders to act on climate*. Edited version of DAVOS speech, p7, The Guardian, 25 January 2019 [online]. https://www.theguardian.com/environment/2019/jan/25/our-house-is-on-fire-greta-thunberg16-urges-leaders-to-act-on-climate
14. Berry, W. (1971). *The unforeseen wilderness: An essay on Kentucky's Red River Gorge* (p. 26). The University Press of Kentucky.
15. Robinson, M. (2018). *Climate justice: Hope, resilience and the fight for a sustainable future.* Bloomsbury Publishing.
16. Heglar, M. A. The big lie we're told about climate change is that it's our own fault. *Vox*, Volume 27 Nov 2018 [online]. Available at: https://www.vox.com/first-person/2018/10/11/17963772/climate-change-global-warming-natural-disasters
17. Editorial, N. (2011). Microbiology by numbers. *Nature Reviews Microbiology, 9*(9), 628.
18. McFarland, B. (2016). *A world from dust: How the periodic table shaped life* (p. 25). Oxford University Press.
19. Denton, M. (1986). *Evolution: Still a theory in crisis*. Discovery Institute Press.
20. Armstrong, R. (2022). *Safe as houses: The more-than-human home*. Lund Humphries.
21. Mott, N. ,*What makes us human? Cooking, study says*. National Geographic. 26 October 2012. [online]. https://www.nationalgeographic.com/news/2012/10/121026-human-cooking-evolution-raw-food-health-science/
22. Negarestani, R. (2008). *Cyclonopedia: Complicity with anonymous materials*. Re.Press.
23. Bock, G. *The history of old stoves. Old house*. 3 Aug 2012. [online]. Available at: https://www.oldhouseonline.com/kitchens-and-baths-articles/history-of-the-kitchen-stove.
24. Atwood, M. (2006). *The tent*. Bloomsbury.
25. Ieropoulos, I., Pasternak, G., & Greenman, J. (2017). Urine disinfection and in situ pathogen killing using a microbial fuel cell cascade system. *PLoS One*. https://doi.org/10.1371/journal.pone.0176475

Brains for Buildings to Achieve Net Zero

Wim Zeiler

People need buildings to protect them against the environmental conditions to be able to work and live in comfortable and healthy indoor air conditions. Architects shaped the built environment since the early beginning of civilization. Building services engineers make it possible to provide comfort and an acceptable indoor air quality for building occupants.

In the last 50 years the world has changed enormously: instead of 3.5 billion there are now living more than 7 billion people on earth with more than 50% in cities with an enormous increased standard of living. Collectively, buildings in the EU are responsible for 40% of our energy consumption and 36% of greenhouse gas emissions, which mainly stem from construction, usage, renovation and demolition [17]. To meet climate goals and comply with the 2030 climate and energy framework, one of the goals is to gain 32.5% improvement in energy efficiency with at present almost 75% of building stock operating inefficiently; see Fig. 1.

There is a need to change the way how architects think about their role in the building design process. We cannot try to solve the problems using the same kind of approach that caused them.

> Until the mechanization of building is in service of creative architects and not creative architecture in service of mechanization we will have no great architecture. –Frank Lloyd [45]

Traditionally a designer of HVAC systems was based on known mechanical systems and techniques. This has consequences for the direction in which architecture has to move towards a more sustainable future, a direction in which technology is used to guide architecture. However, there is a gap between technology and architecture and the research as the architect still takes a major leading role in designing both the

---

W. Zeiler (✉)
Department of the Built Environment, TU Eindhoven, Eindhoven, Netherlands
e-mail: w.zeiler@bwk.tue.nl

© The Author(s), under exclusive license to Springer Nature Switzerland AG 2023
A. Sayigh (ed.), *Towards Net Zero Carbon Emissions in the Building Industry*, Innovative Renewable Energy, https://doi.org/10.1007/978-3-031-15218-4_4

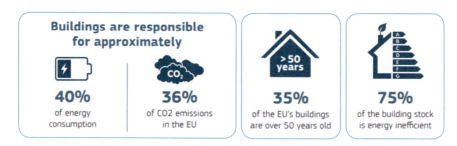

Fig. 1 The current energy situation of the building stock

indoor environment and the energy efficiency of buildings, with the role of the HVAC engineer as a traditional supporting role of the other consulting engineers during the process. The concept, the basic design, is conceived by the architect first, and then there is room for other disciplines. However, the design of a highly sustainable building, due to the increased complexity of building design, inevitably calls for more design collaboration in the conceptual design phase as well. Only the early open collaboration of architects and engineers can facilitate the creation of the necessary new knowledge and solutions beyond the specific scope of each individual discipline [27, 48]. According to the Royal Institute of British Architects (RIBA) president Jane Duncan, architects, engineers and builders must collaborate [12]. To fulfil the demand for zero energy buildings, there is an urgent need for synergy between the architectural and engineering domain. CIBSE, along with the RIBA and other partners, promote more effective assessment of expected and realized energy performance.

A good design is important but also maintaining performance and condition in operation over the years. Most buildings have many problems with comfort and indoor air quality while using much more energy than expected. The maintenance of the installations is more action oriented than performance oriented, which means that the costs are higher and the number of malfunctions and nuisances for the user is higher. Therefore, it is important to detect deviations as soon as possible, so that there is constant analysis of all circumstances and fault detection and diagnosis.

The energy transition requires more optimally functioning installations that use less energy while users want healthier and more productive climate conditions. The complexity of the installations increases sharply and therefore the necessary experience and knowledge to solve problems. However, there is a growing shortage of experienced people who are able to analyse these processes and their data. Therefore, it is becoming increasingly important to develop systems to automate the continuous monitoring, fault detection and diagnostic functions. It is important to improve and safeguard the methods for data analysis and control related to BEMS and measurement and control systems of installations and to develop suitable algorithms based on big data analytics and machine learning.

This chapter shows the way to add more brains to the design of buildings by making use of an integral design approach, as well as show the possibility to incorporate additional artificial brains to buildings to safeguard their optimal operation.

Combined Brains for Building Design: Integral Design

> Architecture will become more informed by the wind, by the sun, by the earth, by the water, and so on. This does not mean that we will not use technology. On the contrary, we will use technology even more because technology is the way to optimize and minimize the use of natural resources. –Richard Rogers

Integral design is a necessity for nature-assisted air conditioning, the basis for net zero buildings, where architect and consulting engineers have to truly collaborate in the conceptual phase of building design process. What is needed is an optimal exchange of interpretations of the design brief as well as an exchange of ideas on possible solutions; see Fig. 2.

Norman Foster and the design board at Foster + Partners are strong supporters of sustainable design and are keen to interpret and integrate engineering principles within design concepts [40]. Their philosophy is that the best projects arise from a totally integrated approach to the design process, where the core disciplines work together to conceive and design a project from its earliest inception [19]. Clearly building design is a team effort, and teamwork is key; therefore, it is necessary to create a place for the needed innovation. The benefits of integrated design are better decisions, higher speed of response, improved ability to iterate and thus reduction of complexity. Early engagement is essential within building design teams. In line with these developments in practice, building design education has moved towards a collaborative practice where designers work in teams [24] and with other disciplines to solve the unstructured problems of design [25]. However, just putting all disciplines together is not enough; there is a clear need for design support to facilitate collaboration between the various design team members from different

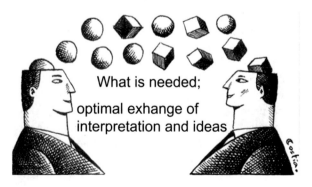

Fig. 2 The needs with the conceptual design phase

disciplines. Design problems are wicked as the information to start with is often very limited and there may be many ways of solving them [24]. This poses difficulties for design teams and highlights the requirement to reach consensus on a variety of matters. Arriving at consensus can be challenging for teams and is affected by cognitive diversity [24].

To cope with this complexity, architects need more support from specialized engineers. The different expertise of engineers must be used more effectively especially in the conceptual design phase to reach for new solutions. This has consequences for the role of the engineers involved; they have to operate early in the conceptual building design process and act more as designers and less as traditional calculating engineers. As a consequence, engineers have to develop new skills. Also the architect has to learn, to not only share his ideas in the conceptual design phase but to really open up his mind and to truly design together with the engineers. Important is that no longer the architect is the one that leads the design process but that the team of architect and engineers leads the design process: designing must become a team effort already in the conceptual phase of design.

Design Methodology

> You never change things by fighting the existing reality. To change something build a new model that makes the existing model obsolete. –Buckminster Fuller

Due to problems resulting from the lack of quality of products and projects, in the early 1960s researchers and practitioners began to investigate new design methods as a way to improve the outcome of design processes [14]. Since then, there has been a period of expansion through the 1990s right up to the present day [3, 31]. Moreover, many of the design methodologies were developed at universities and are rarely applied in industrial applications [15].

In 1999, the professional Dutch organization for architects and consulting engineers together with the University of Technology Delft and the Building Services Society started a research to develop an integral design method to improve the conceptual building design process. Since 2003 this research has continued at the University of Technology Eindhoven and led to a design method based on intensive use of morphological charts [28, 47], and its outcome was evaluated in a situation as close as possible to practice among professionals. The design method has a distinctive feature, the step pattern of activities (generating, synthesizing, selecting and shaping that occurs within the design process; see Fig. 3.

A morphological chart is a kind of matrix with columns and rows which contain the aspects and functions to be fulfilled (see Fig. 4 step 1) and the possible solutions connected to them (see Fig. 4 step 2). These functions and aspects are derived from the program of demands. In principle, overall solutions can be created by combining various sub-solutions to form a complete system solution combination [36]. Morphological chart structures the solution space and encourages creativity.

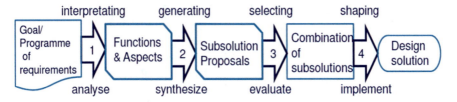

Fig. 3 The four-step pattern of integral design

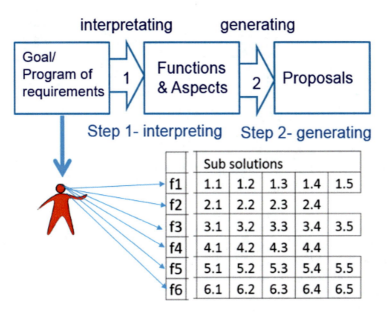

Fig. 4 Concept of a morphological chart

Morphological charts are essentially tools for information processing; it is not confined to technical problems but can also be used in the development of management systems and in other fields [37].

The use of the morphological charts and morphological overview is an excellent way to improve the design process communication procedure. It makes it possible to record information about the solutions for the relevant functions and aids the cognitive process of understanding, sharing and collaboration [38].

In the first step of the integral design method, the individual designer has to make a list of what he thinks are the most important functions that has to be fulfilled based on the design brief. This is derived from their own specialist perspective. The morphological charts are formed as each designer translates the main goals of the design task, derived from the program of demands, into functions and aspects and is then put into the first column of the morphological chart; see Fig. 5, step 3. In the second step of the process, the designers add the possible part solutions to the related rows of the functions/aspects of the first column. Based on the given design task, each

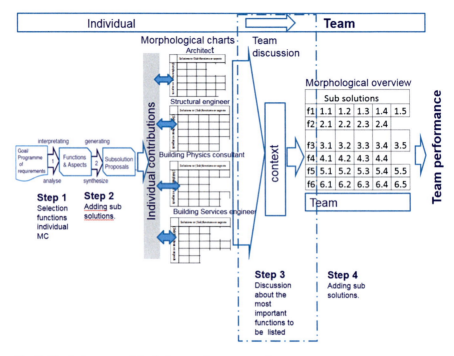

Fig. 5 The design steps of the design team's process cycle

design team member perceives reality due to his/her active perception, memory, knowledge and needs. The morphological charts represent the individual interpretation of reality, leading to active perception, stimulation of memory, activation of knowledge and definition of needs. These individual morphological charts can be combined by the design team to form one morphological overview; see Fig. 5, step 4.

Integral design requires, besides a particular composition of the team, also team building. Only selecting people and putting them together is not enough. It is important to invest sufficient time in the formation of a team. Assuming a minimum quality, the individual qualities of participants are less important than team performance. Usually a team goes through four phases before it is really tuned to each other. [50] has described the development of cooperation within groups. Groups develop themselves in a certain order into a team: forming, storming, norming, performing and adjourning [35]; see also Fig. 6:

- Framing (Orientation Phase) – There is no team spirit yet. Individual positions and roles are not taken yet. Group members take a cautious approach and are in need of guidance. In this first phase, the "forming" phase, you get to know your teammates, and there is an enthusiasm about the new project.
- Storming (Power Phase) – In this phase, the group members try to take their position in the group. This process often leads to conflict when people have different ideas. In the second phase, the "storming" phase, you discover that there is

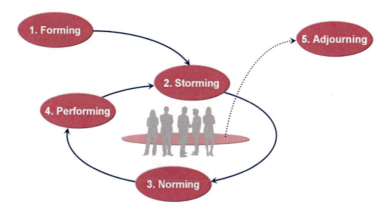

Fig. 6 Stages of team development [35]

difference of opinion in some respects and that not all the tasks and responsibilities fit well with each other.
- Norming (Affection/Standardization Phase) – Group members come closer together. The rules and methods of cooperation are determined. The common team goals are shared and determined.
- Performing (Performance Phase) – In this phase there really is a solid team. Team members complement each other and work together harmoniously to a common team goal. The team is able to work independently.
- Adjourning (Goodbye Phase) – This team is eliminated, and the design task has been carried out. Features of the phase include dissolution, withdrawal and increased independence.
- It is a good model for the promotion of cooperation within a team [51], which can be used to illustrate the steps of morphological approach (see Fig. 7).

Researchers in several disciplines have applied the construct of mental models to understand how designers perform tasks based on their knowledge, experience and expectation [4]. Mental models are often seen as critical indicators of team success [22]. Figure 8 depicts McComb's [33] three-phase convergence process framework indicating a directional mental model convergence process with feedback loops [22]. First, the team members orient themselves by capturing information pertinent to the task. Second, the team members differentiate among the information gathered to discover similarities, differences or irrationalities in their individual approaches. Third, the information becomes integrated into the team members' views: the individuals' internal representations of the design task from an individual perspective change into a team perspective [22]. Each team member can only be analysed from the exchange of communication acts [9, 10]. As we wanted to analyse the process within the design team, we looked for ways to make the communication explicit so that it would be possible to analyse the process. Using the transition from individual morphological charts towards the team's morphological overview enables to illustrate the results of the communication especially during the differentiation and

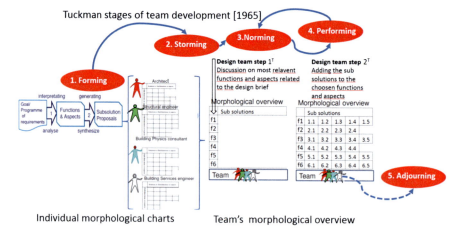

Fig. 7 The second phase generating the morphological overview

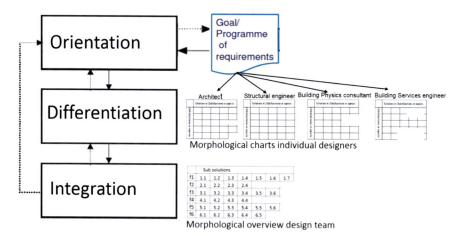

Fig. 8 The process of orientation, differentiation and integration within the process of creating a morphological overview [33]

integration that takes place in the group's discussion to form the morphological overview.

Putting the morphological charts together enables "the individual perspectives from each discipline to be put on the table", which in turn highlights the implications of design choices for each discipline. This approach supports and stimulates the discussion on and the selection of functions and aspects of importance for the specific design task. Important is the keeping of a phase of individual creativity during the morphological chart.

By structuring design (activities) with morphological overviews as the basis for reflection on the design results stimulates communication between design team members and helps the understanding within design teams. It stimulates

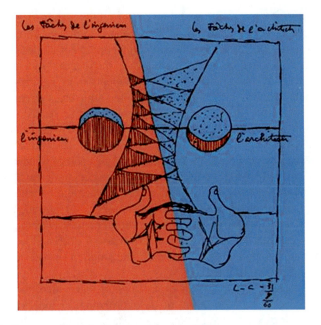

Fig. 9 Symbolic composition of architect and engineer [30]

collaboration as it makes it easier to come forward with new design propositions. Through visualizing the contributions, morphological overviews stimulate the understanding of the different perspectives among design team members; see Fig. 9.

Experiments

> Under the symbolic composition I have placed two clasped hands, the fingers enlaced horizontally, demonstrating the friendly solidarity of both architect and engineer engaged, on the same level, in building the civilization of the machine age. [30]

Since the year 2000 we, together with the Royal Society of Architects (BNA), the Association of Consulting Engineers (NLIngenieurs) and the Society of Building Services Engineers, organized series of workshops in the Netherlands. More than 200 professionals, with at least 12 years of experience, from different professional organizations voluntarily participated in these workshops. After extensively experimenting with different setups for the workshop, a 2-day workshop setting was selected [39]. The 2-days workshop was organized as part of a professional training program for architects and consulting engineers (structural engineers, building services engineers and building physics engineers).

In connection with the integral design research project for professional in the Dutch building industry, we developed an educational project, the master project integral design. The concept of the integral design workshop for professionals was implemented within the start-up workshop of our multidisciplinary masters' project. The different design assignments all were related to the design of zero energy buildings. These complex tasks require early collaboration of all design disciplines involved in the conceptual building design and as such let the students experience the added value of the design method. Master students from architecture, building physics, building services, building technology and structural engineering participated in these projects. The basis of this project, which serves as a learning-by-doing start-up workshop for master students, is a method with extensive use of morphological charts combined to a morphological overview of the design team. The master project integral design was initiated by the chair of building services in the 2005/2006 academic year. During the start-up workshop professionals participated in the student's design teams, and this specific intervention within the design process has been investigated. Having a tested framework for introducing the design method allowed us to investigate the effects of different interventions as well as the analysis of several aspects, such as the effectiveness of different designers or the effect of communication in words or sketches. The program and set-up of the workshop is presented in Fig. 10.

All the assignments had a similar level of complexity which made the results comparable. To investigate the effect of the morphological tools of the integral design approach, they were used in similar workshops setting for different types of students, professionals and practitioners;

- Bachelor Students 2015–2021

The students of the course in which the workshop was held were second and third year bachelor students, age around 20–22, all Dutch. The students were from the Faculty of the Built Environment and of the Faculty of Psychology and Technology.

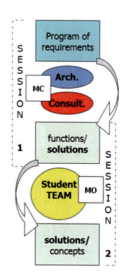

Fig. 10 Program and set-up of start-up workshop

- Master Students 2011–2018

  These were fourth year students (architectural, structural, building physics and building services) all from the Faculty of the Built Environment, age around 22–24.

- Architectural Master Students

  One workshop was held for students of architecture all working in a master thesis project design atelier as part of their MSc graduation project. So they were fifth year students who nearly had finished their studies, age around 23–25. This was the only mono-disciplinary group in the comparison.

- PDEng students 2012–2013

  The students from the postdoctoral engineering (PDEng) program Smart Energy Buildings and Cities (SEB & C) were from all different International MSc discipline backgrounds, age 24–26.

- Professionals 2009

  In the research of Savanovic (2009), the concept of working with morphological overviews was tested in different series of workshops for professionals, with at least 12 years of experience. There were 4 series of workshops with in total 96 participants for testing different set-ups. Here only the results of the fourth workshop are included.

- Professionals 2015

  In 2015, the researchers participated in the start-up of a real professional project for the design of a nearly zero energy building [6]. The professionals had around 20 years of experience.

- Practitioners 2019

  The Dutch Society for Building Services Engineers (TVVL), together with the TU Eindhoven organized a master call. There was no restriction towards the participants, unlike the workshops for professionals in the research of Savanovic (2009) where the participants should have a least 12 years of experience.

Results Design Approach

> Architectural intent, function and structural, services and environmental intent – all need to be brought together at an early stage to achieve the aesthetic, functional and civic integrity of a good building. –Mike [52]

Central element of the integral design process is the use of morphological charts by individual designers which were combined into one morphological overview by the design team. During all experiments the design teams consisted of different disciplines. Unfortunately in the conceptual phase of the design, it is not possible to accurately evaluate the quality of the mentioned functions/aspects or sub-solution. Only a quantitative analysis is possible by counting the number of mentioned functions/aspects and sub-solutions. The number of functions and sub-solutions mentioned by the designers in their morphological charts and the design team's morphological overview were counted. The average increase of functions and solutions as mentioned by the design teams in their morphological charts and morphological overview is represented in Fig. 11.

Results show that the group interaction is of great importance during the conceptual design phase and has a clear positive effect on the number of functions and aspects discussed as well as on the number of generated part solutions. This was founded by the original research with professionals [Savanovic 2009] as well as in the educational setting with different types of students. Given the number of involved design teams in the series of workshops, with more than 300 students and well over hundred professionals as participants, there is a sound quantitative basis for conclusions.

So this is a good way to get the brains of the design team together; however, it does not solve the problem with the operation phase of the building. Fortunately enough the new developments with the application of artificial intelligence, big data analytics and machine learning offer new possibilities to also add more brains to the building to improve their performance in real life.

Artificial Brains for Building Operation: Data Analytics and Machine Learning

> Artificial Intelligence, deep learning, machine learning — whatever you're doing if you don't understand it — learn it. Because otherwise you're going to be a dinosaur within 3 years.– Mark Cuban, American entrepreneur

A good design is important but also maintaining performance and condition in operation over the years. Buildings are not yet capable of controlling their energy-consuming and energy-producing devices to achieve the user's desired comfort or to respond more flexibly to local demand and (sustainable) energy supply.

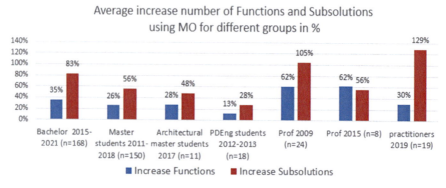

Fig. 11 Average relative increase in the number of functions and sub-solutions in morphological overviews compared to the individual morphological charts

With energy management of technical installations in buildings, substantial savings can be achieved, up to 15–30% [16, 20, 44]. A study conducted in England indicates even that 25–50% savings could be made if the installations all worked without faults or deviations [46]. In reality, buildings use much more energy than assumed during the design. The availability of data from buildings offers new opportunities for improving actual energy performance and closing the so-called energy performance gap. Due to the combination of continuous monitoring, error detection and diagnosis, approximately 25% energy can be saved, which is five times higher than the nationally produced amount of Dutch solar and wind energy in 2016 [11]; see Fig. 12.

However, the main objective of the building services, the perceived comfort of occupants, was not considered. The traditional focus is on the means and not on the goal itself: a healthy, productive and comfortable indoor climate within the buildings. There is a strong relationship between occupants and energy us; see Fig. 13 from IEA Annex 66.

In addition, when using the 4S3F, the number of complaints regarding thermal comfort will decrease sharply, resulting in higher productivity and less absenteeism. This amounts to about 3–5% of labour costs and is actually the biggest saving for building users. Another saving lies in the time saved by automation with building managers and facility managers.

The European Commission has adopted the revised European Energy Performance of Buildings Directive (EPBD III) with the aim of improving the energy efficiency of buildings, thereby reducing energy consumption. The EPBD III prescribes system requirements for improving the energy performance of technical building systems. Utility buildings with heating or air conditioning systems with a power of more than 290 kW must be equipped with a building automation and control system (GACS) from 2026. These systems must be capable of:

- Continuously monitoring, tracking, analysing and adjusting energy consumption.

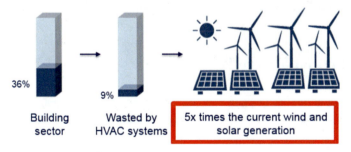

Fig. 12 The benefits of reducing the energy inefficiency of operating buildings

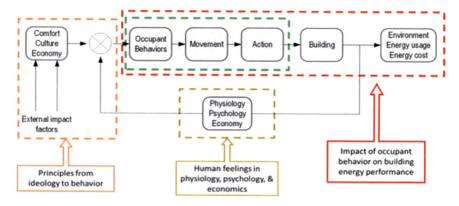

Fig. 13 Relationship between occupants and building's energy usage (IEA Annex 66)

- Assessing the energy efficiency of the building, detecting efficiency losses of technical building systems and informing the manager of the facilities or technical installations about the possibilities to improve this.

The current building management systems cannot comply with this. Data is produced and (sometimes) shown in graphs, but analysis thereof is missing and is not automated; interfaces to inform and support the administrator in his decisions are very limited.

The maintenance of the installations is more action oriented than performance oriented, which means that the costs are higher and the number of malfunctions and nuisances for the user higher. Therefore, it is important to detect deviations as soon as possible with continuous commissioning, so that there is constant monitoring of all circumstances and error detection in combination with a diagnosis.

There are four main methods [8] for energy performance evaluation of climate installations: engineering calculations, simulation, statistical methods and machine learning. The theoretical possibilities of data analysis and machine learning are promising, but in the built environment they have not yet been made sufficiently applicable for the specific requirements of the domain. Although several studies have been carried out into the causes of energy gaps, a structured analysis is still lacking to deal with the enormous amount of data which can be processed leading to the detection and diagnosis of a specific energy gap.

Fault Detection and Diagnosis

In the categorization of fault detection and diagnosis methods (see Fig. 14) (Zhao, 2019), there are two main categories: knowledge-based methods and data-driven methods.

It is more promising to apply combinations of both methods, combining data-driven techniques with physical models, which are in line with knowledge and practice of installation designers and operators.

There is an urgent need for robust energy management methods in practice for continuous remote management of climate installations in terms of energy efficiency, comfort and performance contracts. The main objective of this work package is to develop the necessary insights regarding sensors, data interpretation, trends signalling, continuous monitoring, error detection, energy diagnosis and predictive maintenance. This is in such a way that it can be converted into practical concepts for product development within the installation industry: this means a modular and scalable approach, so not everything at the same time but focused on the most important parts.

Although much of a promise, the reality of data analysis is that about 80% of the time is needed for cleaning up and organizing the data [23]. So it's a relatively inefficient process. In addition, 80% of waste is caused by only 20% of components. To increase the efficiently of the process, two promising methods are being tested and further developed to use building data more efficiently and effectively: the 4S3F framework and Pareto-Lean analyses. The first is promising because it is based on installation technical knowledge (and BEMS data, of course) and the second because it comes from powerful tools for improving processes. In addition, different methods and tools are used for data analysis for error detection and diagnosis. The results will be add-on's modules on top of existing BEMS.

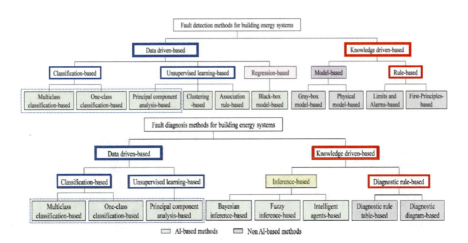

Fig. 14 Classification of faultdetection methods for buildings [49]

Automated Fault Detection by Diagnostic Bayesian Networks 4S3F Method

Automated fault detection greatly increases both indoor environmental quality (IEQ) and energy performance. For this purpose, a generic error detection method has been developed, the 4S3F method that works based on diagnostic Bayesian networks (DBN) with three generic types of errors (component, control and model errors) and four generic types of symptoms (balance, energy performance, operational state and additional symptoms) [41, 42]. Symptoms and errors are linked once in a Bayesian network set-up based on the principle schemes of a specific installation. Figure 15 shows relationships between error and symptom types in these four types of symptom and three types of errors (4S3F).

The application of the 4S3F framework on sensor faults is strongly based on system engineering for both detection and diagnosis of faults, applying energy, mass and pressure conservation laws for detection purposes applied to both aggregated systems and subsystems based on the process and information diagram (P & ID). The symptom detection part tested in a case study demonstrated that the framework can identify symptoms at multiple system and subsystem levels. Sensors are the core of an FDD system, and therefore it is essential to be able to automatically diagnose sensor faults. By considering hard sensors as components of the systems and soft sensors as models, they can be integrated into the 4S3F method. The fault identification part of the 4S3F framework is a DBN that interprets the symptoms to identify the faults. In a DBN, posterior probabilities of faults are estimated from the results of the detection. The DBN must be set up conforming to the P & ID, i.e. using the same structured standard DBN models for common components, controls and models are developed and can be combined in order to represent the complete HVAC system. The Bayesian network is continuously powered by data from the BEMS and is therefore able to continuously identify errors. The 4S3F system is therefore plugged into the BEMS. It is important to improve and safeguard the

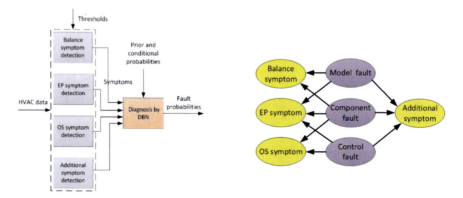

Fig. 15 The 4S3F model: relationships between error and symptom types and the 4S3F structure [41]

methods for data analysis and control related to BEMS and measurement and control systems of installations and to develop suitable algorithms.

This method has been specifically designed to connect to the design process of installations and to the knowledge and skills of installation designers and building managers. It uses the knowledge embedded in basic schemes, unfolded in a DBN, which performs automated analysis based on BEMS data and machine learning and rule-based AI methods. Deviations in an energy, mass or pressure balance for a system, deviations from a performance factor (e.g. coefficient of performance or efficiency) or deviations from state value (e.g. temperature, flow rate, pressure, on-off status of a component) are automatically detected by the status values measured by the BEMS. Additional symptoms based, for example, on inspection or maintenance information or on specific fault detection methods of HVAC components can be added if necessary.

From the measured symptoms, the errors that cause these symptoms are identified. There may be errors in the missing data models, the so-called soft sensors. There may also be HVAC components and systems that do not function properly, for example, too low installed capacity or too low efficiency due to aging or because it is defective. The latter type concerns errors in the control of the HVAC components and the system, for example, the control of supply temperatures and the on-off strategy of components such as the control of the order of energy generators.

Pareto-Lean Analyses

In the current situation the BEMS can generate enormous amount of data. Some people speak of the new gold as they image that you can do smart things with it and earn much money. Unfortunately in reality the data mining process (see Fig. 16) is difficult and time-consuming to really find the gold, the useful knowledge, between all the data. It is just like gold digging in the past; it is hard work, and nothing is got for free. In data analysis, data preparation is known to take up to 80% or even more of the project time [23], and so a small portion is left on the analysis and optimization itself. So instead of starting collecting all data, it is important to think which data you really need, which parameters are the most important and during which periods of the year.

The Pareto analysis, also known as the 80/20 rule, assumes that the majority of problems (80%) can be identified by a few major causes (20%), or 80% of the problems can be solved with 20% of the effort. This analysis method is often used in decision-making issues or in solving complex problems in, for example, industrial engineering. An example of a proven applicability of the Pareto analysis is found in a lamp production process where the stem making is responsible for more than 87% of the total defects [2]. The Pareto analysis is a practical way of identifying causes of problems, and it encourages analysing and organizing. It is proven to be a successful systematic approach in, for example, economic aspects [21]. Figure 17 illustrates this Pareto analysis, in which the required effort (causes) is plotted against the

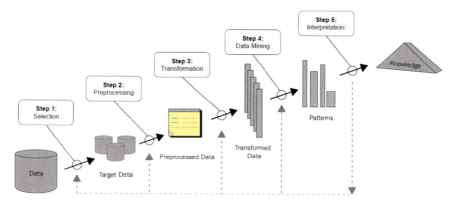

Fig. 16 Data mining process to come from data to useful knowledge. (Modified from [29])

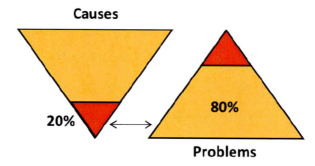

Fig. 17 Methodology based on the Pareto analysis. Majority of problems (80%) identified by a few major causes (20%) [5]

solutions (problems) [5]. According to this definition of the Pareto analysis, the hypothesis has been formed that the Pareto analysis was also applicable to energy reduction. The Pareto analysis identifies the few causes that result in the majority of energy consumption problems. The analysis identifies problems and rates the influencing parameters, resulting in the most important parameter to focus on first. "It is normally easier to reduce a tall bar by half than to reduce a short bar to zero. Significantly reduce one big problem, and then hop to the next", as cited by [7]. Figure 18 illustrates a Pareto diagram.

It is therefore important to be focused and try to make a selection of the most important aspects and associated datasets in order to discover the deviations and causes. Especially useful are the Pareto analysis (Barlett, 2015) and the LEAN energy analysis (L.E.A.) [1, 26]. These are intended to focus on the data to be used so that not all data needs to be cleaned up and used which would take a lot of time. The combination of both methods can also be used to carry out that analysis [18], and thus the strengths of both methods can be used. Both methods for energy analysis, pareto analysis and LEAN, can be combined into an eight-step approach [13]:

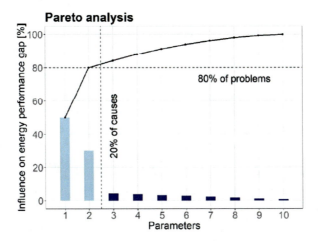

Fig. 18 Pareto diagram projected on the energy problem [54])

1. Pareto analysis, step 1: Identification of problems. The building is modelled and energy performance simulated in MATLAB or EnergyPlus and compared to the measured energy consumption, to identify the energy gap.
2. Pareto analysis, step 2: Identify the underlying causes of each problem. Although all institutions, control strategies and properties in buildings and building systems together determine the behaviour of energy consumption, they are reduced to a small amount of identified "main parameters".
3. Pareto analysis, steps 3/4/5: Rank, score and group problems and causes. The impact of the parameters on annual energy consumption is assessed on the basis of a sensitivity analysis to determine which ones are most important in assessing energy performance gaps.
4. Pareto analysis, step 6: Evaluation of the energy gap by critical parameters. This step investigates whether the energy performance differences in the case studies can be assessed/explained by the selected parameters from the steps before.
5. LEAN, step 1: Collect weather and energy usage data. Collect the required measurement data for a linear comparison with simulated energy consumption.
6. LEAN, step 2: Create basic/benchmark models. Identification of characteristic correlations in energy performance of the case study building and the creation of benchmark models, which can be used to assess measured energy efficiency.
7. LEAN, step 3: Identify energy gaps with regression coefficients of benchmark regression models. The identification of the energy gap consists of assessment with coefficients of multi-parameter regression models.
8. LEAN energy analysis, step 4: Assessment of the remaining energy gaps (which could not be explained by the Pareto analysis). The results are used to assess the energy performance differences, in addition to the previous results of the Pareto analysis.

Through (smart) data analysis of sensor or energy data from main meters or coarse submeters, the problem of the performance gap of building operations can be determined by the analysis of the data. This can then be applied to a large number of buildings. This involves the use of enormous amounts of data; however, the application of Pareto-LEAN method can reduce the number of data points required. This will increase the efficiency and effectiveness. Using the data gives the possibility to add brains to the building and thus let them perform much more efficient than currently possible. In this way, it is possible to focus on those data from the entire mountain of data, which is really the most important to detect deviations and identify causes. This allows targeted choice of data and specific datasets to be used for the machine learning algorithms. This greatly increases efficiency and effectiveness by only having to work on data with a limited learning set. However, obtaining accurate energy predictions is still difficult due to the changes in weather and user behaviour. Data mining methods can provide a better understanding of the interactions between user behaviour and future energy needs that have significant impact asset management and the utilization of energy flexibility. But this requires a good and effective approach. A preliminary study examined a methodology for the identification, characterization and evaluation of robust typical energy consumption patterns. A cross-validation has been created to illustrate the added value of the evaluation using variable distance metrics and clustering algorithms. By using identified and characterized patterns, reliable evaluation of the energy use forecast can be made [32].

For the purpose of energy reduction of the built environment, this task is based on analysis using (physical and/or/or/plus data-driven) models at building and main component level, in addition to tightening rule-based controls at component level. This involves a combination of top-down and bottom-up with a middle out approach; see Fig. 19. This approach fits in well with the 4S3F approach through the classification into systems, subsystems and components, and synergy will therefore be sought.

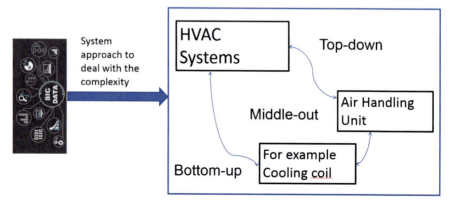

Fig. 19 Combined approach to top-down, bottom-up and middle out

Towards Machine Learning

Models of the building and building services are also used. Input for the model is the actual behaviour of users, for example, control lighting, presence or thermostat settings. The models can partly be implemented as a virtual sensor for parameters that are also measured in reality, but for which there is often no sensor in practice. The difference between parameter values measured in reality and the parameter values calculated by the model (e.g. energy consumption, indoor temperature or cooling capacity) serves as error indicators.

Through (smart) data analysis of energy data from main meters or coarse submeters, a large part of the problem or impact of sustainability measures can be determined by the analysis of the data. Due to the amount of sensors/data required, this can then be applied to a large number of buildings. This involves the application of Pareto-LEAN method to reduce the number of sensors required.

A number of machine learning algorithms were evaluated to predict electricity demand at individual construction level in hour intervals [43]. Hourly prediction is important to understand short-term dynamics, but most studies are limited to annual, monthly, weekly or daily data resolutions. Two years of data was used in training the model, and the prediction was carried out using another year of untrained data. Learning algorithms such as boosted-tree, random forest, SVM linear, quadratic, cubic, fine-Gaussian as well as ANN were analysed and tested to predict the electricity demand of individual buildings; see Fig. 20. The results showed that boosted-tree, random forest and ANN are the best results for hourly prediction when comparing algorithms based on computer time and error accuracy [43].

These results are based on the total energy consumption; it is important to now look at the energy use of specific user groups of electricity in a building, as well as also to look at gas use or possibly district heating.

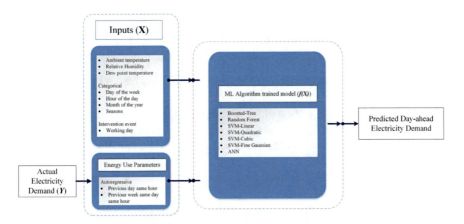

Fig. 20 Development opportunities data based model [43]

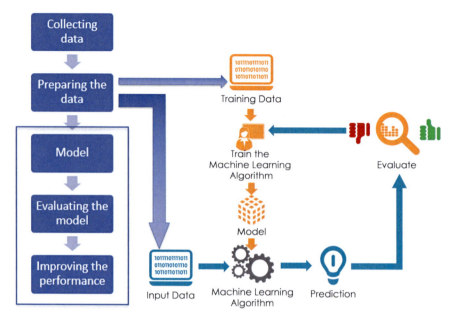

Fig. 21 Principle application machine learning

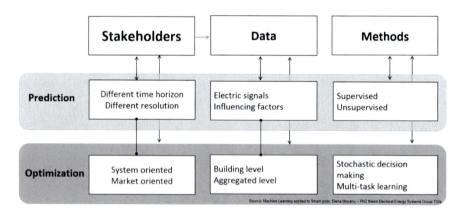

Fig. 22 Relationship between stakeholders, data and machine learning methods

Although only the application of specific machine learning algorithms is promising in theory, in practice it turns out that using all the data leads to a huge amount of work. The data collection and the preparation of the data are time-intensive; see principle design application machine learning, Fig. 21.

An important innovative element is the linking of prediction and optimization of the machine learning method that are applied. Depending on the specific interest of the stakeholder (user, administrator, installer, etc.), the most suitable method or combination of methods is used. Research has shown that for the optimization of the process and the reliability of the prediction, a combination often gives the best result [34]; see Fig. 22.

Discussion

The proposed design method had a major positive effect on the number of proposed sub-solutions and also on the amount of functions and aspects considered in the conceptual phase of the design process by the design team members. This indicates that the effectiveness and productivity of design teams were largely improved by adding structure to the process. As such integral design is a necessity for nature-assisted HVAC systems for net zero energy buildings where architect and consulting engineers have to truly collaborate and make optimally use of their combined brains in the conceptual phase of building design process.

The energy transition requires more optimally functioning installations that use less energy. Users want healthier and more productive climate conditions. The complexity of the installations increases sharply and therefore the necessary experience and knowledge to solve problems. There is a growing shortage of experienced people who are able to analyse this data. Therefore, it is becoming increasingly important to develop systems to automate these continuous monitoring, error detection and diagnostic functions.

Conclusions

To achieve net zero buildings in practice, we need to use all the brains of the stake holders involved especially architects and engineers. In chapter "Thermal Behavior of Exterior Coating Texture and Its Effect on Building Thermal Performance", different and complementary approaches are presented, one aimed at het design phase and the other at the operation phase:

A break with the traditional line of thoughts of architects as well as consulting engineers is therefore needed. A new design model, integral design, was developed to support interaction between all the disciplines involved in the conceptual building design process by structuring the communication and solution generation process in steps. By structuring the information flow about the tasks and solutions of the other disciplines, the method forms a design within the design process and enables a structured approach even in the conceptual design phase. The use of the morphological overview based on the individual morphological charts creates a way to share interpretations and ideas for solutions forming a basis for synergy leading to more and innovative designs; see Fig. 23.

A new design model, integral design, was developed to support interaction between all the disciplines involved in the conceptual building design process by structuring the communication and solution generation process in steps, thus stimulating a break with the traditional line of thoughts of architects as well as consulting engineers. By structuring the information flow about the tasks and solutions of the other disciplines enables a structured approach even in the conceptual design phase. This would really activate more brains to the building design. As such it is a good

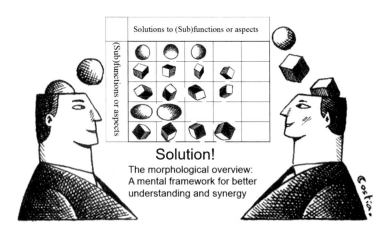

Fig. 23 The morphological overview to connect the minds of the design team

method for supporting architects and engineers in the highly complex tasks of designing sustainable net zero or even plus energy buildings.

The application of data analysis using machine learning for continuous monitoring is new in this area, as is the application of machine learning combined with models for error detection and diagnosis. It is essential to find the right techniques for the applications in this specific domain of HVAC, e.g. filters with their resulting indoor air quality level or energy wastage due to shifted setpoint. The modular approach to the development of a continuous monitoring system through data analysis by machine learning and supported by thinking methods such as 4S3F or Pareto-LEAN offers new possibilities to solve some of the problems that have so far proved too complex.

The modular approach to developing a fault detection and diagnosis system through data analysis and machine learning offers new possibilities to actually bring multiple perspectives and focus together. The result is an approach to automatically and continuously determine the performance of climate installations in buildings, to identify undesirable deviations or trends and to diagnose them. This leads to more effective and efficient maintenance and management of the installation (also lower costs) as well as to a long-term lower energy consumption, better experienced comfort and indoor air quality. This is important for the operator of the building (owner or a third party responsible).

In addition to being a better guarantee of comfort, more reliable systems are also a better guarantee of balance. A big issue with flexibility is the reliability of delivery; often flexibility is offered in advance. This flexibility can be crucial for, for example, a grid operator. If at the moment it is not possible to deliver due to malfunctions or deteriorated performance, problems may arise for the grid operator, a risk for the provider. If it is clear in advance that an error is coming, maintenance can be planned, and flexibility may not be offered. This can even prevent problems in the energy grid in the long term.

The combination of the approaches presented in this chapter offers a real possibility to improve the current situation of the built environment and will make it possible to make an important step to reach for truly net zero buildings.

References

1. Abels, B., Server, F., Kissock, K., & Ayele, D. (2011). *Understanding industrial energy use through Lean energy analysis, mechanical and aerospace engineering faculty publications* (Paper 164). http://ecommons.udayton.edu/mee_fac_pub/164
2. Ahmed, M., & Ahmed, N. (2011). An application of Pareto analysis and Cause-and-Effect diagram (CED) for minimizing rejection of raw materials in lamp production process. *CS Canada Management Science and Engineering, 5*(3), 87–95.
3. Atkinson, H., & Opperheimer, M. R. (2016). Design research – History, theory, practice: Histories for future-focussed thinking. *Proceedings DRS 2016*. Brighton.
4. Badke-Schaub, P., Neumann, A., Lauche, K., & Mohammed, S. (2007). Mental models in design teams: A valid approach to performance in design collaboration? *CoDesign, 3*(1), 5–20.
5. Bartlett B.M., (2015). Powerful Strategic Planning Rules That Could Revolutionize Your Business And Life, Available at: www.benmbartlett.com
6. Bartlett, B. M. (2013). Here's a powerful strategic analysis and decision making tool. *High Performance Strategy*. www.benmbartlett.com
7. de Bont, K., Zeiler, W., & van der Velden J. (2016). Integral design method to support nZEB design: A real project experiment. In *Proceedings Clima 2016*, Aalborg.
8. Bonacorsi, S. (2014). *Step-by-step guide to using Pareto analysis*. Process Excellence Tools process excellence network.
9. Borgstein, E. H., Lamberts, R., & Hensen, J. L. M. (2016). Evaluating energy performance in non-domestic buildings: A review. *Energy and Buildings, 128*, 734–755.
10. Casakin, H., & Badke-Schaub, P. (2013). Measuring sharedness of mental models in architectural and engineering design teams. In *Proceedings ICED13*, Seoul.
11. Casakin, H., & Badke-Schaub, P. (2014). Mental models and creativity in engineering and architectural design teams. In *Proceedings design computing and cognition*, University College London.
12. CBS. (2017). Aandeel hernieuwbare energie 5,9 procent in 2016. https://www.cbs.nl/nl-nl/nieuws/2017/22/aandeel-hernieuwbare-energie-5-9-procent-in-2016
13. CIBSE. (2016, January). RIBA president wants 'frank debate' on supply chain conflict. *CIBSE Journal*.
14. Corten K., Willems E., Walker S., Zeiler W., (2019), Energy performance optimization of buildings using data mining techniques, Proceedings Clima, Bucharest, Romania.
15. Corten, K., & Zeiler, W. (2020). The Pareto LEAN energy analysis for identification of energy performance gaps. In *CIBSE ASRAE Technical Symposium*, September virtually held.
16. Cross, N. (2007). Editorial forty years of design research. *Design Studies, 28*(1), 1–4.
17. Davies, M. (2014). I cannot think of a better career than services engineering, CIBSE Journal April, https://www.cibsejournal.com/general/i-cannot-think-of-a-better-career-than-servicesengineering/.
18. Dorst, K. (2016). Design practice and design research: finally together? In *Proceedings DRS 2016*, Brighton.
19. van Dronkelaar, C., Dowson, M., Spataru, C., & Mumovic, D. (2016). A review of the regulatory energy performance gap and its underlying causes in non-domestic buildings. *Frontiers in Mechanical Engineering, 1*, 1–14.
20. EU. (2020). *Energy efficiency in buildings*. https://ec.europa.eu/info/news/focus-energy-efficiency-buildings-2020-feb-17_en

21. Huls, A. J., Lops, B., & Zeiler, W. (2018). Time series analysis for re-commissioning of building service installations. In *Proceedings International Conference on Time Series and Forecasting*, September 19–21, Granada.
22. Jackon, A., & Heywood, M. (2019, September 18). Development of integrated design. *CIBSE-ASHRAE Group seminar and webinar, Foster+Partners*, London.
23. Jing, R., Wang, M., Zhang, R., Li, N., & Zhao, Y. (2017). A study on energy performance of 30 commercial office buildings in Hong Kong. *Energy and Buildings, 144*, 117–128. https://doi.org/10.1016/j.enbuild.2017.03.042
24. Juran, J. M. (1992). *The new steps for planning quality into goods and services. Quality by design*. Simon and Schuster.
25. Kennedy, D. M., & McComb, S. A. (2010). Merging internal and external processes: Examining the mental model convergence process through team communication. *Theoretical Issues in Ergonomics Science, 11*(4), 340–358.
26. Keogh, E., & Kasetty, S. (2004). *On the need for time series data mining benchmarks: A survey and empirical demonstration*. https://www.cs.ucr.edu/~eamonn/Data_Mining_Journal_Keogh.pdf
27. Kiernan, L., Ledwith, A., & Lynch, R. (2017). How design education can support collaboration in teams. In *Proceedings E & PDE*, Oslo.
28. Kiernan, L., Ledwith, A., & Lynch, R. (2019). Comparing the dialogue of experts and novices in interdisciplinary teams to inform design education. *International Journal of Technology and Design Education*, published online January 23th. https://doi.org/10.1007/s10798-019-089495-8
29. Kissock, K., Abels, B., Sever, F., & Ayele, D. (2015). *Understanding industrial energy use through Lean energy analysis*. Department of Mechanical and Aerospace Engineering, University of Dayton.
30. Kovacic, I., & Filzmoser, M. (2014). Designing and evaluation procedures for interdisciplinary BIM use – An explorative study. In *Engineering Project Organization Conference Devil's Thumb Ranch*, Colorado July 29–31.
31. van den Kroonenberg, H. H.. (1988). Stimulating creativity and innovations by methodical design, In *Creativity and innovation: Towards a European network*. Springer Netherlands.
32. Lara, J. A., Lizcano, D., Martínez, M. A., & Pazos, J. (2014). Data preparation for KDD through automatic reasoning based on description logic. *Information Systems, 44*, 54–72.
33. Le Corbusier. (1960). Science et Vie.
34. Le Masson, P., Hatchuel, A., & Weil, B. (2012). How design theories support creativity – An historical perspective. In *Proceedings 2nd International Conference on Design Creativity*, ICDC2012, Glasgow.
35. Leprince, J., & Zeiler, W. (2020). A robust building energy pattern mining method and its application to demand forecasting. In *Proceedings 3rd International Conference on Smart Energy Systems and Technologies (SEST 2020)*, 7–9 September, held virtually.
36. McComb, S. (2007). Mental model convergence: The shift from an individual to being a team member. *Research in Multi-Level Issues, 6*, 95–147.
37. Mocanu, E. (2017). *Machine learning applied to smart grids* (PhD thesis). TU Eindhoven, Eindhoven.
38. Nieuwenhuis, M. A. (2010). The Art of Management. http://123management.nl/0/030_cultuur/a300_cultuur_11a_teamontwikkeling_tuckman.html
39. Nieuwenhuis, M. A. (2012). *The art of management 2003-2010*. http://123management.nl/0/030_cultuur/a300_cultuur_11a_teamontwikkeling_tuckman.html
40. Ölvander, J., Lundén, B., & Gavel, H. (2008). A computerized optimization framework for the morphological matrix applied to aircraft conceptual design. *Computer Aided Design, 41*(2), 187–196.
41. Pahl, G., Beitz, W., Feldhusen, J., & Grote, K. H. (2006). *Engineering design, a systematic approach* (3th ed.), Ken Wallace, K., & Blessing, L., translators. Springer.

42. Ritchey, T. (2010). *Wicked problems social messes, decision support modelling with morphological analysis*. Swedish Morphological Society.
43. Savanovic, P. (2009). Integral design method in the context of sustainable building design, PhD thesis, TU Eindhoven, Eindhoven, Netherlands.
44. Smith, A. (2019, September). Piers' review, integrated design at Foster+Partners. *CIBSE Journal*, 40–42. https://www.cibsejournal.com/general/piers-review-integrated-design-at-foster-partners/
45. Taal, A. C. (2021). *A new approach to automated energy performance and fault detection and diagnosis of HVAC systems, Development of the 4S3F method*. PhD Thesis TU Eindhoven, Bouwstenen 325, Eindhoven.
46. Taal, A. C., Itard, L. C. M., & Zeiler, W. (2018). A reference architecture for the integration of automated energy performance fault diagnosis into HVAC systems. *Energy and Buildings, 179*, 144–155.
47. Tuckman, B. (1965). Developmental sequence in small groups. *Psychological Bulletin, 63*(6), 384–399
48. Walker, S., Khan, W., Katic, K., Maassen, W., & Zeiler, W. (2020). Accuracy of different machine learning algorithms and added-value of predicting aggregated-level energy performance of commercial buildings. *Energy and Buildings, 209*, 1–14.
49. de Wilde, P. (2014). The gap between predicted and measured energy performance of buildings: A framework for investigation. *Automation in Construction, 41*, 40–49.
50. Wright, F. L. (1953, May). The language of organic architecture. *The Magazine of Building, Architectural Forum*, 106–107.
51. Yu, Y., Woradechjumroen, D., & Yu, D. (2014). A review of fault detection and diagnosis ethodologies on air-handling units. *Energy and Buildings, 82*, 550–562. https://doi.org/10.1016/j.enbuild.2014.06.042
52. Zeiler, W. (2017). Morphology in conceptual building design. *Technological Forecasting and Social Change*. https://doi.org/10.1016/j.techfore.2017.06.012
53. Zeiler, W. (2019). Methodological design: Effects of a morphological approach for different students and professionals. In *Proceedings ICED19, International Conference on Engineering Design*, 5-8 August, Delft.
54. Zhao, Y., Li, T., Zhang, X., & Zhang, C. (2019). Artificial intelligence-based fault detection and diagnosis methods for building energy systems: Advantages, challenges and the future. *Renewable and Sustainable Energy Reviews, 109*, 85–101.

Using Building Integrated Photovoltaic Thermal (BIPV/T) Systems to Achieve Net Zero Goal: Current Trends and Future Perspectives

Ali Sohani, Cristina Cornaro, Mohammad Hassan Shahverdian, Saman Samiezadeh, Siamak Hoseinzadeh, Alireza Dehghani-Sanij, Marco Pierro, and David Moser

Introduction

According to some estimates, by 2035, the global energy consumption rate will have undergone a 50% rise from 1990 levels. This predicted rise will stem from large increases in population growth and urbanization and will influence the energy consumption of buildings. Today, buildings account for 40% of global energy consumption [1–3], with nonrenewable energy being consumed for their cooling, heating, and lighting [4].

In large cities, for instance, Tokyo, San Francisco, Hong Kong, and New York, buildings account for significantly greater greenhouse gas (GHG) emission and energy consumption than transportation does [5]. In response, global roadmaps are

---

A. Sohani · C. Cornaro
Department of Enterprise Engineering, University of Rome Tor Vergata, Rome, Italy
e-mail: ali.sohani@uniroma2.it; cornaro@uniroma2.it

M. H. Shahverdian · S. Samiezadeh
Optimization of Energy Systems' Installations Lab., Faculty of Mechanical Engineering-Energy Division, K.N. Toosi University of Technology, Tehran, Iran

S. Hoseinzadeh
Department of Planning, Design, and Technology of Architecture, Sapienza University of Rome, Rome, Italy
e-mail: siamak.hosseinzadeh@uniroma1.it

A. Dehghani-Sanij (✉)
Waterloo Institute for Sustainable Energy (WISE), University of Waterloo, Waterloo, ON, Canada
e-mail: a7dehgha@uwaterloo.ca; alireza.dehghanisanij@uwaterloo.ca; ads485@mun.ca

M. Pierro · D. Moser
EURAC Research, Bolzano, Italy
e-mail: marco.pierro@eurac.edu; david.moser@eurac.edu

seeking to replace the fossil fuels used to power buildings with renewable, clean energy resources, with the goal of changing buildings with high energy consumption into net zero energy ones [6, 7]. Therefore, decarbonization has become a great environmental priority, and new buildings designed to be net zero are a focus in the European Union (EU). The UK is hoping to achieve an 80% decline in national emissions by 2050 and then to modify the plan for zero emission systems [8]. Japan intends for its public and residential buildings to be net zero energy buildings (NZEBs) by 2020 and 2030, respectively [9]. Furthermore, after 2030, new US commercial buildings will be required to be net zero. Meeting these goals will require the reduction of primary building energy consumption through energy-efficient envelopes [10].

Photovoltaic (PV) systems are a promising alternative for harnessing clean, inexhaustible, and abundant solar power for the generation of environmentally benign energy [11]. In Italy, attempts are being made to achieve the goal of using 55% renewable energy resources for the country's power supply. The power production of PV technologies is planned to increase from 25 TWh per year at the present time to 72 TWh per year in 2030 [12]. Moreover, German PV installations have reached a total energy capacity of 50 GW and are expected to generate up to 413 GW by 2050 [13]. The global market for PV solutions is also expected to enjoy a 1.7% growth per year, suggesting a rise from approximately 42,000 million USD in 2019 to around 47,000 million USD in 2024 [14]. The permanent load of a building, including concrete rooftops and walls, can be replaced with PV systems, leading to fossil energy-free buildings and a pollution-free environment [14].

Building integrated photovoltaics (BIPVs) are solar-generating components that can be used to replace traditional construction materials and envelopes (e.g., shading, atria, window, and roof components) with PV and so support clean power generation. Hence, BIPV provides a building envelope and supplies electricity. New buildings can be constructed using BIPV, and existing buildings can have it retrofitted. In light of its dual functionality, BIPV is an efficient and effective approach to decreasing construction labor and material costs. In addition, BIPV helps maintain the decorative features of buildings. Thus, the implementation of BIPV has grown significantly in recent years. Semitransparent BIPV structures (e.g., glass-on-glass) have become an appealing choice for architects as they can provide an excellent exterior appearance while allowing daylight into buildings and controlling solar gain. They are also appealing for facade glazing [15]. In a BIPV, the system produces electricity, while no heat from the panel is recovered. When there is heat recovery from the panel, the system is called BIPV/T [16]. BIPV/T technologies have better energy efficiency in comparison to BIPV systems. Furthermore, due to saving more fossil fuels, they enjoy the better environmental performance as well [17].

Having briefly introduced BIPV/T systems, this chapter next discusses their working principles, their use in different parts of a building, and mathematical modeling of BIPV/T units in both free and forced convection conditions. The net zero goal is described as is the role of BIPV/T systems in reaching this goal.

BIPV/T Systems: How They Work

A BIPV/T system is schematically depicted in Fig. 1. In such a system, on the one hand, a portion of the radiation gained from the sun by PV supports electricity generation. On the other hand, air flows through the channel between the PV and building wall and absorbs a part of the irradiance that is not transformed into electricity. The inlet air can be totally supplied from the atmosphere, or it be a mixture of rooms' return air and ambient air. The outlet air can also be utilized for space heating in cold seasons and desiccant regeneration in hot seasons. These options mean energy, and consequently money, is saved, and fossil fuels need not be consumed for space heating or desiccant regeneration, so GHG emissions are negligible [18].

In addition to installation on the sidewalls of buildings, BIPV/T systems can be installed on roofs [19]. Over time, BIPV/T systems have gradually come to be used in other parts of buildings as well. In fact, it is now possible to use these systems in all parts of buildings' exterior shells, such as roofs, facades, walls, skylights, or special structures such as ledges and awnings [16]. Therefore, it is not too unlikely that BIPV/T systems will soon reach cover the entire exterior of the buildings and can be designed to be architecturally attractive as well as functional [20]. Moreover, PV cells can be employed in the form of dye-sensitized solar cells, offering the opportunity of using them in windows [21].

Mathematical Modeling

This section focuses on system modeling. A BIPV/T system, in addition to the air duct behind the panels, consists of five layers: glass, top EVA (ethylene vinyl acetate), silicon, bottom EVA, and Tedlar, which are further modeled. The sketch of

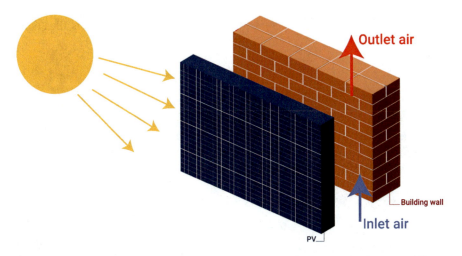

Fig. 1 Working principle of a building integrated photovoltaic thermal (BIPV/T) system [17]

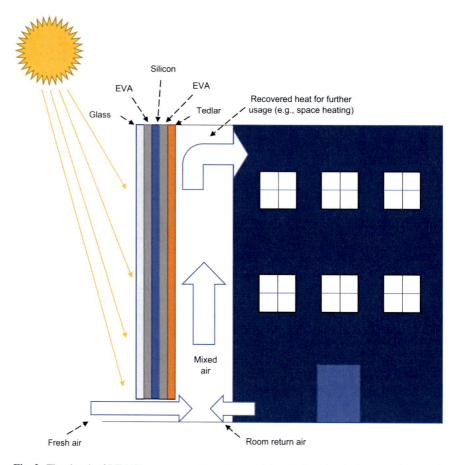

Fig. 2 The sketch of BIPV/T system considered for modeling by the explained approach in part 3

BIPV/T system considered for modeling by the explained approach is illustrated in Fig. 2.

Glass

Energy enters the glass layer through the transfer of conductive heat between the top layer of EVA and the glass as well as the sun's radiation. The energy output of the glass layer is through heat transfer and radiation between the glass and the ambient air, as seen in Eq. (1):

$$c_{p,g} \delta_g A \rho_g \frac{dT_g}{dt} = \alpha_g GA + Q_{\text{cond}-EVA1,g} - Q_{\text{conv}-g,a} - Q_{\text{rad}-g,\text{sky}} \tag{1}$$

where the subtitle g symbolizes glass, a is the ambient air, and c_p, δ, A, ρ, T, t, α, G, Q_{cond}, Q_{conv}, and Q_{rad} represent the specific heat capacity, thickness, area, density, temperature, time, absorption coefficient, solar radiation, conductive heat transfer, convective heat transfer, and radiant heat transfer, respectively.

In Eq. (1), the terms from the left are:

- Glass energy changes
- Heat absorbed by receiving solar radiation
- Conductive heat transfer between the glass layer and the top layer of EVA
- Convective heat transfer between the glass layer and the ambient air
- Radiation from glass to the sky

The coefficient of conductive heat transfer is calculated from Eq. (2):

$$Q_{cond-EVA1,g} = \frac{T_{EVA1} - T_g}{R_{EVA1,g}} \quad (2)$$

where $R_{EVA1,g}$ is the thermal resistance between the glass and EVA layers and can be calculated from Eq. (3):

$$R_{EVA1,g} = \frac{\delta_{EVA1}}{2k_{EVA1}A_{EVA1}} + \frac{\delta_g}{2k_g A_g} \quad (3)$$

where thermal conductivity is shown by k.

The convective heat transfer between the glass layer and the ambient air is obtained from Eq. (4):

$$Q_{conv-g,a} = \frac{T_g - T_a}{R_{conv-g,a}} \quad (4)$$

where the thermal resistance and convective heat transfer coefficient are obtained from Eqs. (5) and (6) [22]:

$$R_{conv-g,a} = \frac{1}{h_{conv-g,a} A} \quad (5)$$

$$h_{conv-g,a} = 2.8 + 3U \quad (6)$$

The wind speed is shown by U.

The relationship in the radiant heat transfer between the glass layer and the sky is as follows:

$$Q_{rad-g,sky} = \frac{T_g - T_{sky}}{R_{rad-g,sky}} \quad (7)$$

Thermal resistance and the sky temperature can be calculated by Eqs. (8) and (9) [23]:

$$R_{rad-g,sky} = \frac{1}{\sigma \varepsilon_g A \left(T_g^2 + T_{sky}^2\right)\left(T_g + T_{sky}\right)} \tag{8}$$

$$T_{sky} = 0.0552 T_a^{1.5} \tag{9}$$

In Eq. (8), σ is the Stefan-Boltzmann coefficient, and ε is the emission coefficient.

Top EVA

The input energy arrives at the top EVA layer through conduction between the top EVA and silicon; output energy from the top EVA layer is through the conduction heat transfer between the EVA and glass layers, as shown in Eq. (10). In the following relations, the PV caption symbolizes the silicon layer.

$$c_{p,EVA1} \delta_{EVA1} A \rho_{EVA1} \frac{dT_{EVA1}}{dt} = Q_{cond-PV,EVA1} - Q_{cond-EVA1,g} \tag{10}$$

In Eq. (10), the terms are as follows:
- Top EVA layer changes
- Conduction between top EVA and silicon
- Conduction between glass and the top layer of EVA

The conduction between top EVA and silicon can be calculated from the following equations:

$$Q_{cond-PV,EVA1} = \frac{T_{PV} - T_{EVA1}}{R_{PV,EVA1}} \tag{11}$$

$$R_{PV,EVA1} = \frac{\delta_{PV}}{2k_{PV} A_{PV}} + \frac{\delta_{EVA1}}{2k_{EVA1} A_{EVA1}} \tag{12}$$

Silicon

The energy input to the silicon layer arrives through solar radiation, and the output energy is the product of power generation and heat conduction between the top EVA and silicon and heat conduction between the bottom EVA and silicon, according to

Using Building Integrated Photovoltaic Thermal (BIPV/T) Systems to Achieve Net Zero... 97

Eq. (13). In the following relation, τ is the transmissivity, and P_{ele} is the production power.

$$c_{p,PV} \delta_{PV} A \rho_{PV} \frac{dT_{PV}}{dt} = \alpha_{PV} \tau_g GA - P_{ele} - Q_{cond-PV,EVA1} - Q_{cond-EVA2,PV} \quad (13)$$

The left-to-right terms in Eq. (13) are

- Silicon layer energy changes
- Solar radiation received by the silicon layer
- Power generation of solar cells
- Conduction between top EVA and silicon
- Conduction between glass and bottom EVA

The conduction between bottom EVA and silicon, as well as the thermal resistance, is calculated from Eqs. (14) and (15):

$$Q_{cond-EVA2,PV} = \frac{T_{PV} - T_{EVA2}}{R_{EVA2,PV}} \quad (14)$$

$$R_{PV,EVA1} = \frac{\delta_{EVA2}}{2 k_{EVA2} A_{EVA2}} + \frac{\delta_{PV}}{2 k_{PV} A_{PV}} \quad (15)$$

Bottom EVA

The input energy to the bottom EVA layer is the conduction between the bottom EVA and silicon, and the output energy is the conduction between the bottom EVA and Tedlar, which can be calculated according to Eq. (16). The Td subtitle represents the Tedlar layer.

$$c_{p,EVA2} \delta_{EVA2} A \rho_{EVA2} \frac{dT_{EVA2}}{dt} = Q_{cond-EVA2,PV} - Q_{cond-Td,EVA2} \quad (16)$$

The terms of Eq. (16) from left-to-right are

- Energy changes of bottom EVA layer
- Conduction between bottom EVA layer and silicon layers
- Conduction between Tedlar and bottom EVA layers

The conduction between the bottom EVA and silicon and the thermal resistance is calculated from Eqs. (17) and (18).

$$Q_{cond-Td,EVA2} = \frac{T_{EVA2} - T_{Td}}{R_{EVA1,Td}} \quad (17)$$

$$R_{PV,EVA1} = \frac{\delta_{PV}}{2k_{PV}A_{PV}} + \frac{\delta_{EVA1}}{2k_{EVA1}A_{EVA1}} \tag{18}$$

Tedlar

The input energy to the Tedlar layer is through the conductive heat transfer between the bottom EVA layers and the Tedlar, and the output energy is through the transfer of radiative and convective heat between the Tedlar and the ambient air, shown in Eq. (19):

$$c_{p,Td}\delta_{Td}A\rho_{Td}\frac{dT_{Td}}{dt} = Q_{cond-Td,EVA2} - Q_{conv-Td,a} - Q_{rad-Td,a} \tag{19}$$

The terms from left-to-right in Eq. (19) are

- Tedlar layer energy changes
- Conduction between Tedlar and bottom EVA
- Convection between Tedlar and air
- Radiation between Tedlar and surroundings

The method of calculating radiant heat transfer rate and thermal resistance is mentioned in Eqs. (20) and (21):

$$Q_{rad-Td,a} = \frac{T_{Td} - T_a}{R_{rad-Td,a}} \tag{20}$$

$$R_{rad-Td,a} = \frac{1}{\sigma\varepsilon_{Td}A(T_{Td}^2 + T_a^2)(T_{Td} + T_a)} \tag{21}$$

In addition, the convective heat transfer rate between Tedlar layer and air stream is obtained according to Eqs. (22) and (23):

$$Q_{conv-Td,a} = \frac{T_{Td} - T_a}{R_{conv-Td,a}} \tag{22}$$

$$R_{rad-Td,a} = \frac{1}{h_{conv-Td,a}A} \tag{23}$$

To obtain the convective heat transfer coefficient—which is between the BIPV/T system and the wall of the building—one must consider whether the air channel of the BIPV/T system has forced or free convection. The governing equations for each of these two conditions are introduced in the subsequent part.

The Air Stream Between BIPV/T System and Wall of Building

In this part, first the governing equations for free convection are presented. It follows by presenting the governing equation for forced convection. For both conditions, the heat transfer rate between panel and air ($\dot{Q}_{PV,air}$) can be determined from Eq. (24):

$$\dot{Q}_{PV,air} = \dot{m}_{air} c_{P,air} \left(T_{air,out} - T_{air,in} \right) \tag{24}$$

Here, $\dot{m}_{air}$, $c_{P,air}$, $T_{air,out}$, and $T_{air,in}$ stand for the mass flow rate, isobaric heat capacity, outlet air stream temperature, and inlet air stream temperature, respectively. In addition to Eq. (24), $\dot{Q}_{PV,air}$ can also be defined from Eq. (25):

$$\dot{Q}_{PV,air} = h_{conv-Td,a} A_{PV,air} \left(T_{Td} - T_{air,mean} \right) \tag{25}$$

where $A_{PV,air}$ denotes the heat transfer area between PV and air stream. $T_{air,mean}$ is the mean temperature of air stream, which can be considered as the average of the values of inlet and outlet temperature of air stream [24]:

$$T_{air,mean} = \frac{T_{air,out} + T_{air,in}}{2} \tag{26}$$

Free Convection

When the convective heat transfer is free, the convective coefficient can be obtained from Eqs. (27) to (29) [25]:

$$\mathrm{Nu}_X = 0.60 \left(\mathrm{Pr}.Gr_X^* \right)^{1/5} \tag{27}$$

$$\mathrm{Nu}_m = 1.25 \left(\mathrm{Nu}_X \right)_{X=L} \tag{28}$$

$$Gr_X^* = Gr_X . \mathrm{Nu}_X = \frac{\beta . g . q_w . X^4}{k . \gamma^2}, \quad \mathrm{Nu}_X = \frac{X . h_{conv-Td,a}}{k} \tag{29}$$

where X denotes the characteristic length. Moreover, Nu, Pr, Gr, and Gr* represent the Nusselt number, the Prandtl number, the Grashof number, and the modified Grashof number, respectively. The average Nusselt number is expressed as Nu$_m$ and β, g, q_w, and γ are the temperature constant of the convective coefficient, the acceleration of gravity, the heat flux (energy per unit area) transferred from the wall to the air stream, and the heat capacity ratio (the ratio of isobaric heat capacity to constant volume heat capacity, which is known as the isentropic expansion factor).

Forced Convection

The forced convective heat transfer coefficient between the airflow and the BIPV/T system is given by Eqs. (30) to (32) [24]:

$$h_c = \frac{k}{D_H}\left\{0.0182\,\mathrm{Re}^{0.8}\,\mathrm{Pr}^{0.4}\left[1+j\frac{D_H}{L}\right]\right\} \quad (30)$$

$$j = 14.3\log\left(\frac{L}{D_H}\right) - 7.9 \quad \text{for } 0 < \frac{L}{D_H} \leq 60 \quad (31)$$

$$j = 17.5 \quad \text{for } \frac{L}{D_H} > 60 \quad (32)$$

where D_H, Re, and L are the hydraulic diameter, Reynolds number, and Length of channel.

The Electrical Model

The primary purpose of PV systems is to generate power, so modeling the electrical power of solar PV systems is very important. There are different ways to predict the electrical performance of PV systems. One of the relations in which the efficiency of the solar system is determined based on the reference conditions is presented in Eq. (33):

$$\eta = \eta_{ref}\left(1 - \beta_{ref}\left(T_{PV} - T_{ref}\right)\right) \quad (33)$$

where η_{ref}, β_{ref}, and T_{ref} represent the efficiency, temperature coefficient, and reference temperature. The reference condition for this equation is solar radiation of 800 W.m^{-2} and temperature of 20 °C, respectively. The reference condition should not be confused with the standard test condition (STC), in which the irradiance and temperature values are 1000 W.m^{-2} and 25 °C, respectively [26]. These coefficients are provided in the customer catalogs of solar panel manufacturers.

The power of the solar system is calculated by Eq. (34):

$$P_{ele} = \eta_{ref}\left(1 - \beta_{ref}\left(T_{PV} - T_{ref}\right)\right)GA \quad (34)$$

Coupling the Two Models

The discussion provided in the previous parts has shown that determining the electrical power requires the PV temperature, and obtaining the PV temperature necessitates knowledge of the electrical power. Therefore, the thermal and electrical stimulation approaches should be coupled together, and the system performance values should be obtained by trial and error. A detailed explanation is available in previous studies by the authors, such as [27]. The trial-and-error process for finding the temperature and electricity production of the system is depicted in Fig. 3.

It is worth mentioning that despite changing the meteorological characteristics by time, by following the same fashion as the published research works in the field such as [28], the equations have been solved in the steady-state condition at each time step. For this purpose, the computer codes developed by a software like MATLAB could be utilized.

Moreover, obtaining the temperature of the outlet air stream is done by following the stages introduced in Fig. 4. As seen, obtaining temperature of PV using the given flowchart of Fig. 3 is a part of that.

Achieving the Net Zero Goal Using BIPV/T Systems

The term "net zero" may refer to two concepts. One is "net zero energy building (NZEB)," referring to a building equipped with renewable energy sources, such as PV panels, to produce the energy required during the year. In cases in which renewable energy sources are not able to meet demand, energy is supplied from the grid. If there is surplus in the generation, the excess is transferred to the network. The net amount of energy transferred from the building to the grid is zero in NZEBs [29]. When referring to NZEBs, the term "net zero" is accompanied by the word "buildings" [30].

The second probable condition the term "net zero" might refer to is "net zero emission (NZE)." Similar to NZEB, NZE means a condition in which the amount of emissions released into the atmosphere is equal to the amount of emissions removed. In order to remove emissions, a variety of solutions can be employed. For instance, CO_2 emissions in the atmosphere can be removed using trees [31].

Using BIPV/T systems can help achieve both NZEB and NZE goals at the same time. The PV panels used in a BIPV/T system represent a type of less fuels like natural gas for heat provision. Moreover, using BIPV/T systems means burning less fossil fuels in furnaces and thermal power plants.

Among related studies, Uygun et al. [32] evaluated the potential of using BIPV systems to achieve NZEB goals in three cities in Turkey: Çanakkale, Antalya, and Rize. The EnergyPlus software program was utilized, while the values of the heating and cooling demand, as well as electricity production, were considered.

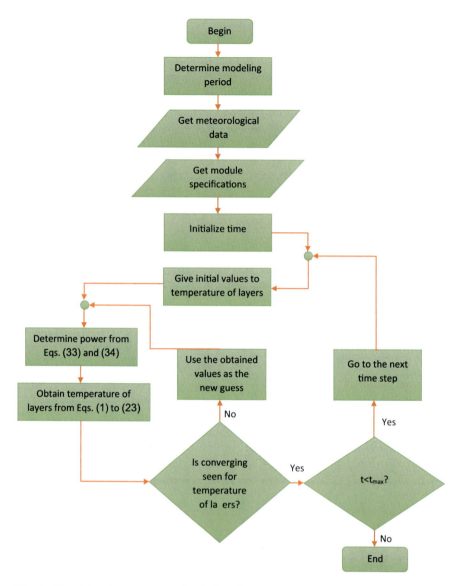

Fig. 3 The trial-and-error process for finding the temperature and electricity production of the system

According to the results, BIPV systems were found to have the potential of meeting the demands for all three locations, which lie in coastal regions.

In another investigation, Nallapaneni and Chopra [33] studied the performance of algal PV for integration with a building face and fulfilling net zero targets. Hong Kong was considered as the case study, and the application in high-rise buildings was investigated. A conceptual design was provided, and a sensitivity analysis was

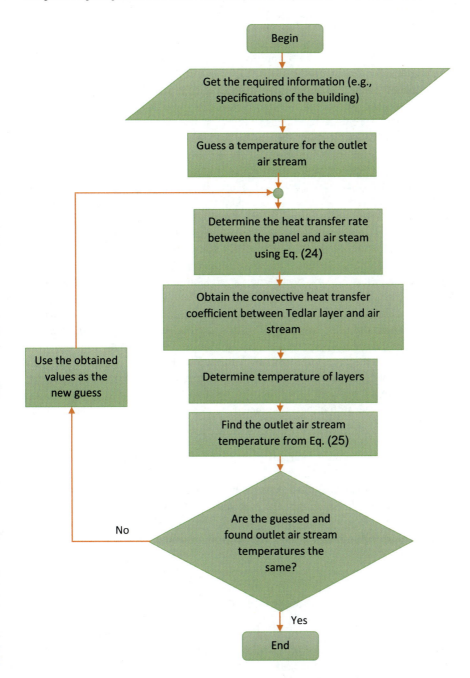

Fig. 4 The trial-and-error process for obtaining the temperature of outlet air stream

conducted to acquire information about the impact of changing solar radiation on system performance. López and Mobiglia [34] chose a heritage building in Switzerland and analyzed the feasibility of employing BIPV systems to reach net zero and positive zero goals.

For a BIPV system in Montreal, Canada, Yip et al. [35] explored the impacts of nine performance criteria, including the shape of the plan, window to wall ratio, and tilt angle. The net energy use intensity in a year was chosen as the system performance indicator. According to the results, the impact of some parameters, such as orientation, was higher than that of others. Another investigation with almost the same research team was conducted in [36].

In addition to the abovementioned studies, in which the net zero concept has been studied directly, a number of other studies have investigated it indirectly. The word indirect refers to investigating the energy and environmental criteria of the system and finding ways for enhancing them, usually by either multi-objective optimization or a parametric study. For instance, Sohani et al. [37] determined the best value of phase change material for integration with a BIPV/T system in Tehran, Iran, by taking a number of objective functions, including environmental and energy ones. The energy generated by the system throughout the year was representative of the energy side, while CO_2 saving was considered by taking the environmental impact into account. A novel optimization method called the dynamic multi-objective optimization approach was employed to gain a better outcome.

Multi-objective optimization was also utilized by Wijeratne et al. [38] with the aim of identifying the best condition for roof sheets and skylights in a BIPV system. The study introduced 7 and 14 solutions for the two aforementioned characteristics, respectively, in the form of non-dominated answers. A passive intelligence concept was proposed by Yoo [5] for a BIPV system, to harvest more irradiance from the sun, as well. According to the paper's discussion, it had a better energy performance, producing more electricity, and was able to reduce the cooling load significantly.

In addition to the research works already mentioned, a number of review studies have also been published during the past years on the topic of achieving net zero goals. Those of Gholami et al. [20], Rosa [39], Awuku et al. [40], Pugsley et al. [41], Ghosh [14], and Wei and Skye [42] can be given as examples.

Conclusion

This work has provided an overview of using both the BIPV and BIPV/T systems to achieve net zero targets. For this purpose, in addition to introducing the BIPV/T concept and the simulation method under both free and forced convection conditions, a number of recent studies have been reviewed in the field of both BIPV and BIPV/T technologies. As noted in our discussion, in a number of research works, sensitivity analysis or a parametric study has been conducted to find the impact of different performance indicators. In another group, different locations have been considered to identify more suitable places for applying BIPV and BIPV/T technologies. In addition to the indicated groups, there has been one more category in which multi-objective optimization was conducted to acquire the foremost condition for the system decision variables.

According to our discussion, both paths can be recommended for future research. One is to consider both NZEB and NZE targets as suitable for study, with the addition of economic perspectives, like the reduction in levelized cost of energy (LCOE) for renewable energies, to give a broader horizon. Another idea is to employ decision-making tools like AHP, LINMAP, and TOPSIS to consider the priorities of policy-makers and customers and not give the same importance to other performance criteria, which is the current approach studies take.

References

1. Dehghani-Sanij, A. R., & Bahadori, M. N. (2021). *Ice-houses: Energy, architecture, and sustainability*. Elsevier, Imprint by Academic Press.
2. M.R.Khani, S., Bahadori, M. N., Dehghani-Sanij, A. R., & Nourbakhsh, A. (2017). Performance evaluation of a modular design of wind tower with wetted surfaces. *Energies, 10*(7), 845.
3. Dehghani-Sanij, A. R. (2022). Providing thermal comfort for buildings' inhabitants through natural cooling and ventilation systems: Wind towers. In A. Sayigh (Ed.), *Achieving building comfort by natural means*. Springer Nature.
4. Boccalatte, A., Fossa, M., & Ménézo, C. (2020). Best arrangement of BIPV/T surfaces for future NZEB districts while considering urban heat island effects and the reduction of reflected radiation from solar facades. *Renewable Energy, 160*, 686–697.
5. Yoo, S.-H. (2019). Optimization of a BIPV/T system to mitigate greenhouse gas and indoor environment. *Solar Energy, 188*, 875–882.
6. Bauer, A., & Menrad, K. (2019). Standing up for the Paris Agreement: Do global climate targets influence individuals' greenhouse gas emissions? *Environmental Science & Policy, 99*, 72–79.
7. Taveres-Cachat, E., Grynning, S., Thomsen, J., & Selkowitz, S. (2019). Responsive building envelope concepts in zero emission neighborhoods and smart cities-a roadmap to implementation. *Building and Environment, 149*, 446–457.
8. Lancet, T. (2019). Net zero by 2050 in the UK. London, UK. 1911.
9. Hu, M., & Qiu, Y. (2019). A comparison of building energy codes and policies in the USA, Germany, and China: Progress toward the net-zero building goal in three countries. *Clean Technologies and Environmental Policy, 21*(2), 291–305.
10. Ballif, C., Perret-Aebi, L.-E., Lufkin, S., & Rey, E. (2018). Integrated thinking for photovoltaics in buildings. *Nature Energy, 3*(6), 438–442.
11. Jäger-Waldau, A., Kougias, I., Taylor, N., & Thiel, C. (2020). How photovoltaics can contribute to GHG emission reductions of 55% in the EU by 2030. *Renewable and Sustainable Energy Reviews, 126*, 109836.
12. Pierro, M., Perez, R., Perez, M., Moser, D., & Cornaro, C. (2021). Imbalance mitigation strategy via flexible PV ancillary services: The Italian case study. *Renewable Energy, 179*, 1694–1705.
13. Kuhn, T. E., Erban, C., Heinrich, M., Eisenlohr, J., Ensslen, F., & Neuhaus, D. H. (2021). Review of technological design options for building integrated photovoltaics (BIPV/T). *Energy and Buildings, 231*, 110381.

14. Ghosh, A. (2020). Potential of building integrated and attached/applied photovoltaic (BIPV/T/BAPV) for adaptive less energy-hungry building's skin: A comprehensive review. *Journal of Cleaner Production, 276*, 123343.
15. Reddy, P., Gupta, M., Nundy, S., Karthick, A., & Ghosh, A. (2020). Status of BIPV/T and BAPV system for less energy-hungry building in India—A review. *Applied Sciences, 10*(7), 2337.
16. Rounis, E. D., Athienitis, A., & Stathopoulos, T. (2021). Review of air-based PV/T and BIPV/T systems-performance and modelling. *Renewable Energy, 163*, 1729–1753.
17. Debbarma, M., Sudhakar, K., & Baredar, P. (2017). Comparison of BIPV and BIPVT: A review. *Resource-Efficient Technologies, 3*(3), 263–271.
18. Sohani, A., Naderi, S., & Pignatta, G. (2021). 4E advancement of heat recovery during hot seasons for a building integrated photovoltaic thermal (BIPV/T) system. *Environmental Sciences Proceedings, 12*(1).
19. Shakouri, M., Ghadamian, H., Hoseinzadeh, S., & Sohani, A. (2022). Multi-objective 4E analysis for a building integrated photovoltaic thermal double skin Façade system. *Solar Energy, 233*, 408–420.
20. Gholami, H., Nils Røstvik, H., & Steemers, K. (2021). The contribution of building-integrated photovoltaics (BIPV/T) to the concept of nearly zero-energy cities in Europe: Potential and challenges ahead. *Energies, 14*(19), 6015.
21. Cornaro, C., Bartocci, S., Musella, D., Strati, C., Lanuti, A., Mastroianni, S., et al. (2015). Comparative analysis of the outdoor performance of a dye solar cell mini-panel for building integrated photovoltaics applications. *Progress in Photovoltaics: Research and Applications, 23*(2), 215–225.
22. Shahverdian, M. H., Sohani, A., & Sayyaadi, H. (2021). Water-energy nexus performance investigation of water flow cooling as a clean way to enhance the productivity of solar photovoltaic modules. *Journal of Cleaner Production, 312*, 127641.
23. Shahverdian, M. H., Sohani, A., Sayyaadi, H., Samiezadeh, S., Doranehgard, M. H., Karimi, N., et al. (2021). A dynamic multi-objective optimization procedure for water cooling of a photovoltaic module. *Sustainable Energy Technologies and Assessments, 45*, 101111.
24. Shahsavar, A., & Rajabi, Y. (2018). Exergoeconomic and enviroeconomic study of an air based building integrated photovoltaic/thermal (BIPV/T) system. *Energy, 144*, 877–886.
25. Ghadamian, H., Ghadimi, M., Shakouri, M., Moghadasi, M., & Moghadasi, M. (2012). Analytical solution for energy modeling of double skin facades building. *Energy and Buildings, 50*, 158–165.
26. Sohani, A., & Sayyaadi, H. (2020). Providing an accurate method for obtaining the efficiency of a photovoltaic solar module. *Renewable Energy, 156*, 395–406.
27. Sohani, A., Sayyaadi, H., Doranehgard, M. H., Nizetic, S., & Li, L. K. B. (2021). A method for improving the accuracy of numerical simulations of a photovoltaic panel. *Sustainable Energy Technologies and Assessments, 47*, 101433.
28. Gu, W., Ma, T., Shen, L., Li, M., Zhang, Y., & Zhang, W. (2019). Coupled electrical-thermal modelling of photovoltaic modules under dynamic conditions. *Energy, 188*, 116043.
29. The U.S. Deparment of Energy, Zero Energy Buildings. 2022. https://www.energy.gov/eere/buildings/zero-energy-buildings. Accessed on 31 Mar 2022.
30. Abdou, N., El Mghouchi, Y., Hamdaoui, S., El Asri, N., & Mouqallid, M. (2021). Multi-objective optimization of passive energy efficiency measures for net-zero energy building in Morocco. *Building and Environment, 204*, 108141.
31. National Grid ESO, What is net zero and zero carbon?, 2022. https://www.nationalgrideso.com/future-energy/net-zero-explained/net-zero-zero-carbon. Accessed on 31 Mar 2022.
32. Uygun, U., Akgül, Ç.M., Dino, İ.G., & Akinoglu, B.G. (2018). Approaching net-zero energy building through utilization of building-integrated photovoltaics for three cities in Turkey-preliminary calculations. In *2018 international conference on photovoltaic science and technologies (PVCon)* (pp. 1–5). IEEE.
33. Nallapaneni, M. K., & Chopra, S. S. (2021). Algal photobioreactor facades coupled with BAPV/BIPV/T in high-rise urban buildings improves indoor air quality and enables energy

resilience in race to net-zero. In *3rd international conference on renewable energy, sustainable environmental and agricultural technologies.* i-RESEAT.

34. López, C. S. P., & Mobiglia, M. (2021). Swiss case studies examples of solar energy compatible BIPV/T solutions to energy efficiency revamp of historic heritage buildings, In IOP conference series: Earth and environmental science (Vol. 863, No. 1, pp. 012006). IOP Publishing.

35. Yip, S., Athienitis, A. K., & Lee, B. (2021). Early stage design for an institutional net zero energy archetype building, part 1: Methodology, form and sensitivity analysis. *Solar Energy, 224,* 516–530.

36. Yip, S., Athienitis, A., & Lee, B. (2019). Sensitivity analysis of building form and BIPV/T energy performance for net-zero energy early-design stage consideration, In IOP conference series: Earth and environmental science (Vol. 238, No. 1, pp. 012065). IOP Publishing.

37. Sohani, A., Dehnavi, A., Sayyaadi, H., Hoseinzadeh, S., Goodarzi, E., Garcia, D. A., et al. (2022). The real-time dynamic multi-objective optimization of a building integrated photovoltaic thermal (BIPV/T) system enhanced by phase change materials. *Journal of Energy Storage, 46,* 103777.

38. Wijeratne, W. M. P. U., Samarasinghalage, T. I., Yang, R. J., & Wakefield, R. (2022). Multi-objective optimisation for building integrated photovoltaics (BIPV/T) roof projects in early design phase. *Applied Energy, 309,* 118476.

39. Rosa, F. (2020). Building-integrated photovoltaics (BIPV/T) in historical buildings: Opportunities and constraints. *Energies, 13*(14), 3628.

40. Awuku, S. A., Bennadji, A., Muhammad-Sukki, F., & Sellami, N. (2021). A blend of traditional visual symbols in BIPV/T application: Any prospects? *Academia Letters,* 4029.

41. Pugsley, A., Zacharopoulos, A., Mondol, J. D., & Smyth, M. (2020). BIPV/T facades–A new opportunity for integrated collector-storage solar water heaters? Part 1: State-of-the-art, theory and potential. *Solar Energy, 207,* 317–335.

42. Wu, W., & Skye, H. M. (2021). Residential net-zero energy buildings: Review and perspective. *Renewable and Sustainable Energy Reviews, 142,* 110859.

Simulated Versus Monitored Building Behaviours: Sample Demo Applications of a Perfomance Gap Detection Tool in a Northern Italian Climate

Giacomo Chiesa, Francesca Fasano, and Paolo Grasso

Introduction

Sustainable and green energy solutions are progressively growing in consideration in the building sector for both new and retrofitted designs and actions. At a European level, the introduction of the EPBD (Energy Performance of Buildings Directive), since its initial 2002/91 version, has progressively supported Member States in introducing and/or upgrading energy and building regulations including the definition of minimal standards, e.g. U-values, supporting a progressive increase of the energy efficiency of the building stock. Furthermore, the EPBD is not a rigid instrument, since it has been improved over time, including the EPBD recast 2010 version and the 2018 one. Recently, the EU (European Union) funded a series of specific H2020 projects to support the 'next-generation of Energy Performance Assessment and Certification' approaches under the call LS-SC3-EE-5-2018-2019-2020. Among these projects, the E-DYCE (Energy Flexible Dynamic Building Certification) project [1] identified five main open issues connected to EBPD topics. These issues include [2] (i) free-running and passive technologies, (ii) smart readiness vision, (iii) energy metering and district network communication, (iv) dynamic hourly models and performance gap, and (v) renovation and operational roadmap. This paper outlines some initial outcomes of the E-DYCE project related to abovementioned issues i, ii and iv are focussing on two demonstrative buildings localized in Northwest Italy and illustrating a new approach for performance gap detection in semi-real time using a dynamic simulation platform which is under development by the authors.

G. Chiesa (✉) · F. Fasano · P. Grasso
Politecnico di Torino, Department of Architecture and Design, Turin, Italy
e-mail: giacomo.chiesa@polito.it; francesca.fasano@polito.it; paolo.grasso@polito.it

Chapter Objectives and Contents

As mentioned above, this chapter focusses on illustrating initial results of the application of a new underdevelopment dynamic simulation platform to detect building performance gap comparing simulated and monitored building behaviours under free-running conditions. Simulated buildings are based on verified models, while the simulation's operational inputs for performance gap are inputted by current standards, e.g. EN 16798-1:2019 [3]. Monitored data are based on a smart cloud-connected monitoring system, while weather data are retrieved from a cloud-connected meteorological station installed for the project. Tests were performed during the extended summer of 2021, also considering transitional periods from the late-spring to the beginning of autumn – from May to October. Additionally, the above-mentioned dynamic simulation platform is based on a Python tool named PREDYCE (Python semi-Realtime Energy DYnamics and Climate Evaluation) that is under implementation on the basis of different development actions, including the 'DYCE' action, based on the mentioned E-DYCE project, and a 'PRE' action, based on another project and adding additional functionalities. The 'DYCE' action includes a larger set of actions with respect to those presented in this chapter. The platform is based on EnergyPlus [4] that, among building energy dynamic simulation engines, is one of the most widely used and recognized [5]: its white box model algorithm to model building dynamics can give very accurate results in terms of both consumption and environmental variable trends, considering it is also used to support the validity of other software used for energy labelling, e.g [6]. This motivates the use of EnergyPlus in this chapter to detect the performance gap between simulated and monitored building behaviours and the increasing interest for both professionals and researchers in the past few years in developing libraries and tools supporting EnergyPlus input model editing and output analysis in a parametric and automatic vision, e.g [7, 8]. This chapter focusses on PREDYCE application scenarios that treat simulation and monitoring results together; see also section "Methodology and PREDYCE". The chapter is organized as follows: Section "Methodology and PREDYCE" shortly introduces the mentioned PREDYCE tool focussing on the 'DYCE' developing action contents and details the methodological pipeline used in this chapter. Section "Demo Building Applications and Results" focusses on two real-project demo applications of this methodology including model verification results and performance gap detection samples. Finally, Section "Conclusions" shortly concludes the chapter by mentioning main results, limitations, and future planned development steps.

Methodology and PREDYCE

PREDYCE Introduction and Use Scenarios

PREDYCE is a newly developed Python library composed of three main modules able to manage EnergyPlus input files (IDFs) in a parametric and automatic mode, executing multiple parallel simulation runs and handling the obtained results. Moreover, additional modules have been developed to manage other important aspects linked to EnergyPlus simulations, e.g. to compile EPW input weather files starting from monitored data from weather stations. A detailed PREDYCE library scheme is illustrated in [9] and in [10, 11]: its architecture is based on the previously mentioned main modules, i.e. (i) an IDF editor module, (ii) an EnergyPlus running module, and (iii) a KPI calculator module. Each module has been built to work harmoniously with the others but also independently in tailored scripts, thus guaranteeing high flexibility and modularity in terms of, for example, data sources for a KPI calculator module, whose input can accept both simulation results and structured monitored data, allowing the development and testing of new methodologies. Similarly, the IDF automatic editing module is able to modify numerous building aspects such as activities, simplified HVAC systems, and envelope materials. The provided set of Python methods for IDF editing and KPIs computation, combined with the integrated EnergyPlus launcher, can help in performing different tasks like sensitivity analysis, retrofitting suggestions, performance gap analysis, or model verification, either automatically or semi-automatically.

The different tasks, which exploit all PREDYCE functionalities, are organized in separate scripts (herein referred to as use scenarios), easily executable by command line or also through a dedicated web service. Thanks to future actions, each script could be treated as a pre-built use scenario through a common application. In particular, the basic PREDYCE scenario is devoted to perform sensitivity analysis allowing parametrization of numerous building characteristics and computation of many possible KPIs according to the most recent European standards. Based on this more general use scenario, two other scenarios have been developed to introduce monitored data in the automatic loop: one is devoted to help in model verification phases, while the other aims to compare KPIs computed on calibrated building models and on monitored data, in order to highlight potential gaps in performance. Each scenario takes in input files and generates output files that are structured in the same way, as shown in Fig. 1. In particular, the main inputs are the building model in IDF format and the weather data in EPW format, which are needed to feed an EnergyPlus simulation; an input JSON file used to personalize the parametric request and apply preliminary modifications to the building model if needed; and finally, a CSV of environmental monitored data if required by the chosen application scenario, e.g. the performance gap one. Main outputs, instead, include a CSV file named *data_res* containing aggregated KPI results for the considered time period; a CSV file named *data_res_timeseries* containing timeseries KPI results with definable timestep resolution (hourly by default) for each performed

Fig. 1 PREDYCE scenario generic input/output workflow

```
{
    "building_name": "MainBlock",

    "preliminary_actions": {
        "change_runperiod": {
            "start": "01-05",
            "end": "30-09",
            "fmt": "%d-%m"}},

    "actions": {
        "change_ach": {
            "ach": [0, 0.2, 0.4],
            "ach_type": ["infiltration"]},
        "add_ceiling_insulation_tilted_roof": {
            "ins_data": ["MW Glass Wool (rolls)"],
            "Thickness": [0.3]}},

    "kpi": {
        "Q_c": {},
        "Q_h": {}}
}
```

Fig. 2 Example of input JSON file for PREDYCE

simulation; and finally, plots (e.g. carpet plots, energy signature) allowing us to deepen the meaning of KPIs.

Figure 2 shows a simple example of a standard PREDYCE JSON input file, in order to explain its structure and general potential. It is made by keywords, which are later used by the scenario scripts to understand how to execute a simulation, and values, which can contain names of IDF objects to be added or modified in the model or values to be set. IDF editing actions, in accordance with specific materials or object names, are made possible by the presence of internal databases of IDF objects that are hidden to the final user. Looking at the main keywords in Fig. 2, the *building name* is the name of the main block of the IDF which is utilized by the tool to know which zone elements need to be edited and then perform calculations on it; the *preliminary actions* are the actions which are executed only once before running the simulations such that all simulated buildings have in common the same modifications listed in this section (e.g. changing the run period or also activating/deactivating the HVAC system); the *actions* are the parametric modifications that have to be applied to the building (e.g. changing infiltration through windows or adding an insulation layer to the ceiling, eventually filtering based on the building's thermal zones); and finally, the *KPI* section includes the key performance indicators that are computed at the end of each simulation (in the example, Q_c and Q_h are the primary energy needs for cooling and heating in the building).

Considering the PREDYCE scenarios that involve comparisons between monitored and simulated data, a structured nomenclature of both sensors located in the building and the IDF model thermal zones is necessary to obtain a correct and automatic spatial association within the tool without the need of an intermediate translator, such that spatial aggregations for KPI analysis correspond. Consequently, the following nomenclature has been adopted for sensors: *building name_block name_ thermal zone name_sensor identifier_type of variable*. The naming part preceding the sensor identifier (e.g. MAC address) follows the IDF model naming structure, which always includes the building name, the block name, and the thermal zone name. To apply this nomenclature, the mentioned naming scheme must be used both within the building model – when initially creating it through an interface, e.g. DesignBuilder or OpenStudio – and on sensor ID. This coherence allows a strict spatial correspondence, making it possible to aggregate analyses and results at both the building and block level. Moreover, at the end of the naming structure, the name of the measured variable must be included. The variable name is then used inside each KPI calculator methodology to recognize which CSV columns need to be included in the computation. The proposed scheme leaves freedom to build the IDF model as desired (e.g. following a multi-zone or a mono-zone approach), allowing different thermal zone aggregations, without impacting the matching. In this chapter, two sample applications of both the performance gap scenario and the model verification scenario are shown assuming preliminary data of two demonstration buildings of the E-DYCE project; see section "Demo Building Applications and Results".

PREDYCE Model Verification Scenario

The PREDYCE semi-automatic model verification scenario has been used previously to the performance gap scenario execution in order to adjust the considered building models to the real indoor air temperature trend, speeding up the manual procedures usually adopted for this purpose. Temperature is adopted as a target verification variable, given that this chapter focused on summer free-running conditions. The following IDF editing values were varied to try aligning the simulated trend to actual building behaviour: U-value of the walls and roof; U-value and SHGC (solar heat gain coefficient) of the windows; internal mass and equipment gains in each thermal zone; and ACH (air changes per hour) ventilation and infiltration. Model verification is made possible by PREDYCE's ability to handle both simulation results and monitored data. The adopted model verification methodology is inspired by [12] and consists in optimizing a combined error measure which includes RMSE (root mean square error) and MBE (mean bias error) (see Eq. (1))

on a given variable or combination of variables, which is in this case represented by the indoor dry bulb temperature.

$$\text{Error}_{tot} = \sqrt{\text{RMSE}^2 + \text{MBE}^2} \quad (1)$$

The calibration signature described in [12] is computed according to Eq. (2), considering indoor dry bulb temperature an objective variable.

$$\text{Calibration signature} = \frac{\text{measured} T_{db}^i - \text{simulated} T_{db}^i}{\text{max measured} T_{db}^i} \cdot 100\% \quad (2)$$

The different IDF editing actions allow the user to shift the curve (e.g. acting on ACH, equipment gains), change coefficient and inclination, and modify amplitude variations – e.g. acting on internal mass – thus reaching a flat line within a 5% error range, which corresponds to reference suggestions for model calibration (see also ASHRAE Guideline 14-2014 for calibration criteria) [13]. Figure 3 shows an example of calibration signature plots before and after the calibration of one of the considered buildings.

The model verification scenario is currently considered to be semi-automatic since. In order to minimize the number of performed simulations, it requires observation of calibration signature plots and at the CSV of aggregated total error results to better choose which parameters to vary and in which ranges. This procedure may be further automatized in future actions, but the current potential of PREDYCE to simultaneously test multiple building parameters and automatically edit the model provides a considerable improvement in terms of effort and time with respect to traditional manual procedures, taking few hours to reach results such as the one shown in Fig. 3.

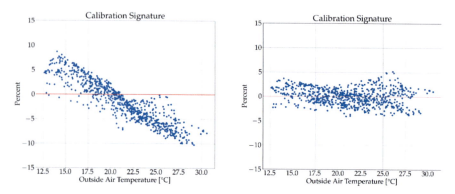

Fig. 3 Example of calibration signatures: (**a**) starting point and (**b**) after model verification on the residential demo building

PREDYCE Performance Gap Scenario

After the model verification phase, updated IDF files are saved as new simulation starting points, and the PREDYCE performance gap scenario is run on the two sample demo cases, considering different building settings. Figure 4 shows the performance gap scenario input/output workflow: among main inputs there is the CSV file containing monitored environmental data; the EPW file that should be also built from monitored data, necessarily an actual weather file; and the input JSON file which allows the user to define schedules, setpoints, and other building activities-related fields as simulation parameters in order to test the parameter impact on the gap against actual behaviour. Moreover, an optional weather input in CSV format is also available, giving the possibility to exploit EPW compiler module functionalities within PREDYCE instead of providing a previously built EPW: in this case, weather station coordinates should be passed through the input JSON file, such that eventually missing weather variables can be computed by the compiler exploiting well-known meteorological formulas. Outputs are instead composed by a zip folder containing the CSV file of aggregated results in which each row represents a considered EnergyPlus run (e.g. the different simulation settings recognizable by keywords *simulated_x*, monitored condition, and finally the delta between monitored KPI and *simulated_x* KPI), the CSV file containing timeseries KPIs with default hourly resolution for both monitored, simulated and delta KPIs (computed as monitored minus simulated results), and finally the optionally required plots.

The input JSON file is used to define both standard and standard modified conditions in which the building model is simulated: particularly, standard modified models were adapted considering a more realistic building usage concerning occupancy, ventilation, and schedules by taking advantage of an inspection-based approach, e.g. see [14]. Moreover, the KPIs to be computed are listed in the input file. In particular, since it focuses on free-running building conditions, distribution of

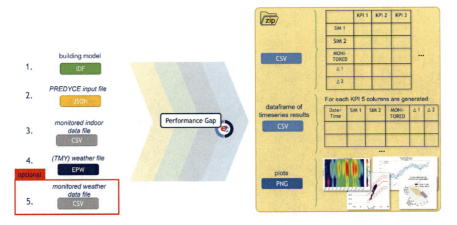

Fig. 4 PREDYCE performance gap scenario input/output workflow

datapoints in adaptive comfort model (ACM) categories is calculated assuming the adaptive thermal comfort model of EN 16798-1:2019 [3]. Additionally, the percentage outside the range (POR) is calculated, returning the percentage of cumulated hours when thermal comfort is not reached, considering them discomfort hours outside cat. II boundaries. Moreover, CO_2 concentration is considered one of the main symptoms of under- and overventilation, returning the number of hours above the threshold of 1000 ppm and under the threshold of 600 ppm, as suggested in Ref. [15]. Besides aggregated KPI results – which are computed on a weekly basis in order to recurringly inform users – when data are not too old to be detached from operational choices while sufficient to describe building phenomena, even timeseries results are returned for CO_2 and indoor dry bulb temperature with an hourly timestep, allowing the user to better identify where a potential problem could be located.

Figure 5 shows part of the input file used for the performance gap analysis in the residential unit: the list of KPIs can be seen, together with the spatial aggregations on which each KPI has to be computed (the different activities refer to different rooms in the house, while r01 refers to the entire unit); inside the *preliminary actions* field a list of two JSON structures can be seen, the first referring to building modifications needed to reach standard conditions, the second to reach standard modified usage conditions (e.g. changing ventilation rate or the occupancy). Keywords used within the *preliminary_actions* field refer to methods in the PREDYCE IDF editor module, while KPIs name the methods within the PREDYCE KPI calculator module.

The described methodology can be summarized by the pipeline shown in Fig. 6, including EnergyPlus building model development, model verification adopting monitored data, and the PREDYCE scenario of the same name, followed by the

```
{
    "scenario": "performance_gap",
    "building_name": "r01",
    "start_date": "2021-10-25",
    "end_date": "2021-11-01",
    "kpi": {
        "adaptive_comfort_model": {}, "n_co2_aIII": {},
        "n_co2_bI": {}, "timeseries_t_db_i": {}, "timeseries_co2": {}
    },
    "aggregations": {
        "adaptive_comfort_model": ["act105aa", "act104aa", "act103aa", "r01"],
        "timeseries_co2": ["act104aa"],
        "n_co2_aIII": ["act104aa"],
        "n_co2_bI": ["act104aa"],
        "timeseries_t_db_i": ["act105aa", "act104aa", "act103aa", "r01"]
    },
    "preliminary_actions": [
        {"<to_standard>" :{}},
        {"change_ach": {
            "ach": 3,
            "Schedule_Name": "_residenziale 16798-1",
            "filter_by": "r0", "relative": false},
        "change_occupancy": {"value": 0.011, "filter_by": "r0", "relative": false},
        "..." : {}
        }]}
```

Fig. 5 Example of input JSON file

Simulated Versus Monitored Building Behaviours: Sample Demo Applications... 117

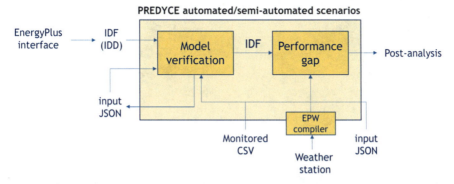

Fig. 6 The methodological pipeline adopted by this chapter

application of the performance gap PREDYCE scenario adopting standard and standard modified IDF input data. Standard scenario is based on given EU input data from standards and norms, e.g. EN 16798-1:2019, while the standard modified scenario refers to adapted input data upon collecting inspection data from the real building including regional and national adaptation, e.g. adapting set points and occupation scheduling.

Demo Building Applications and Results

Demo Buildings General Description

Two demo buildings are adopted in this chapter to support application testing. Both buildings are participating as demonstrations of the EU H2020 project E-DYCE, and they represent two different building typologies: a single-family building and a public school. These buildings are located in Torre Pellice, a small city located in the Turin metropolitan area, in the Pellice Valley, in Northwest Italy. Even if positioned at the bottom of the valley, the climate is cold and influenced by the Alps. It is classified in the Italian climate zone F, reaching 3128 heating degree-days$_{20}$. According to the Italian Presidential Decree n° 412/93 et seq., the climate zone F does not have any specific limitations to the heating activation period, although, for the purposes of this paper, the heating systems were considered active from 5 October through 22 April on the basis of interviews. Torre Pellice is a very representative demo city for small municipalities in Northern Italy and in the Piedmont Region, and it is positioned 5357th in terms of population among the 7978 Italian municipalities (ISTAT 2018). Nevertheless, with 4545 inhabitants Torre Pellice is in line with most small cities, considering that the Italian average is 7980, but the median is 2457. The single-family house typology is also very representative as a demo case considering that, typically, small cities are mainly composed of small houses rather than the multi-block buildings that characterize medium-to-large

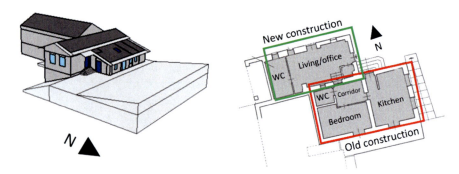

Fig. 7 The considered residential building: (**a**) comprehensive view and (**b**) internal view

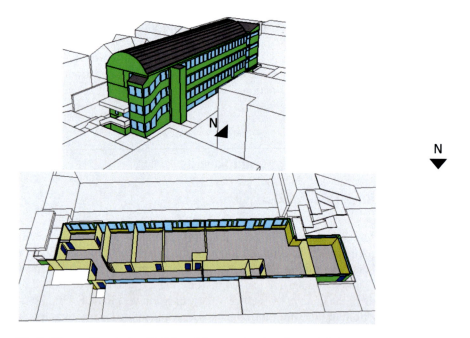

Fig. 8 The considered school building: (**a**) comprehensive view and (**b**) basement floor

cities; see related ISTAT data. Moreover, the considered municipality school is also representative of Italian public school constructions, as it was built in 1975 and features reinforced concrete pillars and external walls with a double layer of bricks and minor infilled insulation due to the cold climate.

Focussing on the selected buildings, base geometrical models are shown in Figs. 7 and 8, showing the residential building and the school building, respectively. Considering the single-family house, it has a 93-m^2 surface area subdivided into three main rooms, two bathrooms, a corridor, and a technical space. The house is on a single floor, with a minor change in elevation between the northern – recently adapted to a residential space – and southern part. Considering, instead, the municipality school, it is composed of four floors similarly organized with a long corridor on the north façade and teaching areas facing south. Three of the four floors (ground to second) are used as a middle school, while the basement floor is used as a kindergarten, directly facing an outside recreational area on the north side. Since the school is a complex building, characterized by different usages and consequently schedules and standard requirements, this chapter only focusses on the kindergarten floor. Unlike other floors, the kindergarten is characterized by larger teaching areas: in particular, the area at the end of the corridor is used mainly for lunch and as a sleeping area in the afternoon, while the two rooms at the beginning of the corridor are divided by a movable panel and kept often open as a single, bigger playroom or also used partially as a sleeping area; finally, the central room is used for daily activities.

Sensors have been installed for environmental data acquisition since April 2021 and allow temperature monitoring in all rooms, relative humidity in most rooms, CO_2 in the most representative spaces, and extra parameters in limited rooms, such as TVOC and illuminance, although the latter are not investigated in this chapter. Sensors and monitoring gateways are based on the Capetti WineCap system [16]. The solution allows to access monitored data remotely and in almost real time by developing a SOAP-based API [17] or by using the provided interface. Additionally, a meteorological station has been installed in order to collect weather variables to feed simulation with the same real boundary conditions of monitored data. The station includes a Thies US climate sensor that monitors temperature, relative humidity, wind (direction and velocity), precipitation data, illuminance, atmospheric pressure, and correlated data, plus a delta ohm pyranometer (class 1) for collecting global horizontal irradiation. Among split irradiation models, the well-known Boland-Ridley-Lauret model [18] has been applied to retrieve diffuse and direct components.

Application of the Model Verification PREDYCE Scenario

Regarding the residential house, the month of June 2021 was chosen as model verification time period, without having any information of actual house occupancy but assuming it was occupied.

Figure 9 compares initial and verified model results vs. monitored data (measured) of the single-family house case. The two graphs clearly demonstrate that the model verification scenario may support an improvement in building behaviours with respect to monitored data thanks to the adaptation of boundary modelling conditions. Figure 10 plots monitored and simulated internal temperatures (building average values of both sensors and simulated thermal zones) considering initial and verified models. These graphs help to better understand improvements in the model supported by the PREDYCE tool. Generally speaking, these changes are obtained by manually performing several simulations, i.e. via an EnergyPlus interface, but

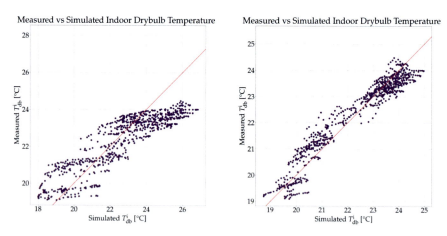

Fig. 9 Indoor temperature measured vs monitored before and after residential house calibration process

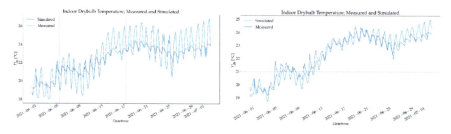

Fig. 10 Indoor temperature trend over time before and after residential house calibration process

the new developed tool allows to automatically compare the two series (without requiring post-production) and to support automatic changes of IDF parameters in given ranges, avoiding to manually perform this task. It is hence possible to verify models in a quicker and more productive way by testing both statistical discrepancies between simulated and monitored data and potentially also analysing the impact of different parameters on these differences. As it is visible from Fig. 10, at the end of the month larger discrepancies occur between verified model and monitored data. Consequently, a drastic change in building usage can be supposed in those days, for example, because of the beginning of a holiday period with consequent occupancy and ventilation going to zero values.

The peculiar construction of the house, made of a newly renovated area and an older uninsulated part, increased the complexity of the process, since both boundary walls and ceiling for the two areas were calibrated as separate parameters, e.g. concerning vertical walls, the best-found values led to a reduction of the model U-value by 90% in the renovated insulated part of the house and an increase of 15% in the old non-renovated area. To reduce the amplitude of differences in temperature between monitored and simulated data, the most effective action was the increase of internal mass, probably because of the massive structure of the mountain house

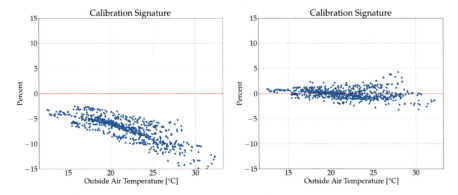

Fig. 11 Kindergarten calibration signatures, before and after the calibration process

considered old. Even internal gains increased, while infiltration through windows drastically dropped. The obtained values are consistent with the building materials and technical elements identified during the inspection phase that followed this analysis.

Regarding the kindergarten building model verification, the chosen time period was from 21 June to 21 July 2021, corresponding to the school closure. Consequently, ventilation was considered inactive and was not used as a variable in the calibration process.

Figure 11 compares original and verified model calibration signatures. Even in this case, the original model shows evident discrepancies, while the verified model presents an error perfectly fitting suggested calibration error thresholds; see ASHRAE Guideline 14-2014. Original errors in calibration signatures shifts from a [−2 to −15] range to a [+4, −4] one. The other two following graphs (see Fig. 12) plot internal temperatures over time including in the same graph monitored and simulated results. The latter model (verified) shows a very good correlation and is able to represent real building behaviours under actual weather conditions. The best values found allowed an increase in the U-value of the walls to 1.25 W/(m² K), coherently with results found during a subsequent inspection. Roof U-value, instead, fell from 1.79 W/(m² K) to 1.2 W/(m² K), and infiltration through windows increased. Also, internal mass increased by 20%, while internal equipment gains dropped.

Application of the PREDYCE Performance Gap Scenario

Standard and Standard Modified Scenarios

The verified models retrieved in the previous section "Application of the Model Verification PREDYCE Scenario" are adopted here to run the PREDYCE performance gap scenario. The input JSON file is used to impose to the models (i)

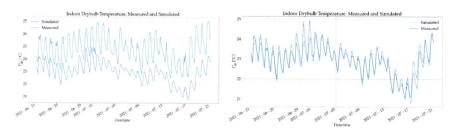

Fig. 12 Indoor temperature trend over time before and after the kindergarten calibration process

standard settings according to Annex C of EN ISO 16798-1:2019, overwriting certain values previously calibrated (e.g. ventilation rate), and (ii) standard modified settings, considering a more realistic building use defined after an inspection. Regarding the residential case, considering its dimension and adjacency (despite through only one wall and floors) to other residential units, the 'residential apartment standard case' of the EN 16798 standard was considered. Concerning internal gains, 28.3 m^2/person are used as a standard, while in the standard modified case, the knowledge that the house is inhabited by only one person is used, leading to the total 93 m^2/person. With the units being scheduled for occupancy, appliances and lighting are not modified in the standard modified setting with respect to the standard, since the house is not actually used with a specific home-office pattern, and it was not possible to structure a proper schedule. Ventilation in standard modified conditions was increased from the 0.5 l/(m^2/s) considered as the standard with 3 ACH (air changes per hour), a high value able to perform ventilative cooling, since during the inspection a high usage of natural ventilation during the summertime was underlined.

Concerning, instead, the kindergarten, standard conditions consider a continuous building usage over the year without including any holiday: weekends are considered unoccupied, while a typical weekday is considered occupied from 7 a.m. to 7 p.m., for a total of 12 h per day. Regarding internal gains, 3.8 m^2/person are considered to lead to 4.92 l/(m^2/h) as a CO_2 generation rate. Standard air flow for ventilation is supposed to be 4.5 l/(m^2/s), and the outdoor temperature setpoint for its activation was set to 17.5 °C, corresponding to the standard heating setpoint. Those values were modified considering a more realistic building usage in the considered kindergarten: schedules were changed, since around 16:30 all children leave the school and no after-school service is provided in the rooms; also, holidays were considered assuming the traditional Italian calendar. Moreover, child presence in the main activity areas was reduced with respect to the standard, considering that some of them go home after lunch and that even the outside area is used, reducing the overall indoor presence. Also, ventilation was increased to 2 ACH always active, considering the COVID-19 government advice to keep windows open as much as possible.

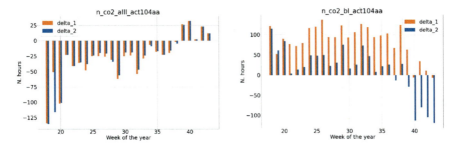

Fig. 13 CO$_2$ KPIs in residential unit kitchen and living room area representing the difference in number of hours between the simulated (standard = 1; standard modified = 2) and monitored values. The graph on the left shows hours above 1000 ppm, while the graph on the right the number of hours below 600 ppm

Performance Gap Results

Concerning the residential unit, a single CO$_2$ sensor was installed in a room used both as a kitchen and living room; thus the CO$_2$ analysis was performed for this specific room, called *act104aa*. Figure 13 shows weekly aggregated results of numbers of hours below the threshold of 600 ppm and above 1000 ppm. Since the performance gap with the standard simulated buildings is negative, it means that there are more hours above 1000 ppm of CO$_2$ concentration with a standard building behaviour than considering the actual one, except in the beginning of the autumn season. Oppositely, the number of hours below the 600-ppm threshold is greater in the actual monitored behaviour than the standard one. However, it can be seen that, especially from the late spring to the early autumn, the standard modified behaviour results were much more similar to the actual ones than the standard ones, which show a greater gap. This means that the ventilation was probably higher than in colder weeks, with the occupant behaviour causing ventilative cooling, while in later weeks the cold season implied a different usage of window openings to prevent heat dissipation, i.e. the difference between the standard modified and real behaviour become negative for hours below 600 ppm.

Looking at the timeseries of CO$_2$ in the room (Fig. 14), it can be seen that monitored data follows a random behaviour, which is difficultly represented by simulated trends that follow a fixed schedule in each weekday. Consequently, aggregated results are more useful to highlight potential behavioural gaps in the residential unit. Moreover, looking at the graph on the right in Fig. 14, it can be seen that in late June monitored data flattened, reinforcing the hypothesis of a holiday made upon observing the model verification results in Fig. 10.

Figure 15, instead, shows differences in weekly hour distribution in ACM categories, giving an idea of indoor thermal comfort monitored with respect to standard conditions. Unshown categories resulted to be empty for both simulated and monitored data, meaning that the residential unit is maintained quite cold in the whole period considered. Particularly, monitored data turn out to be, on average, colder

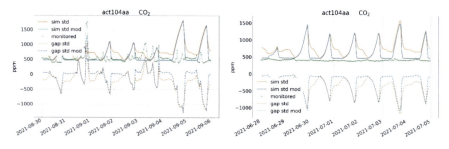

Fig. 14 Timeseries CO_2 values in residential unit kitchen and living room area for different periods

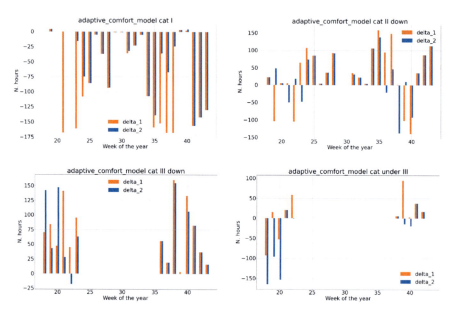

Fig. 15 Distribution of hours in ACM categories, averaged on all residential unit zones

than what is expected in standard conditions, except for the first weeks in May, when there are more simulated hours below comfort cat. III than monitored ones.

Looking at indoor temperature trend over time, the average behaviour shown in Fig. 16 is quite different depending on specific thermal zones, as shown in Fig. 17 (the kitchen) and in Fig. 18 (the second living room located in the newly renovated part of the building). In fact, despite the average monitored behaviour is almost in line with both standard and standard modified simulated conditions, the kitchen area shows several peaks at a higher temperature during the entire month of May and then gradually flattening at the end of the month. The newly renovated area, instead, shows a colder trend with respect to simulated standard conditions, but quite in line with the standard modified in which occupancy is more realistically set. Kitchen peaks are explainable because of a wood stove located in the room, which was used despite the end of the heating season to face the last colder weeks of the

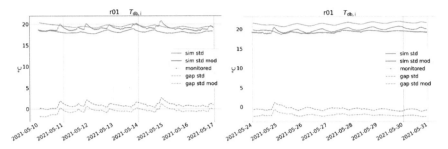

Fig. 16 Timeseries of average air temperature values in all residential unit zones

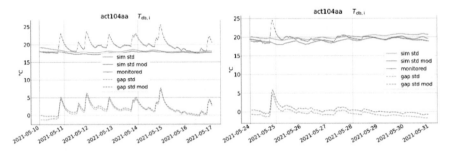

Fig. 17 Timeseries air temperature values in residential unit kitchen (left) and second living room area (right)

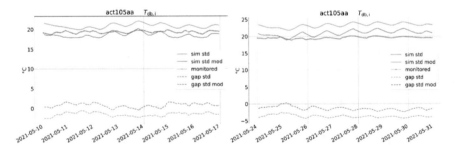

Fig. 18 Timeseries air temperature values in newly renovated residential zone

year. However, the more the other rooms are far from the kitchen, the less they can benefit from the impact of the stove, resulting in colder monitored data even with respect to the standard. The same observations can be made by looking at the POR in Fig. 19, considering both the kitchen and the zone average: during the first simulated weeks in May, simulated behaviour is worse than monitored behaviour, especially in the kitchen, where a higher temperature is maintained, and then during the proper summer season comfort is maintained with almost all hours in cat. II, resulting in POR = 0. Finally, with the beginning of the autumn season, simulated behaviour was recorded to be slightly better than monitored, perhaps because – waiting

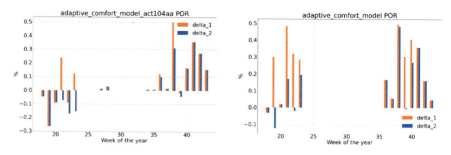

Fig. 19 ACM POR in the kitchen/living room (left) and average in all residential unit zones (right)

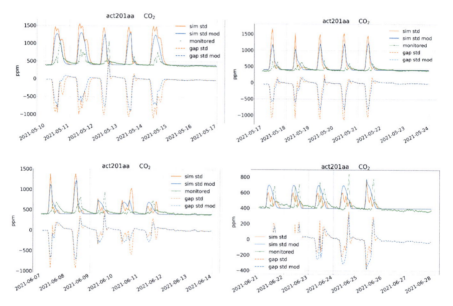

Fig. 20 Timeseries CO_2 values in kindergarten act201aa thermal zone

for the beginning of the heating period – the stove was not used even in the first, colder days.

Concerning the kindergarten results, Figs. 20, 21, and 22 show CO_2 trends over time in the three main activity areas of the floor (*act201aa/ab/ac*) during 4 weeks in May and June 2021. The three rooms show a more regular behaviour with respect to the residential case, because of the cyclic schedule followed by children in the rooms. However, the three areas show quite different trends, underlining the need to analyse different spatial aggregations to investigate the average behaviour that could be affected by room values distribution and to better localize potential problems. Particularly, the three rooms seem to be used in different moments of the day, suggesting the need of even more detailed schedules to better simulate a standard modified behaviour. *Act201aa* is mainly used in the afternoon, which corresponds to lunchtime and the afternoon nap. Differently, *act201ab* is mostly used in the

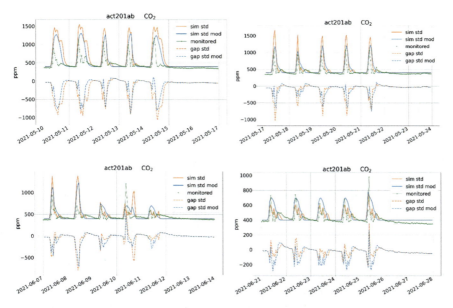

Fig. 21 Timeseries CO_2 values in kindergarten act201ab thermal zone

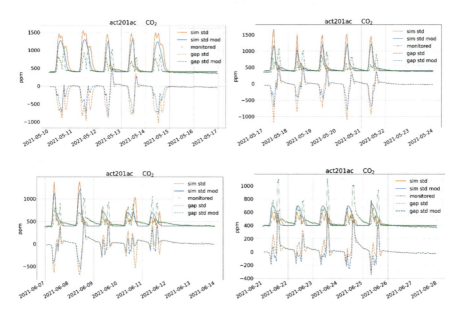

Fig. 22 Timeseries CO_2 values in kindergarten act201ac thermal zone

morning, while *act201ac* usually shows two peaks, one in the morning and the other in the afternoon, corresponding to the double use as playroom in the morning and sleeping area in the afternoon. Peaks of CO_2 concentration seem to increase towards

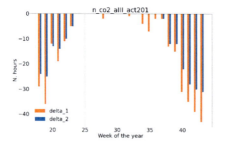
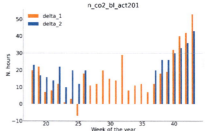

Fig. 23 CO_2 KPIs aggregate results of all kindergarten teaching areas

the warmer summer periods, as if natural ventilation were drastically decreased not to overheat the rooms in the afternoon or reduce airflows or noise during the nap. Moreover, unlike the first weeks of May, when CO_2 trends seem to reach 1000 ppm peaks given the outdoor temperature is still too cold to allow high ventilation rates, in late May all rooms show lower CO_2 peaks, usually below 700 ppm, suggesting an increased natural ventilation usage. In general, CO_2 peaks can be affected by the unnatural ventilation approach forced by covid-19 rules, which can lead to the opposite undesired result of overventilating an area. The latter risk is mainly impacting during the winter months due to higher heating requirements but may also lead to unwanted draft during nap time.

In fact, looking at aggregated weekly results in Fig. 23, both standard and standard modified building simulation settings tend to overestimate the number of weekly hours above the threshold of 1000 ppm despite the performance gap is lower with respect to standard modified behaviour, because of the adapted occupancy schedule which avoids a late-afternoon CO_2 drop and a natural ventilation strategy. For the same reason, the number of hours below the threshold of 600 ppm is greater in the monitored data. Particularly, standard behaviour looks more incongruous with the standard modified than actual monitored behaviour, because of the better balance between occupancy and ventilation and, especially in the summer weeks, the appropriate holiday schedules.

Similar observations can be made also by looking at each room's temperature trend. Particularly, Fig. 24 shows the average teaching area temperature in the same weeks analysed for CO_2 concentration. During the month of May, monitored data show very low temperatures due to the end of the heating season and to the still cold outdoor temperature. Overventilative tendencies could explain the very low peaks in early morning that can reach 15 °C. In June, instead, monitored indoor temperature gets closer to the simulated standard profile, suggesting that the adopted ventilative strategies are also closer to the standard ones. Standard and standard modified behaviours turn out to be very similar in terms of temperature trends, given the input differences are also limited with respect to the residential demo case.

The same results can also be highlighted by the weekly aggregated results for the considered time period. Figure 25 shows the distribution of the identified gaps in ACM categories, together with the POR. Since the POR shows mainly positive

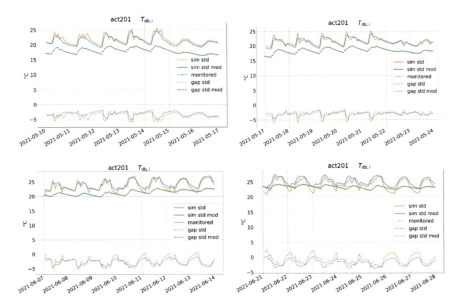

Fig. 24 Timeseries of average air temperature values in all kindergarten teaching areas

results, it means that overall, the kindergarten behaved slightly worse than expected in standard conditions. Particularly, standard behaviours show more hours in adaptive thermal comfort categories I, II up, and III up, while monitored data tend to represent colder temperatures with more data in categories II, III down, and even below cat. III. A similar behaviour can be seen in the central summer weeks, especially in late June and early September, when monitored data are more present in cat. I, while standard behaviours tend to show hotter temperatures than the actual building trend. The central summer weeks, instead, in July and August, are affected by an incorrect standard occupancy, which does not consider holidays.

Considering that the two considered buildings are located in a mountain region, the main problem to reach the best free-running mode in terms of both thermal comfort and indoor air quality was recorded to be, even in the late spring and late summer, finding a good balance between natural ventilation strategies and consequent natural cooling. Especially in the school, the application of a non-optimized ventilation strategy could result in very cold days and even in increased heating consumption in the corresponding season. Results show that, despite trying to apply the best strategies to maintain indoor thermal comfort and air changes (also considering the pandemic), it is difficult without the aid of visual supports and eventual suggestions to understand when the pollutant concentration is increasing above a certain level and when it is low enough to not require additional ventilation. Also, this suggests that mechanical ventilation machines could be of great aid in maintaining indoor comfort in a school building, especially during colder hours and periods, when overcooling may represent a risk. For this reason, three mechanical ventilation units have been recently installed in another floor of the school building to support further tests and verifications.

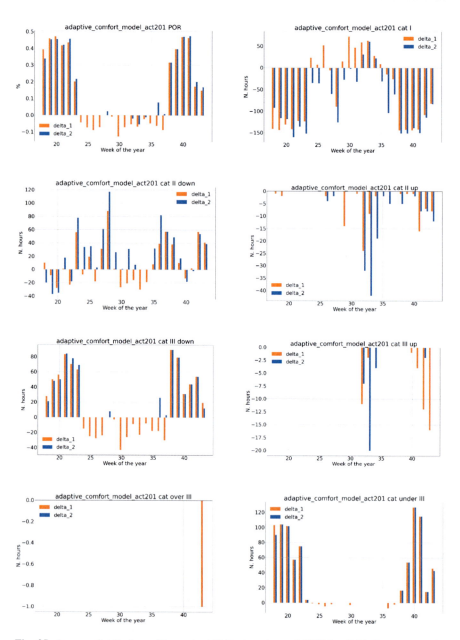

Fig. 25 Average distribution of hours in ACM categories and POR in all kindergarten teaching areas

Results also showed the relevance of both aggregated and timeseries data in understanding potential problems inside the considered space: aggregated results are indispensable in case of random trends, much like the CO_2 concentration in the residential unit, while timeseries data resulted to be of great aid in understanding the more regular school behaviour. Also, considering different levels of spatial aggregation, such as specific rooms or thematic areas (e.g. all the teaching areas together), resulted to be useful in understanding localized problems that could disappear in an average behaviour. Furthermore, the adoption of single zone-specific analyses may help to underline and justify peculiar effects like those given by the wood stove in the residential unit.

Conclusions

The chapter describes initial applications of a new Python library tool able to manage EnergyPlus simulations for different purposes. The current version of the PREDYCE tool is described by detailing the needed inputs and potential outputs of its main scenarios with special regards to the model verification scenario, able to support calibration processes comparing monitored and simulated building data and allowing parametric changes to suggest model error reduction. Furthermore, the performance gap scenario is described to support the identification till real time of discrepancies between monitored and model/simulated building KPIs. The methodology is applied to real demonstration buildings, showing how the proposed pipeline may be used in practice. Results clearly show the potential of the new underdevelopment tool, suggesting several development lines, including the integration on a large building management middleware to support multi-data source integration, taking advantage of the PREDYCE simulation flexibility. This work is part of a larger project that aims to suggest new paths and issues for the next generation of building energy performance certification visions. Currently, PREDYCE faces some limitations: it is developed to work with EnergyPlus version 8.9 (8.x in general), although it may be upgraded to v9.x in future. Additionally, the run of several simulations may benefit from a server facility but may be easily managed through remote REST API services or future middleware solutions. Finally, larger tests on additional demos, considering different building typologies, climates, and national backgrounds, are planned to be performed throughout the coming months.

Acknowledgements Dr. Ahmadi is thankfully acknowledged for having worked on demo inspections and in situ surveys and for developing initial drafts of DesignBuilder models.

This project has received funding from the European Union's Horizon 2020 research and innovation programme under grant agreement No 893945 (E-DYCE).

References

1. E-DYCE. (2020). *Energy flexible DYnamic building CErtification - EU H2020 project No 893945*. https://edyce.eu/
2. Chiesa, G. (2021). *D1.2 Operational dynamic Energy Perfomance Certificate (EPC) specifications*. PoliTO. https://edyce.eu/reports-and-results/
3. European Committee for Standardization. (2019). EN 16798-1:2019 - Energy performance of buildings - Part 1: Indoor environmental input parameters for design and assessment of energy performance of buildings addressing indoor air quality, thermal environment, lighting and acoustics. Brussels. https://standards.cen.eu/dyn/www/f?p=204:110:0::::FSP_PROJECT,FSP_ORG_ID:41425,6138&cs=11EDD0CE838BCEF1A1EFA39A24B6C9890
4. DOE, NREL. (2020). *EnergyPlus™*. DOE, BTO, NREL. https://energyplus.net/
5. de Wilde, P. (2018). *Building performance analysis*. John Wiley & Sons, Ltd. https://doi.org/10.1002/9781119341901
6. De Luca, G., Bianco Mauthe Degerfeld, F., Ballarini, I., & Corrado, V. (2021). Accuracy of simplified modelling assumptions on external and internal driving forces in the building energy performance simulation. *Energies, 14*, 6841. https://doi.org/10.3390/en14206841
7. Energy In Cities Group, University of Victoria. (2020). *BESOS - Building and Energy Systems Optimization and Surrogate-modelling*. https://besos.uvic.ca/
8. Faure, G., Christiaanse, T., Evins, R., & Baasch, G. M. (2019). BESOS: A collaborative building and energy simulation platform. In *Proceedings of the 6th ACM International Conference on Systems for Energy-Efficient Buildings, Cities, and Transportation*, ACM, pp. 350–351. https://doi.org/10.1145/3360322.3360995
9. Chiesa, G., Fasano, F., & Grasso, P. (2021). A new tool for building energy optimization: First round of successful dynamic model simulations. *Energies, 14*, 6429. https://doi.org/10.3390/en14196429
10. Chiesa, G. (2022). *E-DYCE - D3.1 - Dynamic simulation platform*. Turin. https://edyce.eu/wp-content/uploads/2022/03/E-DYCE_D3.1_Dynamic-simulation-platform_28.02.2022_Final.pdf
11. Chiesa, G. (2022). *E-DYCE - D3-2 - Free running module*. Turin. https://edyce.eu/wp-content/uploads/2022/03/E-DYCE_D3.2_Free-running-module_28.01.2022_Final.pdf
12. Claridge, D., & Paulus, M. (2019). Building simulation of practical operational optimisation. In J. Hensen & R. Lamberts (Eds.), *Building performance simulation for design and operation* (2nd ed., pp. 399–453). Routledge.
13. ASHRAE, ASHRAE Guideline 14-2014. (2014). *Measurement of energy, demand, and water savings*. https://www.techstreet.com/standards/guideline-14-2014-measurement-of-energy-demand-and-water-savings?product_id=1888937
14. de Kerchove, T. (2022). *E-DYCE - D2.2 - inspection process*. Losanne. https://edyce.eu/wp-content/uploads/2022/03/E-DYCE_D2.2_Inspection-process_02.03.2022_Final.pdf.
15. Pomianowski, M., & Kalyanova Larsen, O. (2022). *E-DYCE - D2.4 -E-DYCE protocol*. Aalborg. https://edyce.eu/wp-content/uploads/2022/03/E-DYCE_D2.4_Protocol_18.02.22_Final.pdf
16. Capetti Elettronica. (2021). *Capetti Winecap system*. http://www.capetti.it/index.php/winecap

17. Snell, J., Tidwell, D., & Kulchenko, P. (2002). *Programming web services with SOAP* (1st ed.). O'Reilly & Associates.
18. Boland, J., Huang, J., & Ridley, B. (2013). Decomposing global solar radiation into its direct and diffuse components. *Renewable and Sustainable Energy Reviews, 28*, 749–756. https://doi.org/10.1016/j.rser.2013.08.023

Open Access   This chapter is licensed under the terms of the Creative Commons Attribution 4.0 International License (http://creativecommons.org/licenses/by/4.0/), which permits use, sharing, adaptation, distribution and reproduction in any medium or format, as long as you give appropriate credit to the original author(s) and the source, provide a link to the Creative Commons license and indicate if changes were made.

The images or other third party material in this chapter are included in the chapter's Creative Commons license, unless indicated otherwise in a credit line to the material. If material is not included in the chapter's Creative Commons license and your intended use is not permitted by statutory regulation or exceeds the permitted use, you will need to obtain permission directly from the copyright holder.

Dynamic Simulations of High-Energy Performance Buildings: The Role of Climatic Data and the Consideration of Climate Change

Stella Tsoka

Introduction

During the last decades, there is a growing concern towards the reduction of buildings' energy needs for heating and cooling purposes; given that residential and commercial buildings account for the 40% of the final energy use in European Union, the recent Energy Performance Buildings Directive 2010/31/EC [1] has obliged all members of the EU to adopt many measures orientated towards innovations and practices to respond to the growing energy demand of the building sector. In parallel, the Directive imposed regulations foreseeing minimum energy performance requirements for new and existing building under major renovation, both for the residential and the tertiary sector. Considering the rising concerns about the excessive energy consumption and its negative impact on the environment due to high CO_2 emissions, there is an increasing interest in the direction of more accurate building energy performance estimations as well as the establishment of energy saving strategies [2].

In this vein, over the past 50 years, remarkable research has been performed, and several methods of different levels of complexity have been established for the estimation of buildings' energy requirements. The significant computational advances of the last 50 years have contributed on the development of the dynamic building energy performance simulation models (BEPS) for the detailed analysis of the buildings' thermal behavior. The application of such tools enables the assessment of the energy performance of a building based on the location, the construction type, the internal gains, the occupancy schedules, and the weather parameters at study

---

S. Tsoka (✉)
Faculty of Civil Engineering, University of Patras, Patras, Greece

Faculty of Civil Engineering, Aristotle University of Thessaloniki, Thessaloniki, Greece
e-mail: stsoka@civil.auth.gr

area [3]. Undeniably, their application on the analysis of the buildings' energy performance instead of models that are based on the simplified procedure of the EN ISO 13790 [4] or the degree-day method/degree-hour method [5, 6] is an important step towards the acquisition of more reliable estimates of the buildings energy performance. To date, the most commonly applied building energy simulation programs [7, 8] are (a) the DOE-2 [9, 10], (b) the TRNSYS [11, 12], (c) the ESP-r [13], and (d) the EnergyPlus [14, 15]. Each one of the abovementioned tools has specific capabilities and features, a detailed discussion of which is performed in a recent scientific review of Coakley et al. [7]. In all the respective models, the major simulation input data involve (a) the hourly weather dataset, reflecting the climatic conditions of the location of the study area to which the examined building is subjected, (b) the occupancy schedules, (c) the internal gains by electrical equipment, occupancy, and lighting, (d) the ventilation and infiltration flow rates, (e) the temperature set points inside the building zones, and (f) the geometrical model of the building along with the thermal and optical properties of the building elements.

Generally, when configuring the energy performance simulations, the attention is primarily given on the accurate representation of the building model and the precise definition of the operation schedules including lighting, equipment, thermostatic control, ventilation, and occupancy; several studies have evaluated the effect of the occupant's behavior and the respective occupancy schedules on the obtained energy performance simulation results [16–18], whereas others have analyzed the role of the energy management systems (EMS) on the estimated buildings' energy performance [19–21]. Yet, as underlined by the IEA Annex 53 [22], the annual weather dataset, comprising of 8760 hourly values of various climatic variables such as dry bulb temperature, dew point temperature, wind speed, solar radiation, etc., will also strongly influence heating and cooling loads calculations, systems' dimensioning, energy production from solar panels, etc. The quality of the introduced climatic data is thus of crucial importance when high calculations' accuracy is required, as in the case of the net zero energy buildings.

To date, the traditional techniques that aim to establish typical weather datasets for dynamic building energy performance simulations at specific locations (see section "Typical Weather Years as an Input Parameter for BEPS") are mainly based on long-term observations issued from weather stations in the peripheral zones of cities, outside the dense urban centers; as a result, complex interactions between solar radiation, wind speed, and the increased urban densities are ignored. The generated weather datasets could be thus considered more consistent for energy performance calculations of buildings located in new residential areas in the peripheral zones, characterized by lower building densities and higher vegetation coverage, rather than for dense urban areas [23].

Yet the road towards the precise investigation of the energy demand of high-energy performance buildings such as the net zero energy buildings would require a holistic selection of suitable weather data that accurately represent the climatic conditions of the area under investigation. Moreover, given that the life cycle of a building extends to 50–70 years, many scientific studies also underline the necessity to consider the climate change effect on energy performance simulations [24]. In other words, the determination of the buildings heating and cooling energy demand should entail both the consideration of the urban warming and the urban heat island effect and also the higher Tair values due to climate change [25].

Based on the abovementioned remarks, the current chapter aims to systematically review the results of previous studies that investigate the effect of the hourly climatic files, introduced as an input boundary condition on the dynamic building energy performance simulation. More precisely, the review focuses on studies that have been based (a) on traditional methods for the acquisition of the required weather datasets, (b) on methods for the consideration of the site specific microclimatic conditions, and (c) on the results of other scientific studies that develop future weather datasets for building energy performance simulations, incorporating the future climate change.

The chapter is organized in the following way: section "Typical Weather Years as an Input Parameter for BEPS" initially describes the main methods for the generation of the so-called typical weather years (TWY) (or typical weather datasets (TWD)) mainly based on the statistical analysis of multiyear climatic records of fixed weather stations. The results of previous studies, assessing the variation of the annual building energy needs as a function of the provided TWY, are then presented and discussed, whereas the last part of the section summarizes the advantages and limitations of the use of TWYs for dynamic BEPS. To continue, section "Urban Microclimate Data as an Input Parameter for BEPS Simulation" focuses on previous studies, using local microclimatic data of the area under study as an input boundary condition for the dynamic BEPS; the different procedures towards this direction are described, along with key issues arising in terms of the dynamic energy performance simulation results. Finally, the results of previous studies creating future weather datasets and investigating the role of climate change on buildings energy demand are described in section "Climate Change and Generation of Future Weather Datasets for Dynamic Building Energy Performance Simulations", while section "Synopsis and Conclusions" presents the summary and the conclusions of the chapter.

Typical Weather Years as an Input Parameter for BEPS

As previously mentioned, a key input parameter for buildings dynamic energy performance simulations is the hourly weather file, comprising of 8760 hourly values of major climatic variables. Given that weather data can present important interannual variations, several years of climatic records are needed so as to obtain an average performance [26, 27]; however, long-term real weather data, measured at a close distance from the study area, are not always available, and, thus, a typical practice for most of the corresponding BEPS models is to adopt the so-called typical weather year files (TWY) [28], the creation of which is based on the statistical analysis of multiyear observations, issued by weather stations outside the urban areas such as airports, university campuses etc. It could be thus claimed that the use of a "typical weather year" files for BEPS has arisen from the need to compromise accuracy and computationally efficient simulations. In other words, the TWYs are artificial years, generated to be representative of the long-term climatic averages of a specific reference location, such as the one of a meteorological station. The "typical year" is created by defining the 12 most representative months from multiyear climatic records. The 12 "typical meteorological months" (TMM) [29], selected from different years, are then concatenated so as to form a single year. To date, several procedures regarding the generation of typical weather data have been reported in the literature, including the "typical meteorological year," the "test reference year," the "weather year for energy calculation," and the "example weather year" techniques.

The "Typical Meteorological Year" Technique

One of the most commonly used hourly data format files for BEPS is the typical meteorological year format, created by the US National Renewable Energy Laboratory in 1978 for 248 locations using long-term observations of solar radiation and weather data from 1952 to 1975 [30]. Later updates, in the early 1990s, introduced the TMY2 format derived from measurements during 1961–1990 [31], while the current TMY3 datasets cover 1020 sites across the USA using data from 1976 to 2005 or 1991 to 2005 [32]. The "typical year" is composed of 12 "typical meteorological months" (TMM) [29], selected from different years and concatenated so as to form a single year. The selection of the 12 TMMs, concatenated so as to form a single year, is known as "Sandia method" [33]. It involves a two-step procedure, based on nine indices with different weights: daily global radiation, daily maximum, minimum and mean air temperature, and relative humidity as well as daily maximum and mean wind velocity. During the first stage, five candidate months, the cumulative distribution functions of which have the smallest deviation from the respective CDF of the long-term distribution, are defined. The comparison between short-term and long-term distribution is performed for each month and each one of the nine parameters mentioned before, using the Finkelstein and Schafer (FS) statistics [34]. The FS statistics are calculated for each month of every year and for all nine climatic parameters, and a weighted sum is then produced, with different weights attributed to each parameter. In the second stage, the final selection of the TMM is performed by examining the months with the lower weighted sums, the smaller deviation of the monthly mean, and median from the corresponding long-term mean and median [35]. A detailed description of the creating process of a TMY is presented in [33].

The "Test Reference Year" Technique

A similar compilation procedure, based on long-term measurements (usually 20 years), is applied for the test reference year (TRY), a technique developed by the Chartered Institution of Building Services Engineers (CIBSE) [36]. Finkelstein-Schafer statistics are estimated for each month and each climatic variable in order to define the typical months which are then aggregated to form a complete year, yet there are two major differences with the TMY technique: (a) only the mean value of dry bulb temperature, wind speed, and the global solar radiation are taken into account, instead of the nine variables considered in the TMY method, and (b) all three variables are equally weighted whereas on the TMY technique, the global radiation is considered to be more critical, having the highest weight factor.

The "Weather Year for Energy Calculation" Technique

Another technique for generating typical hourly weather datasets is the weather year for energy calculation (WYEC) method, initially proposed by Crow [37]. It relies again on the definition of 12 representative months to form a complete year; still, there are two major differences in comparison with the previous techniques: individual months are only selected if the average monthly dry bulb temperature has a difference up to 0.20 °C of the respective long-term monthly average. After the initial selection, in case abnormalities or extreme events are found, individual days or hours can be adjusted so as for the monthly mean values to come closer to the respective long-term values. The complete procedure of creating hourly weather values with this technique along with the corresponding further improvements was described by [38, 39].

The "Example Weather Year" Technique

Finally, a different approach is proposed in the example weather year (EWY) method, introduced by Holmes et al. [40]. A complete year is now defined, the monthly mean weather values of which contain the least abnormalities in comparison with the long-term observations. Thus, a whole representative year instead of representative months is sought. More precisely, monthly mean values of global and diffuse radiation, daily mean wind speed, mean, maximum and minimum dry bulb temperature, and their standard deviation from the long-term mean are estimated. The years that contain monthly means that differentiate more than standard deviation from the corresponding long-term mean are rejected, and the last remaining year would become the example year chosen [41].

Stochastic Generation of Synthetic Weather Years for Building Energy Simulation

According to the previous analysis, the creation of typical weather datasets requires the statistical analysis of various multiyear climatic records. Nevertheless, as long-term, hourly time resolution weather data are rarely available for every site, it is a very common practice for engineers to use stochastic weather generators, so as to create complete climate datasets for any desired geographic location [42–45]; the so-called synthetic weather years [46] can be thus used in areas that long-term hourly resolution data are not available. More precisely, stochastic models provide the user the possibility (a) to use the software's database for the TWY generation (i.e., long-term measurements in various meteorological stations spread all over the world) or (b) to import his own monthly values of the necessary climatic variables, such as dry bulb temperature, solar radiation, relative humidity, and wind speed. In the latter case, the stochastic generation of each climatic variable is based on random choice using its statistical daily and hourly distribution autocorrelation but also interactions with the other variables [47]; intermediate data will also have the same statistical properties as the monthly imported data, i.e., average value, variance, and characteristic sequence [48, 49].

In case the generation of a SWY concerns an unobserved location, the creation of the annual weather dataset will be based on long-term data from the nearest meteorological stations; more precisely, the observed climatic parameters will be interpolated, and the necessary for the building energy simulation time series will be generated using the interpolated values of the parameters [50]. As underlined by Semenov et al. [51], stochastic weather generators can only be considered as an important scientific tool for creating synthetic time series of weather parameters, *statistically identical* to the observation rather than predictive tools for weather forecasting.

To date, several stochastic weather generators for the creation of hourly resolution annual weather data have been developed. An extended list of relevant tools is provided by Skeiker [52]. In the field of building energy analysis, the most commonly encountered generators, when conducting the research of relevant scientific studies are RUNEOLE [47], UCKP09 [53], ENER-WIN [54], and Meteonorm [48], with the latter one being the most widely applied; its database covers more than 8000 meteorological stations, on the basis of which, interpolation models can provide climatic results at any location worldwide. Apart from the climate database and the spatial interpolation tool, the model also consists a stochastic weather generator for the creation of synthetic years [55]. Apart from the field of building energy performance calculations, stochastic models have been also broadly used for the acquisition of yearly precipitation data or storm parameters, necessary for the design of hydraulic structures, ground erosion analysis, crop management policies, etc. [56–59]. A detailed review on the use of stochastic weather models for agricultural, ecosystem, and hydrological impact studies is provided by Wilks and Wilby [60].

The Use of Typical Weather Years on Dynamic BEPS

As mentioned in the previous subsections, the existing techniques for the determination of TMMs involve different statistical procedures, and as a result, the aggregated TWYs may vary significantly [61]. Given that the simulation output is strongly related to the weather parameters, potential differences on the individual hourly weather variables of the climatic files can induce high deviations on the output of the building energy simulations [62] (see Table 1).

In a previous study of Huang and Crawley [63], the effect of using different methods for the creation of hourly weather datasets on the estimation of the energy consumption of a three-story office building has been evaluated. Simulations were conducted with the DOE-2 model and concerned five locations in the USA. The generation of the examined hourly weather files has been based the TMY, TMY2, TRY, WYEC2, and WYEC methods, while the same 30 years climatic records have been used. The results of the analysis suggested that the average variation in the annual energy consumption due to weather file variation is ±5% with absolute maximum and minimum discrepancies of 11% and 1.8%, respectively.

In the same context, Seo et al. [64] have assessed the impact of using different approaches for the creation of typical year weather datasets, on the energy performance calculations of a three-story office building. The total energy needs and the peak electricity demand have been calculated for different types of TWYs (IWEC, TMY, TMY2) and were compared with the simulation results obtained for a 30-year average climatic dataset. The acquired results indicated a maximum deviation of 5% between the simulation outcome using the various TWY datasets and those calculated using the long-term average weather data. Moreover, the authors mentioned that for the ten US climates that have been analyzed, 15 years of recorded data would have been also adequate for typical weather year generation.

In the study of Hong et al. [65], the peak electricity demand and the energy use of three types of office buildings in 17 different locations have been assessed. The simulated results, calculated for the long-term average of 30 actual meteorological years (AMY), were then compared with the respective results for a TMY dataset, reflecting typical weather conditions for the same 17 locations. The analysis revealed significant variations between the weather variables of AMY and TMY datasets, a fact that has also affected the calculated energy use for heating and cooling. Substantial deviations on the calculated HVAC energy use, reaching 37%, were reported for the medium-size office building, whereas for the large office buildings, the HVAC energy needs using the TMY dataset were found 9.2% lower than the respective AMY results.

Moreover, Yoo et al. [66] estimated the energy performance of a generic office building, using the TRY and the TMY methods for the city of Seoul, South Korea. The obtained results were then compared with the respective simulation results obtained for a 20-year long-term dataset of hourly climatic data. The analysis indicated that the established TMMs through the two procedures were generally different with only the months (April and July) being the same, whereas both the TMY

Table 1 Summary of previous studies, using different weather files for dynamic BEPS

Reference	Aim of the study	BES model	Results
Crawleu et al. [67]	Investigate the influence of using different weather datasets, generated with the TMY,TMY2, TRY, WYEC2, and WYEC methods for eight locations, on the simulated annual energy and peak loads	DOE-2	Annual energy consumption varied from −11% to 7% Annual peak cooling loads varied from −11.5% to 30.5% Annual peak heating loads varied from −48.5% to 3.2% TMY2 dataset represented more accurately the long-term observations
Seo et al. [64]	Assess the influence of using typical year weather data, generated with five different techniques and 30 years actual weather data, on the simulation of annual total energy consumption. Simulations are performed for ten locations	DOE-2	Maximum difference on annual energy use, reported at 5% For the ten US climates analyzed, 15 years of recorded data would have been also adequate for typical weather year generation
Hong et al. [65]	Analyze the impact of using 30 years actual meteorological data and TMY data for the simulation of 3 types of office buildings in 17 different locations	Energy+	HVAC energy use differences may vary from −37% to 18% Annual weather variation has a greater impact on the peak electricity demand than on energy use Simulated energy savings and peak demand reduction through energy conservation measures, using the TMY3 weather data and 30 years actual weather data present large discrepancies
Bhandari et al. [68]	Analyze the impact of using three actual weather datasets from different sources for a calendar year, on the simulated energy performance	Energy+	Individual hourly variables varied up to 90% Annual building energy consumption varied up to ±7% Annual peak loads varied up to ±40%
Yang et al. [69]	Compare the energy simulation results obtained when using a TMY weather file, actual weather data from individual years, and weather file issued from long-term means. Simulations performed in five locations for an office building	DOE-2	Monthly loads and energy use results using the TMY follow the long-term means quite well Mean bias error of energy use results between TMY and long-term mean range from −1.7% to 3.6% Root mean square error of energy use results between TMY and long-term mean range from −4.3% to 5.4%

(continued)

Table 1 (continued)

Reference	Aim of the study	BES model	Results
Tsoka et al. [71]	Evaluate the effect of the methodology for the establishment of TMMs, when stochastic methods are used for the creation of TWY has been evaluated	Energy+	Establishment of different TMMs for each applied methodology Variation of the annual heating energy needs may up to 3.2% and 9.3% for a typical insulated and non-insulated apartment, respectively Cooling energy demand variation up to 7.7% and 13.6% for a typical insulated and non-insulated apartment, respectively, depending on the weather file

and TRY simulation results generally followed the respective values for the long-term dataset, yet the TMY method presented a larger underestimation of the energy needs, compared to the TRY method. Peak underestimations of the annual heating and cooling load reached 4.70% and 5.80% correspondingly for the TMY, while the respective values for TRY were found 2.28% and 0.64%.

In the same vein, Aguiar et al. [70] compared synthetic and typical years for Lisbon, Portugal, with regard to the thermal performance of a building test cell. The analysis indicated an overall good performance of the synthetic datasets suggesting that the latter could generally substitute the long-term observed series in a reasonable way.

In a recent study of Tsoka et al. [71], the effect of the methodology for the establishment of TMMs, when stochastic weather generators are used for the creation of TWY, has been evaluated. Using a multiyear dataset of climatic records, TMMs have been defined, according to (a) the weather year for energy calculation method (i.e., TMM_WYEC) and (b) the long-term monthly averages (i.e., TMM_10years average); the corresponding average monthly values of air temperature, wind speed, relative humidity, and solar radiation were then introduced in Meteonorm weather generator, and the two hourly weather datasets were stochastically generated (i.e., TWY_10years average, TWY_WYEC). Along with the two previous datasets, an additional weather dataset was also stochastically created using the climate database of the software (i.e., TWY_Meteonorm default). The analysis revealed that the simulated annual heating energy needs may vary up to 3.2% and 9.3% for a typical insulated and non-insulated apartment, respectively, depending on the weather dataset used; the corresponding deviations on annual cooling requirements range up to 7.7% and 13.6% as a function of the weather file, used as input in the simulation.

Finally, Koci et al. [72] evaluated the energy demand of a residential building in Central Europe, using a TRY dataset for the respective location; the obtained results have been then compared with the corresponding energy demand, estimated for actual measured, weather datasets, for the years 2013–2017. The analysis revealed lower heating energy needs when the actual weather datasets have been applied, ranging by 4–15%; the respective discrepancies for the cooling energy needs ranged between 4% and 20%, indicating the need to update the typical weather datasets, generally used in dynamic BEPS at the specific location.

It can be concluded that the "typical weather years," determined via the previously mentioned techniques, have been widely used by engineers and researchers, to achieve both accuracy and computationally efficient simulations. Yet, despite the significant advantages, the application of the typical weather datasets has some serious drawbacks, affecting the quality of the obtained results. The first important limitation involves the neglection of extreme weather conditions; in all the existing methods for the generation of a TWY, the peak and extreme year conditions are excluded, and the issued TWY files are representative of the average weather conditions over a year and can be therefore only considered suitable for predicting the buildings' average energy demand rather than peak heating and cooling loads [62]. Yet, as extreme weather events such as heat waves or frost winter days are essential when it comes to the estimation of peak energy use, various methods towards the establishment of less typical and extreme weather years have been proposed [55]. The second major limitation entails the neglection of the local microclimatic conditions. More precisely, all the existing techniques are mainly based on long-term observations issued from weather stations outside the city centers; as a result, complex interactions between solar radiation, wind speed, and the increased urban densities are ignored [73, 74]; in this framework, many previous scientific studies have tried to establish various methodologies to assess the magnitude of the energy impact of higher urban Tair, because of the UHI effect, on the buildings' heating and cooling energy needs, and the main applied techniques along with the acquired results from previous relevant studies are presented in the following section.

Urban Microclimate Data as an Input Parameter for BEPS Simulation

Up to the present time, the impact of local microclimate on the buildings' heating and cooling energy needs has been conducted either with the use of onsite measurements or via integrated computational methods. In the first case, simulations are generally conducted with climatic variables, measured in rural and urban stations, and the obtained energy needs are comparatively assessed; in the second case, integrated computational methods between microclimatic models and dynamic BEPS tools are applied so as account of the impact of the urban environment. A summary of relevant studies is given in the following subsections.

Evaluation of the Role of Urban Microclimate on Buildings Energy Performance, Using Urban and Rural Climatic Data

In relevant scientific studies, the analysis initially focuses on the comparison between onsite climatic records in urban and suburban areas, with the latter generally characterized as "reference" locations. From one hand, the comparison of the two datasets permits the calculation of the maximum difference between the urban and the reference station [75], known as the urban heat island intensity [76], and on the other hand, the diurnal variation of the climatic parameters occurring in both sites can be also assessed. At a second step, the measured data of the urban area and the reference station are generally employed to develop sample years that are then used as input boundary condition for the building energy performance simulations. Calculations are mainly performed for reference buildings involving typical residential or office multistory buildings, single family buildings, etc. The comparative assessment of the simulated energy needs for both climatic datasets (i.e., reference and urban area) provides further insight on the energy penalty due to the differentiated urban microclimatic conditions. The results of relevant scientific studies are presented in Table 2.

In the study of Cui et al. [77], the hourly weather data of ten urban and rural weather stations for the calendar year 2000 were used as input for energy performance simulations of a typical building using the DeST software, a dynamic simulation tool frequently used in China [78]. Simulation results suggested substantial differences on the simulated heating and cooling energy demand as result of the high ambient temperatures, recorded in the urban areas. The average annual heating requirements of the rural and the urban stations were 59.4 and 49.9 kWh/m^2, respectively, while the corresponding values of the average annual cooling needs were 56.2 and 62.6 kWh/m^2, respectively.

Similar remarks were also drawn by Magli et al. [79] who have analyzed the hourly air temperature records, issued from an urban and a suburban weather station in Modena, Italy, for the calendar years 2011 and 2012. The results indicated that the urban Tair values are always higher than the respective values at the suburban area, while the obtained hourly climatic records of the urban and the suburban weather station were employed to modify the default climatic dataset of the TRNSYS simulation tool. The simulation output has shown that the annual heating energy needs for the suburban area were 26% and 23% higher than the respective needs of the urban area, for 2011 and 2012 correspondingly. In summer, the annual cooling energy requirements for the urban area were 8% and 8.5% higher than the corresponding needs of the suburban area, for 2011 and 2012, respectively.

The influence of the urban microclimate on the cooling energy needs of typical residential buildings in the city of Rome, Italy, was examined by Zinzi and Carnielo [80]. Onsite hourly Tair and RH values were continuously measured in three urban areas, one peripheral zone, and in one reference, undisturbed location during summer 2015 and 2016. Results from the monitoring campaign indicated higher monthly average temperatures in the historical city center, characterized by increased

Table 2 Summary of previous studies, using onsite climatic records for building energy performance simulations

Reference	Country	Aim of the study	Results
Cui et al. [77]	China	Evaluate the temporal and spatial characteristics of the UHI in Beijing. Assess the effect of urban climate on buildings' energy demand	UHI intensity in winter and summer close to 6 °C and 4 °C, respectively. Urban areas' heating load lower by 16% compared to the rural area; cooling load increased by 11% due to the UHI
Magli et al. [79]	Italy	Compare the hourly Tair records issued from an urban and suburban weather station for 2011 and 2012. Evaluate the effect of local microclimate on buildings' heating and cooling energy needs	Annual heating energy higher by 23–63% for the suburban area. Annual cooling energy needs higher by 8–8.5% 8.5% for the suburban area
Zinzi et al. [80]	Italy	Compare the Tair between urban and rural zones in Rome for 2015 and 2016. Assess the effect of high Tair values in cooling energy needs of urban buildings. Evaluate the impact of UHI on the indoor thermal conditions of buildings	UHI intensity up 10 1.8 °C during the monitoring period. Average annual cooling energy needs of the urban buildings higher by 30–45% than the respective cooling demand at a reference location. Increase of the overheating hours inside the urban buildings due to the UHI
Street et al. [81]	USA	Compare hourly climatic records of an urban and two rural weather stations. Evaluate the energy impact of UHI regarding heating and cooling needs	Maximum urban heat island intensity ranging between 1.30 and 2.80 °C. Heating energy needs of the urban dwellings lower by 2–10% compared to the reference station. Increase on the cooling energy needs of the urban buildings by 20–37% compared to the rural area
Salvati et al. [82]	Spain	Compare Tair records from urban and rural fixed weather stations in Barcelona. Evaluate the UHI intensity during summer. Assess the energy impact of UHI on buildings energy needs	Maximum UHI intensity of 4.3 °C in summer. Average relative increase of the sensible cooling load between 19% and 24% for the urban areas compared to the rural ones
Guattari et al. [83]	Italy	Comparatively assess the hourly climatic records of four urban and two rural weather stations in Rome for 2014–2016. Evaluate the UHI intensity on an annual basis. Assess the effect of urban microclimate on buildings energy needs	Peak Tair differences between urban and rural stations reported in summer and ranging between 2.2 and 4.7 °C. Cooling energy demand increased by 16% in urban areas compared to the rural ones. Heating energy demand decreased up to 15.8% for the urban areas due to the urban warming

building densities and very low presence of vegetation. The monitored Tair data for summer 2015 and 2016 were then used as boundary conditions for energy performance simulations of typical residential buildings. Calculations were performed for both insulated and non-insulated building envelopes with the TRNSYS tool. Simulation results indicated a high-energy penalty for a non-insulated building in the city center; the average annual cooling energy needs for the three urban areas were 30–45% higher than the respective cooling demand at a reference location. Similar trends were reported for the insulated building envelope although the magnitude of difference was lower.

The energy impact of higher ambient urban air temperatures in Boston, USA, was evaluated by Street et al. [81]. At a first step, climatic data from an urban and two rural weather stations were used to evaluate the maximum urban heat island intensity which varied between 1.30 and 2.80 °C, depending on the selected reference station. Onsite Tair measurements from the urban and the two rural weather stations were then used as input boundary condition in the EnergyPlus simulation model to simulate the energy performance of a small office building. It was calculated that the heating energy needs in the central area of the city were reduced by 2–10% depending on the used reference station. As for the cooling energy requirements, they were found to be 20–37% higher in the urban area, than in the reference locations.

Salvati et al. [82] have evaluated the intensity of the UHI phenomenon in Barcelona, Spain, and they have analyzed the effect of higher urban air temperatures and lower wind speeds on the energy performance of typical building units. More precisely, the assessment of the UHI intensity was based on the comparison of hourly air temperature, wind speed, and wind direction records, measured inside two street canyons and a rural, reference weather station; climatic records only concerned single days during summer 2014 and not a complete calendar year. During summer, the maximum differences between urban and rural sites occurred at midnight and in the afternoon, while the wind speed was always lower at the street level, compared to the respective values in the undisturbed rural area, due to wind sheltering by buildings. The variation on the sensible cooling load a building unit as a function of the urban and rural climatic data was also assessed via the EnergyPlus simulation tool. The hourly climatic measurements for the two summer days were used so as to modify the default weather file of the EnergyPlus simulation model. The simulation results suggested an average relative increase of the sensible cooling load, ranging between 19% and the 24%, for the 10th and the 17th of July 2014, correspondingly.

In the same vein, Santamouris has performed a comparative analysis of 13 previous scientific studies in which climatic measurements from urban and reference rural stations were employed for the dynamic BEP simulations of various reference buildings [75]. It was found that the average annual cooling energy penalty as a consequence of higher urban temperatures was close to 13.1%. Evidence from the analyzed scientific studies also suggested that in cooling dominated climates (i.e., areas with an average summer temperature higher than 27.0 °C), the increase of the cooling energy demand as a consequence of the urban heat island phenomenon significantly outweighs the corresponding heating energy conservation. On the other hand, in heating dominated climates (i.e., areas where the average summer Tair is lower than 23.0 °C), the higher ambient urban Tair will lead to a considerable reduction of the heating energy demand, while the increase of the building cooling needs would be of lower importance [75].

It can be concluded that the consideration of the complex interactions between the climatic variables and the urban fabric is of high importance towards the accurate calculation of the energy requirements of urban buildings [84], yet, the abovementioned results highlight that the magnitude of the energy impact of the urban warming is strongly related to the intensity of the urban overheating, the buildings' typology and envelope characteristics, the site specific microclimatic conditions as a function of the respective morphological characteristics of the study area, etc.

Yet, even if experimental data through measurements have always provided the basis for model development and validation, analyzing the effect of the urban microclimate on buildings energy performance using measured hourly climatic records presents some important limitations that should be taken into consideration. Firstly, the collection of onsite urban measurements is a considerably time-consuming process, while the observed climatic parameters generally reflect the actual meteorological conditions of the calendar year they have been conducted rather than the average climatic characteristics of the study area. In parallel, field measurements can be rather prone to errors associated either with the calibration of the respective measuring equipment [85] or the potential near wall phenomena, formed along the building elements, near which the microclimatic sensors are placed [86]. Another important restriction also involves the number of the measured climatic parameters, generally limited to air temperature and relative humidity records, and, thus, the decreased urban wind speeds and the reduced convective heat losses as a consequence of wind sheltering by building volumes is not considered during the simulations [74]. Finally, the study of the energy penalty of the urban warming on buildings using measured urban and rural climatic data does not allow the evaluation of various mitigation strategies since climatic records before and after the application of several mitigation techniques will be performed under different meteorological conditions and as a result, the climatic variables cannot be considered comparable.

In contrast to the field measurements, the use of numerical methods for the urban microclimate analysis and its effect on the buildings' energy needs allows the simulation of various urban morphologies so as to draw general conclusions, recommendations, and policy implications. The respective computational approaches, reported in the literature as coupled/coupling simulations, involve the methodological approaches of integrating models from two different software (i.e., microclimate tools and BEPS models) so as to increase the accuracy of the buildings energy performance analysis. A detailed review of various relevant studies has been performed by Bozonnet et al. [87], Frayssinet et al. [88], and Krebs and Johansson [89]. In the next section, the results of relevant scientific studies are presented.

Evaluating the Effect of the Urban Microclimate on Buildings' Energy Performance, Using Numerical Methods

The dynamic BEPS models have been originally designed to evaluate the performance of stand-alone buildings [74]; the neighboring buildings are only considered as solar radiation obstacles, while their behavior with regard to heat transfer exchange with the examined building is not considered [88]. Yet the existing scientific knowledge suggests that an integrated simulation approach between BEPS codes and microclimate models is necessary towards more efficient buildings energy demand calculations, accounting for the local thermal environment [90, 91]. More precisely, the treatment of the physical phenomena, occurring at the buildings-environment interface, affecting the heating and cooling energy requirements involves (a) the solar radiation fluxes received at the building facades as a function of the near obstructions, albedo values, etc. [92]; (b) long-wave radiation flux between the constructions' surfaces and the sky, as a function of the surfaces' emissivity and the opening to the sky [93, 94]; and (c) the near-wall phenomena and the heat exchange through convention [74].

However, very limited models offer an integrated analysis of both the buildings energy requirements and their surrounding environment. A respective simplified approach is provided by the CitySim [95] and SUNtool [96] model, in which the effect of the solar masks of the urban environment is accounted for during the estimation of the buildings' energy performance; still, airflow effect and the convective heat fluxes cannot be handled by the models. To address this issue, previous scientific studies have proposed integrated simulation approaches to accurately simulate the effect of radiative exchanges, heat transfer, and local airflow on the building's energy performance. The most commonly applied techniques involve coupling (a) microclimate simulation models such as the ENVI-met model with a dynamic BEPS tool or (b) a CFD code with a BEPS tool.

ENVI-met simulation model is one of the most widely applied software for urban climate analysis [97] as it provides an integral modelling of radiation and convention phenomena. Various studies have proposed an integrated simulation approach between the ENVI-met microclimate model and dynamic BEPS tools. In these studies, microclimate simulations of the study area, in which the case study building is located, are initially performed, and at the next step, the ENVI-met simulation outputs (i.e., Tair, RH, Tdew, WS), reflecting the microclimatic conditions of the air layer around the analyzed building, are employed to modify the existing hourly weather file of the dynamic BEPS tool. In this way, the site-specific microclimatic conditions, used as boundary conditions at the BEPS, are accounted during the energy performance simulations, yet the coupling is mainly done for specific days and not for a complete year's period, given the time-consuming microclimate simulation.

In this context, Yang et al. [73] established an integrated simulation approach between the ENVI-met v.4 model and the EnergyPlus tool via the software platform Building Controls Virtual Test Bed (BCVTB). Microclimate simulations were performed for a generic case study area in Guangzhou, China, for two typical summer days; the obtained hourly climatic records were then used to modify the existing epw weather file of the EnergyPlus model. In parallel, aiming at an accurate evaluation of the convective heat losses, the convective heat transfer coefficient for each facade of the building was calculated according to the microclimatic results obtained by ENVI-met and it was sent back to EnergyPlus to replace the default CHTC. To evaluate the role of urban microclimate, the calculated cooling energy needs were then compared with the respective values for an isolated, stand-alone building. It was found that the consideration of the local thermal environment significantly affects the energy performance simulation results as an increase of the sensible cooling load of around 10% was observed when the local thermal conditions have been considered.

Similarly, Gobakis and Kolokotsa [90] established a coupling of the ESP-r dynamic simulation tool with the ENVI-met v.4 model to improve the accuracy of the energy requirements' calculations of a tertiary building in Crete, Greece, while the analysis has been restricted to 4 typical days, due to the increased computational cost of microclimate simulations. The ENVI-met simulation output in front of the investigated building facades was extracted and used to modify the ESP-r hourly weather file for the specific simulated days. In parallel, the hourly values of the air temperature, wind speed, and wind direction on each surface of the investigated buildings were extracted from ENVI-met and introduced in a Python code to calculate dynamically the convective heat transfer coefficient according to three different equations. Heating and cooling energy needs were quantified both following the coupling procedure and for the default Meteonorm hourly weather dataset. The analysis indicated a reduction of 2.5% for the heating requirements during the winter period when the local microclimate was considered, while an increase of 8.60% on the cooling energy needs due to local climatic conditions was reported.

In the research of Sharmin and Steemers [98], a one-way coupling computational approach between the ENVI-met v.4 and the IES-VE dynamic energy simulation model was applied for the energy performance calculations of a typical residential building, whereas four different urban arrangements were considered to account of diverse microclimatic conditions. Microclimate simulations were performed for a single hot, summer day, and the acquired hourly microclimatic variables were used to modify the existing hourly weather file for the respective simulated day, so as to evaluate the corresponding cooling energy requirements. The obtained results suggested higher building cooling energy needs by 6–8%, when the urban microclimate has been accounted for, because of higher outdoor Tair and lower wind speed, resulting in lower convective losses.

Morakinyo et al. [99] have established an one way model coupling procedure to assess the impact of the outdoor microclimate as a function of vegetation, on the indoor summer thermal conditions of a university building in Nigeria. ENVI-met microclimate simulation results were performed for three individual summer days, considering both exposed and shaded buildings; the generated ENVI-met v.4 output, reflecting the microclimatic conditions in front of the investigated buildings, was extracted to modify the weather dataset of the EnergyPlus tool. Further dynamic simulations with the EnergyPlus revealed significant differences in the indoor thermal environment due to the diverse outdoor microclimatic conditions and the corresponding tree shading effect.

A one-way coupling approach between the ENVI-met and the EnergyPlus tool to evaluate the effect of cool materials on buildings energy needs has been also established by Tsoka et al. [100]. In this study, microclimate simulations have been conducted for 12 representative days (one for each month), defined through a detailed statistical analysis of long-term climatic records, while both the existing conditions and high albedo pavements have been evaluated. The major microclimatic parameters, estimated in front of the examined building unit, were then extracted from the ENVI-met model, and their average values have been introduced in Meteonorm weather generator to stochastically create the site-specific, annual climatic datasets, henceforward entitled lurban specific weather datasets" (USWDs). The generated, hourly weather datasets, representative of the microclimatic conditions of the urban site (before and after the high albedo pavements applications), have been then used as an input boundary condition for the building unit's dynamic energy performance simulations with the EnergyPlus model.

Finally, in the study of Santamouris et al. [101]. a one-way coupling of the ENVI-met v.4 model with the EnergyPlus was performed to assess the effect of various UHI mitigation strategies on the cooling energy requirements of a typical residential building in Sydney. Microclimate simulations for various mitigation strategies were done for a typical summer day, and the hourly values of the major outdoor microclimatic variables, extracted from the ENVI-met software, were used to modify the existing hourly weather file of EnergyPlus. The weather file was modified for 10 days, under the assumption that all these days presented identical microclimatic characteristics with the simulated one. The results indicated a reduction of the maximum cooling load of 0.4% for every 10% of ground surfaces' albedo, whereas every 10% of increase of the urban greenery led to a reduction of the peak building cooling load of almost 0.5%.

To continue, another technique for the assessment of the urban microclimate's effect on the buildings heating and cooling energy requirements involves the coupling between BEPS and CFD models. As already described, BEPS tools calculate the buildings' energy requirements and the surfaces' temperatures without however capturing the dynamic effects of airflow near the facade that have a strong impact on the convective heat losses and thus cooling and heating energy requirements. Establishing an integrated computational method between a BEPS tool and a CFD code could be thus considered as an alternative to handle the latter limitation [102].

As mentioned by Bouyer et al. [91], three possibilities are offered: (a) full dynamic coupling, in which iterations between the CFD/BES are performed till the achievement of strict convergence; (b) quasi-dynamic coupling, involving only one iteration for both CFD and BES; and (c) and intermediate coupling in which simplifications on the energy balance equations are performed to reduce the computational cost. As full dynamic coupling is vastly computationally expensive, Malys et al. [92] have compared the three abovementioned levels of coupling to determine the level of detail, required for the accurate estimation of buildings heating and cooling energy requirements. The sensitivity analysis involved the coupling between the Solene-microclimat thermoradiative model and the Code_Saturne CFD tool. The obtained results revealed a rather weak impact of the chosen coupling method, on the assessment of the heating energy needs both for an insulated and non-insulated building, while the selection of CHTC calculation mode was more prominent for the non-insulated envelope.

Allegrini et al. [103] have established a coupling approach between the CitySim BEPS model and the Open FOAM CFD code to examine the effect of the urban microclimate on the buildings energy demand. Energy performance simulations concerned generic urban neighborhoods in Zurich, Switzerland, and the CFD simulations were run for steady-state conditions, assuming four different weather conditions, critical for the space cooling energy requirements. In the same context, Barbason and Reiter [104] have developed a coupling approach between the TRNSYS BEPS program and the FLUENT CFD code to accurately estimate the overheating conditions of a residential building. The results of both the abovementioned studies suggested that the unique application of BEPS (i.e., consideration of stand-alone building) cannot accurately describe the thermal behavior of the entire building, while on the other hand, the coupled approach can considerably increase the simulation accuracy.

In this context, Pandey et al. [105] have proposed a new coupling simulation approach between the EnergyPlus and the Ansys Fluent CFD code to evaluate the performance of phase change materials on the built environment. In the same vein, Liu et al. [106] have evaluated the effect of the urban microclimate on buildings energy performance, focusing on the convective heat losses in areas of different surface densities. More specifically, new correlations of CHTCs, dependent of the plan area density of generic study areas, were initially developed using CFD simulations; they were then implemented in the EnergyPlus simulation tool to investigate their impact on the energy performance of a typical residential building located in Philadelphia, USA. Dynamic simulations were conducted both for the newly generated CHTCs and for the default correlations of the EnergyPlus. The analysis revealed that, for an urban neighborhood with plan area densities ranging from 0.04 (i.e., isolated building) to 0.44 (dense area), the variation on the estimated cooling and heating energy needs was up to 4% and 1.3%, as a function of the applied CHTC correlation. Moreover, the shading by the adjacent buildings was found the most crucial parameter concerning the estimated buildings' energy performance; increasing the plan area density from 0.04 to 0.44 led to reduction of the cooling energy needs by 32%, while the heating energy requirements increased by 24% due to lower heat gains.

Finally, Bouyer et al. [91] have proposed a coupling approach of Fluent CFD code and Solene-microclimat thermoradiative model to assess the impact of the urban microclimate, involving solar shading, short and longwave radiation fluxes, and airflow, on the buildings energy requirements; the obtained results showed that the coupling method vs usual BEPSs (i.e., only solar masks by adjacent buildings are considered) led to a decrease of the calculated heating energy needs by 19% and to an increase of the estimated cooling energy requirements by 30%.

Based on the results of the abovementioned studies, it can be said that the combination of the dynamic building energy simulation tools with urban microclimate models is of crucial importance when high accuracy on the estimation of buildings energy demand is required. However, an important issue that may arise on the coupling techniques concerns the correspondence of the building surfaces between the two programs; since different geometry modelers are employed in each of the codes, special attention should be paid to establish the correct exchange of data between the two models [28, 73]. Another important issue of the coupling of a CFD and a BEPS code concerns the method of the time step control and whether a one step coupling, a quasi-dynamic, or a full dynamic coupling procedure is applied [107]. According to the findings of Malys et al. [92] and Robitu et al. [108], a full dynamic coupling is recommended for the assessment of the buildings' cooling energy demand and also for the evaluation of the effect of various mitigation strategies including trees, cool materials, water ponds, etc. Yet, despite its high resolution and the increased accuracy of the results, the respective method is extremely computationally expensive. An intermediate coupling proposed by Bouyer et al. [91] could be a good compromise to face this limitation, but still, further research would be necessary to assess the sensitivity of the corresponding method on various simulation parameters [92].

Given the differences on the temporal resolution of the BEPS models and the microclimate CFD simulation tools, the previously mentioned coupling approaches were mainly established for limited time periods, varying from 24 h to only a few days. Even if significant knowledge on the effect of the local climate on the building's energy demand can be retrieved from the daily analysis, a complete annual simulation would be necessary to obtain a global view of the buildings' energy performance as a function of its surrounding environment.

In this context, Tsoka [109] aimed on the establishment of a novel computational method to combine dynamic building energy performance simulation (BEPS) tools with microclimatic models, to generate typical hourly weather datasets that are representative of the microclimatic conditions of urban areas. In fact, the one-way coupling method has been based on three tools: (a) the ENVI-met v.4 microclimate model, (b) the Meteonorm weather generator, and the (c) the dynamic building energy simulation tool EnergyPlus. The site-specific typical weather datasets have been then used to evaluate the energy performance of generic insulated and non-insulated building units in Thessaloniki, Greece. The annual heating energy needs of the examined, non-insulated building units were found 8.2–11.5% lower when the site-specific microclimate data have been considered rather than the default TWD for Thessaloniki. On the other hand, the higher T_{air} values in the urban districts, captured in the USWDs, resulted in a rise of the annual cooling energy needs, between 13.4% and 28.2%, depending on the study area.

Finally, the existing scientific evidence suggest that the numerical methods have significantly increased the accuracy of climatic input boundary conditions and, thus, the reliability of the dynamic building energy performance simulations. Nevertheless, in the abovementioned studies, the climate change effect and the continuous rise of air temperature are not accounted for in the simulations. Given that the outside air temperature is strongly correlated with the energy demand of buildings, assuming stable climatic conditions and ignoring the global warming during the buildings' life cycle can lead to significant errors on their energy performance assessment. To this aim, many recent scientific studies, the results of which are presented in the next section, have proposed numerical methods to generate future weather datasets for the dynamic energy performance simulations of contemporary buildings.

Climate Change and Generation of Future Weather Datasets for Dynamic Building Energy Performance Simulations

Climate change has a major impact on the urban built environment, not only as regards the buildings' performance and the formation of the outdoor and indoor thermal comfort conditions but also with regard to the higher probability of confronting natural disasters, such as flashfloods. The impact of these effects, often with severe economic and social extensions, is gradually growing, and, thus, it is of imperative importance to establish high accuracy methods to evaluate the magnitude of climate change and propose suitable adaptation strategies. To this aim, the Intergovernmental Panel on Climate Change (IPCC) developed a series of different scenarios (Special Report on Emissions Scenarios (SRES)) that correspond to the main driving forces of future emissions, depending on economic, technological, and population growth. There are four "families" of scenarios, A1, A2, B1, and B2, and within the A1 "family" there are three different groups (A1FI, A1T, and A1B) according to the development of energy technologies [110]. In the Fifth Assessment Report, IPCC introduced updated emission scenarios, the Representative Concentration Pathways (RCPs), which are used in climate models for the future projections of climate. Four different RCPs are developed including a declining pathway that leads to very low radiative forcing (RCP2.6), two intermediate stabilization pathways (RCP4.5/RCP6), and a high radiative forcing pathway (RCP8.5).

Future emission scenarios are the main input data for general circulation models (GCMs), the models that are used for the evaluation of the future climate change and provide climate data at spatial resolution of 150–300 km [111]. Yet, given the coarse spatial resolution of the GCMs, they have to be temporally and spatially downscaled so as to be compatible with dynamic building energy performance simulation tools, as the latter require information at finer spatial scale and at a temporal resolution of 1 h [112]. To date, there are two main approaches to downscale GCMs: statistical and dynamical downscaling.

The statistical downscaling method develops statistical relationships between measured climatic variables (often issued by historical records at a meteorological station) and larger (GCM)-scale variables, employing either analogue methods regression analysis or stochastic methods [28]. In the existing literature, the most widely applied statistical method involves the "morphing" approach, a method proposed by Beclher et al. [29] to adjust present-day datasets according to future projections, so that they can be further used for dynamic building energy performance simulations. More precisely, the methodology for "morphing" weather data involves three different algorithms depending on the weather parameter to be changed, and it includes (a) a "shift" of a current hourly weather data parameter by adding the predicted absolute monthly mean change as issued by a GCM for a given location, (b) a "stretch" of a present-day hourly weather data parameter by scaling it with the predicted relative monthly mean change, as forecasted by the GCM, and (c) a combination of a "shift" and a "stretch" for current hourly weather data. A detailed overview of the "morphing" downscaling approach is given in [113].

The simplicity and the easiness on the application of the morphing method are the main factors for its wide implementation by the scientific community. As suggested by Robert and Kummert [33], especially for the estimation of the annual energy performance of climate sensitive buildings such as NZEBs, multiyear simulations with weather data that consider global warming and the ongoing climate change should always be performed [114]. Perez-Andreu et al. [115] evaluated the influence of climate change on the energy demand of a residential building in Spain, and the morphing approach was implemented to generate future weather datasets for the periods of 2048–2052 and 2096–2100. The obtained energy performance simulation results indicated a decrease of the heating energy demand, as a function of the outdoor temperature increase, while on the other hand, the demand for cooling and the risk of overheating considerably increased.

In the study of Shen [116], the morphing methodology was employed to downscale the output of GCMs and predict future hourly weather data for the period 2040–2069; the predicted hourly weather data have been then used to evaluate the future energy performance of residential buildings in the USA. The analysis indicated an overall reduction of the annual heating energy needs by 14.7–49% and an increase of the annual cooling energy use by 17.4–36.4% as a result of the higher, future T_{air} values. In the same vein, Chan et al. [117] employed the morphing approach so as to create future hourly weather files that would be further used for the energy performance evaluation of residential and office buildings in Hong Kong. The results indicated a substantial increase on energy consumption of the air-conditioning systems due to the expected higher T_{air} values, ranging by 2.6–14.3% for the office building and by 3.7–24% for the residential flat, depending on the emission scenario.

Another method for the statistical downscaling of GCMs involves the use of stochastic weather generators that employ computer algorithms to generate long time series of climatic parameters with statistical properties similar to existing historical climatic records. The use of these models can be very helpful in cases that there are no complete datasets of all necessary weather variables as they can fill in missing data and permit the generation of long synthetic time series. To date, the future weather generators that have been widely employed are the Meteonorm weather generator [118], the CCWorldWeatherGen [119], and the WeatherShift™ [120]. Indicatively, Rey-Hernández et al. [121] used the CCWorldWeatherGen to create future weather datasets so as to model the long-term effect of climate change on a zero energy building in Spain. Dynamic energy performance simulations were conducted with Design Builder tool, and the results showed that a significant increase on the cooling energy demand should be expected for the years 2050 and 2080, while on the other hand, the heating energy needs will drop.

Undoubtedly, the use of such weather generators is an important asset for the generation both of future time series, especially for locations that lack sufficient historical data. Yet, as underlined by Herrera et al. [55] when it comes to future weather datasets, an important disadvantage relies on the inherent hypothesis that future weather sequences will be similar to those observed historically. To address this issue, other scientific studies have used more sophisticated approaches for the generation of future weather datasets for dynamic BEPS by implementing dynamic downscaling of the GCMs. More precisely, the dynamical downscaling refers to the use of a regional climate model (RegCM), driven by the output of a GCM (used as initial and boundary conditions), to dynamically extrapolate the effects of large-scale climate processes to regional or local scales of interest [122]. RegCMs can thus generate climate information at a much finer resolution than GCMs, down to 2.5–100 km, embracing in a more detailed way the topographical particularities and the climatic dynamical processes of the region of interest. Still, an important disadvantage of the method relies on the considerable amount of computational power required along with the large storage devices for the creation of the datasets [123].

In the study of Berardi et al. [24], both statistical and dynamical downscaling methods have been employed to create future weather files for Canada for the energy performance evaluation of 16 building prototypes. Based on the obtained results, the authors suggested that the higher spatial resolution of the dynamical downscaling compared to the statistical methods resulted in a better representation of the local climate conditions, leading to the higher accuracy of the dynamic buildings' energy performance simulations. Similar results were also mentioned in the study of Tootkaboni et al. [124], who comparatively evaluated the effect of different future weather datasets, created both with statistical and dynamical downscaling approaches, on the dynamic energy performance analysis of typical buildings. Again, the statistical downscaling methods of "morphing" and stochastic approach may provide adequate information to comparatively assess the long-term changes in energy building performance. But still, it is the dynamical downscaling method that provides reliable simulation results given its finer resolution.

However, it should be emphasized that in the existing literature, the impact of climate change on the buildings' energy performance is assessed only for the general governing climatic conditions (at a mesoscale) and not for the microclimatic conditions, characterizing each study area. As a result, complex processes that occur between building blocks, solar radiation, and wind speed, determining the urban climate, are neglected; nonetheless, given that the buildings' energy demand is strongly correlated to climatic conditions of the study area, it becomes obvious that the acquisition of high accuracy building energy simulation results necessitates from the one hand a deep knowledge of microclimatic conditions of each study area and from the other hand the consideration of the occurring future climate change, strongly influencing the buildings' energy demand.

Towards this direction, Tsoka et al. [125] aimed on the evaluation of the combined impact of climate change and urban heat island on the heating and cooling energy demand of a generic building unit in the city of Thessaloniki, Greece. The study focused on the generation of future weather datasets using both statistical and dynamical downscaling methods. For statistical downscaling, the Meteonorm stochastic weather generator was employed, creating a future weather dataset for the year 2050, whereas for the dynamic downscaling, the use of the RegCM4 model allowed the projection of future climate conditions for the future period 2041–2060. The compounding effect of urban heat island on the warming due to climate change is explored through the three-dimensional dynamic microclimatic ENVI-met model, and the input boundary conditions are based on the output of the RegCM4. The energy performance simulation results indicated that the climate change will lead to a substantial increase of the cooling energy needs while reducing the heating energy demand. The acquired simulation results also highlighted the importance of considering the intensifying effect of urban heat island on climate change, when forecasting the future buildings' energy demand, since its neglection may lead to an overestimation of the heating energy demand by 21% and an underestimation of the cooling energy needs by 22.4%.

Synopsis and Conclusions

In this chapter, the effect of the climatic datasets as an input parameter on dynamic building energy performance simulations has been discussed. The existing scientific evidence proposes various methods for the acquisition of suitable weather datasets, corresponding to different levels of complexity and presenting diverse characteristics. Still, the results of the studies, reviewed in this chapter, agree that the hourly weather file, introduced in the BEPS models as a required input boundary condition, will strongly affect the simulation output concerning heating and cooling energy demand.

Given that it is not possible to have multiyear climatic observations at every site, the generation of typical weather datasets for dynamic BEPS has been traditionally based on the statistical analysis of historical records from weather stations in the peripheral zones of cities (i.e., airports, university campuses, etc.). The obtained standardized weather files correspond thus to a single year that reflects the average regional climate conditions. These weather datasets, widely used by the scientific community, fail however to include peak and extreme year conditions; the issued TWY files are thus representative of the average weather conditions over a year, and they are therefore more suitable for predicting the buildings' average energy demand rather than peak heating and cooling loads during extreme events. Moreover, since the creation of TWDs is traditionally based on climatic observations, recorded outside the cities, the complex thermal and energy balance of urban areas cannot be accounted for in the energy performance simulations of urban buildings, a limitation that may lead to significant inaccuracies in the estimation of buildings energy demand, especially for high-performance buildings such as the NZEBs.

To address this restriction, many scientific studies have proposed the use of onsite microclimatic measurements, so as to consider the site-specific climatic conditions to which the examined building is subjected. This method may improve the accuracy of the obtained energy performance simulation results, but there are still some limitations that should not be neglected. Firstly, the collection of onsite urban measurements is a considerably time-consuming process, and climatic records of at least one calendar year are required when the annual building energy requirements are assessed. Besides, the observed climatic parameters will reflect the actual meteorological conditions of single calendar years rather than the average climatic characteristics of the study area. Secondly, field measurements of microclimatic parameters can be rather prone to errors, associated either with the calibration of the respective measuring equipment or with the potential boundary effects formed along the building elements, near which the microclimatic sensors are placed. Finally, due to limitation of resources, most of the relevant studies only perform onsite microclimatic measurements of one or two climatic variables, mainly involving air temperature and relative humidity records; the dynamic simulations of the heating and cooling loads do not account of the complex convective heat transfers, affecting the buildings' thermal balance.

Numerical methods, establishing coupled simulations between BEPS tools and microclimate models, can address the abovementioned limitations and increase the simulation accuracy. Yet a significant difficulty of the integrated computational approach lies on the different spatial and temporal scales of the employed models; the analysis of the buildings' energy performance using dynamic simulation tools is typically conducted for a period of 1 year, whereas microclimate simulations are mainly carried out for only a few days or just for 24 h due to the high computational cost of the latter ones. As a result, in the scientific studies that combined the microclimate CFD models with BEPS tools, the analysis of building energy requirements only concerned 24 h or a very limited number of days.

In all the abovementioned scientific methods for the acquisition of climatic parameters for BEPS (i.e., statistical methods for TWDs generation, onsite measurements, and coupled simulations), the analysis is conducted for the present-day climatic parameters. Nevertheless, due to climate change, the weather data during a building's life cycle may differ considerably. In new buildings that will serve for more than 50 years, climate change is expected to have adverse influence on their life cycle performance regarding energy needs and thermal comfort. They should thus be designed and constructed with effective measures and environmental features to cope with the impact of climate change. Towards this direction, different methods for the generation of future weather datasets have been proposed by the scientific community, including the use of weather generators or more sophisticated approaches such as the downscaling of general circulation models.

References

1. Recast, E. (2010). Directive 2010/31/EU of the European Parliament and of the Council of 19 May 2010 on the energy performance of buildings (recast). *Official Journal of the European Union, 18*(06), 2010.
2. Soares, N., et al. (2017). A review on current advances in the energy and environmental performance of buildings towards a more sustainable built environment. *Renewable and Sustainable Energy Reviews, 77*, 845–860.
3. Cox, R. A., et al. (2015). Simple future weather files for estimating heating and cooling demand. *Building and Environment, 83*, 104–114.
4. Iso, E. (2008). *13790: 2008 Energy performance of buildings–Calculation of energy use for space heating and cooling*. International Standard Organisation.
5. Allen, J. C. (1976). A modified sine wave method for calculating degree days. *Environmental Entomology, 5*(3), 388–396.
6. Fischer, R., et al. (1982). Degree-days method for simplified energy analysis. *ASHRAE Transactions, 88*, 522–571.
7. Coakley, D., Raftery, P., & Keane, M. (2014). A review of methods to match building energy simulation models to measured data. *Renewable and Sustainable Energy Reviews, 37*, 123–141.
8. Sousa, J. (2012). Energy simulation software for buildings: Review and comparison. In *International workshop on Information Technology for Energy Applicatons-IT4Energy, Lisabon*. Citeseer.
9. Birdsall, B., et al. (1990). *Overview of the DOE-2 building energy analysis program, version 2.1 D*.
10. York, D. A., & Cappiello, C. C. (1981). *DOE-2 engineers manual (Version 2. 1A)*. Lawrence Berkeley Lab./Los Alamos National Lab.
11. Beckman, W. A., et al. (1994). TRNSYS the most complete solar energy system modeling and simulation software. *Renewable Energy, 5*(1–4), 486–488.
12. Klein, S. A. (1988). *TRNSYS-A transient system simulation program*. University of Wisconsin-Madison, Engineering Experiment Station Report, pp. 38–12.
13. Strachan, P. (2000). *ESP-r: Summary of validation studies*. Scotland, UK.
14. Crawley, D. B., et al. (2001). EnergyPlus: Creating a new-generation building energy simulation program. *Energy and Buildings, 33*(4), 319–331.
15. Crawley, D. B., et al. (2000). EnergyPlus: Energy simulation program. *ASHRAE Journal, 42*(4), 49.
16. Erickson, V. L., et al. (2009). Energy efficient building environment control strategies using real-time occupancy measurements. In *Proceedings of the first ACM workshop on Embedded Sensing Systems for Energy-Efficiency in Buildings*. ACM.
17. Parys, W., Saelens, D., & Hens, H. J. J. O. B. P. S. (2011). Coupling of dynamic building simulation with stochastic modelling of occupant behaviour in offices–a review-based integrated methodology. *Journal of Building Performance Simulation, 4*(4), 339–358.
18. D'Oca, S., et al. (2014). Effect of thermostat and window opening occupant behavior models on energy use in homes. In *Building simulation*. Springer.
19. Ellis, P. G., Torcellini, P. A., & Crawley, D. (2008). *Simulation of energy management systems in EnergyPlus*. National Renewable Energy Lab. (NREL).
20. Shaikh, P. H., et al. (2014). A review on optimized control systems for building energy and comfort management of smart sustainable buildings. *Renewable and Sustainable Energy Reviews, 34*, 409–429.
21. Huang, W. Z., et al. (2006). Dynamic simulation of energy management control functions for HVAC systems in buildings. *Energy Conversion and Management, 47*(7–8), 926–943.
22. Annex, I. E., 53. *Annex 53 total energy use in buildings: analysis & evaluation methods*.

23. Guattari, C., Evangelisti, L., & Balaras, C. A. (2017). On the assessment of urban heat island phenomenon and its effects on building energy performance: A case study of Rome (Italy). *Energy and Buildings*. https://doi.org/10.1016/j.enbuild.2017.10.050
24. Berardi, U., & Jafarpur, P. (2020). Assessing the impact of climate change on building heating and cooling energy demand in Canada. *Renewable Sustainable Energy Reviews, 121*.
25. Allegrini, J., & Carmeliet, J. (2017). Simulations of local heat islands in Zürich with coupled CFD and building energy models. *Urban Climate, 24*, 340–359.
26. Janjai, S., & Deeyai, P. (2009). Comparison of methods for generating typical meteorological year using meteorological data from a tropical environment. *Applied Energy, 86*(4), 528–537.
27. Oxizidis, S., Dudek, A., & Aquilina, N. (2007). Typical weather years and the effect of urban microclimate on the energy behaviour of buildings and HVAC systems. *Advances in Building Energy Research, 1*(1), 89–103.
28. Kolokotsa, D., et al. (2016). Development of a web based energy management system for University Campuses: The CAMP-IT platform. *Energy and Buildings, 123*, 119–135.
29. McLeod, R. S., Hopfe, C. J., & Rezgui, Y. (2012). A proposed method for generating high resolution current and future climate data for Passivhaus design. *Energy and Buildings, 55*, 481–493.
30. Hall, I. J., et al. (1978). *Generation of a typical meteorological year*. Sandia Labs.
31. William, M., & Urban, K. (1996). *User's manual for TMY2S*. National Renewable Energy Laboratory. Also can be accessed at http://rredc.nrel.gov.solar/old_data/nsrdb/tmy2
32. Wilcox, S., & Marion, W. (2008). *Users manual for TMY3 data sets*. National Renewable Energy Laboratory Golden, CO.
33. Gazela, M., & Mathioulakis, E. (2001). A new method for typical weather data selection to evaluate long-term performance of solar energy systems. *Solar Energy, 70*(4), 339–348.
34. Finkelstein, J. M., & Schafer, R. E. (1971). Improved goodness-of-fit tests. *Biometrika, 58*(3), 641–645.
35. Kalamees, T., & Kurnitski, J. (2006). Estonian test reference year for energy calculations. *Proceedings of the Estonian Academy of Sciences Engineering, 12*(1), 40–58.
36. Ozdenefe, M., & Dewsbury, J. (2016). Simulation and real weather data: A comparison for Cyprus case. *Building Services Engineering Research and Technology, 37*(3), 288–297.
37. Crow, L. W. (1981). Development of hourly data for weather year for energy calculations (WYEC), including solar data, at 21 stations throughout the US. *ASHRAE Transactions, 87*(1), 896–900.
38. Crow, L. W. (1984). Weather year for energy calculations. *ASHRAE Journal (United States), 26*, 42–47.
39. Augustyn, J. R. (1998). WYEC2 user's manual and software toolkit. *ASHRAE Transactions, 104*, 32.
40. Holmes, M., & Hitchin, E. (1978). An example year for the calculation of energy demand in buildings. *Building Services Engineering, 45*(1), 186–189.
41. Wong, W., & Ngan, K. (1993). Selection of an "example weather year" for Hong Kong. *Energy and Buildings, 19*(4), 313–316.
42. Radhi, H. (2009). A comparison of the accuracy of building energy analysis in Bahrain using data from different weather periods. *Renewable Energy, 34*(3), 869–875.
43. de la Flor, F. J. S., et al. (2008). Climatic zoning and its application to Spanish building energy performance regulations. *Energy and Buildings, 40*(10), 1984–1990.
44. Ebrahimpour, A., & Maerefat, M. (2010). A method for generation of typical meteorological year. *Energy Conversion and Management, 51*(3), 410–417.
45. Eames, M., Kershaw, T., & Coley, D. (2012). A comparison of future weather created from morphed observed weather and created by a weather generator. *Building and Environment, 56*, 252–264.
46. Tumini, I., Higueras García, E., & Baereswyl Rada, S. (2016). Urban microclimate and thermal comfort modelling: Strategies for urban renovation. *International Journal of Sustainable Building Technology and Urban Development, 7*(1), 22–37.

47. Adélard, L., et al. (2000). A detailed weather data generator for building simulations. *Energy and Buildings, 31*(1), 75–88.
48. Meteotest, Meteonorm, Global Meteorological Database, Version 7.1. Handbook Part I. (2015).
49. Boland, J. (1995). Time-series analysis of climatic variables. *Solar Energy, 55*(5), 377–388.
50. Hutchinson, M. (1995). Stochastic space-time weather models from ground-based data. *Agricultural and Forest Meteorology, 73*(3–4), 237–264.
51. Semenov, M. A., & Barrow, E. M. (1997). Use of a stochastic weather generator in the development of climate change scenarios. *Climatic Change, 35*(4), 397–414.
52. Skeiker, K. (2006). Mathematical representation of a few chosen weather parameters of the capital zone 'Damascus' in Syria. *Renewable Energy, 31*(9), 1431–1453.
53. Mylona, A. (2012). The use of UKCP09 to produce weather files for building simulation. *Building Services Engineering Research and Technology, 33*(1), 51–62.
54. Degelman, L. O. (1991). A statistically-based hourly weather data generator for driving energy simulation and equipment design software for buildings. In *Proceedings of Building Simulation*.
55. Herrera, M., et al. (2017). A review of current and future weather data for building simulation. *Building Services Engineering Research and Technology, 38*(5), 602–627.
56. Richardson, C. W., & Wright, D. A. (1984). *WGEN: A model for generating daily weather variables*. U.S.D.A. Agricultural Research Service.
57. Wallis, T. W., & Griffiths, J. F. (1995). An assessment of the weather generator (WXGEN) used in the erosion/productivity impact calculator (EPIC). *Agricultural and Forest Meteorology, 73*(1–2), 115–133.
58. Nicks, A., & Gander, G. (1994). CLIGEN: A weather generator for climate inputs to water resource and other models. In *Proceedings of the 5th international conference on Computers in Agriculture*.
59. Dubrovsky, M. (1996). Met&Roll: The stochastic generator of daily weather series for the crop growth model. *Meteorological Bulletin, 49*, 97–105.
60. Wilks, D. S., & Wilby, R. L. (1999). The weather generation game: A review of stochastic weather models. *Progress in Physical Geography, 23*(3), 329–357.
61. Argiriou, A., et al. (1999). Comparison of methodologies for TMY generation using 20 years data for Athens, Greece. *Solar Energy, 66*(1), 33–45.
62. Jentsch, M. F., Bahaj, A. S., & James, P. A. (2008). Climate change future proofing of buildings—Generation and assessment of building simulation weather files. *Energy and Buildings, 40*(12), 2148–2168.
63. Huang, Y. J., & Crawley, D. (1996). *Does it matter which weather data you use in energy simulations?* Lawrence Berkeley National Lab.
64. Seo, D., Huang, Y. J., & Krarti, M. (2010). Impact of typical weather year selection approaches on energy analysis of buildings. *ASHRAE Transactions, 116*(1), 416–427.
65. Hong, T., Chang, W.-K., & Lin, H.-W. (2013). A fresh look at weather impact on peak electricity demand and energy use of buildings using 30-year actual weather data. *Applied Energy, 111*, 333–350.
66. Yoo, H., Lee, K., & Levermore, G. J. (2015). Comparison of heating and cooling energy simulation using multi-years and typical weather data in South Korea. *Building Services Engineering Research and Technology, 36*(1), 18–33.
67. Crawley, D. B. (1998). Which weather data should you use for energy simulations of commercial buildings?/discussion. *ASHRAE Transactions, 104*, 498.
68. Bhandari, M., Shrestha, S., & New, J. (2012). Evaluation of weather datasets for building energy simulation. *Energy and Buildings, 49*, 109–118.
69. Yang, L., et al. (2008). Building energy simulation using multi-years and typical meteorological years in different climates. *Energy Conversion and Management, 49*(1), 113–124.

70. Aguiar, R., Collares-Pereira, M., & Conde, J. (1988). Simple procedure for generating sequences of daily radiation values using a library of Markov transition matrices. *Solar Energy, 40*(3), 269–279.
71. Tsoka, S., et al. (2017). Evaluation of stochastically generated weather datasets for building energy simulation. *Energy Procedia, 122*, 853–858.
72. Kočí, J., et al. (2019). Effect of applied weather data sets in simulation of building energy demands: Comparison of design years with recent weather data. *Renewable and Sustainable Energy Reviews, 100*, 22–32.
73. Yang, X., et al. (2012). An integrated simulation method for building energy performance assessment in urban environments. *Energy and Buildings, 54*, 243–251.
74. Allegrini, J., Dorer, V., & Carmeliet, J. (2012). Influence of the urban microclimate in street canyons on the energy demand for space cooling and heating of buildings. *Energy and Buildings, 55*, 823–832.
75. Santamouris, M. (2014). On the energy impact of urban heat island and global warming on buildings. *Energy and Buildings, 82*, 100–113.
76. Oke, T. R. (1982). The energetic basis of the urban heat island. *Quarterly Journal of the Royal Meteorological Society, 108*(455), 1–24.
77. Cui, Y., et al. (2017). Temporal and spatial characteristics of the urban heat island in Beijing and the impact on building design and energy performance. *Energy, 130*, 286–297.
78. Yan, D., et al. (2008). DeST—An integrated building simulation toolkit Part I: Fundamentals. In *Building Simulation*. Springer.
79. Magli, S., et al. (2015). Analysis of the urban heat island effects on building energy consumption. *International Journal of Energy and Environmental Engineering, 6*(1), 91–99.
80. Zinzi, M., & Carnielo, E. (2017). Impact of urban temperatures on energy performance and thermal comfort in residential buildings. The case of Rome, Italy. *Energy and Buildings, 157*, 20–29.
81. Street, M., et al. (2013). Urban heat island in Boston–An evaluation of urban air-temperature models for predicting building energy use. In *Proceedings of BS2013: 13th conference of International Building Performance Simulation Association*.
82. Salvati, A., Roura, H. C., & Cecere, C. (2017). Assessing the urban heat island and its energy impact on residential buildings in Mediterranean climate: Barcelona case study. *Energy and Buildings, 146*, 38–54.
83. Guattari, C., Evangelisti, L., & Balaras, C. (2018). On the assessment of urban heat island phenomenon and its effects on building energy performance: A case study of Rome (Italy). *Energy and Buildings, 158*, 605–615.
84. Santamouris, M. (2015). Regulating the damaged thermostat of the cities—Status, impacts and mitigation challenges. *Energy and Buildings, 91*, 43–56.
85. Young, A., & Mitchell, N. (1994). Microclimate and vegetation edge effects in a fragmented podocarp-broadleaf forest in New Zealand. *Biological Conservation, 67*(1), 63–72.
86. Niachou, K., Livada, I., & Santamouris, M. (2008). Experimental study of temperature and airflow distribution inside an urban street canyon during hot summer weather conditions—Part I: Air and surface temperatures. *Building and Environment, 43*(8), 1383–1392.
87. Bozonnet, E., et al. (2013). Modeling methods to assess urban fluxes and heat island mitigation measures from street to city scale. *International Journal of Low-Carbon Technologies*, ctt049.
88. Frayssinet, L., et al. (2017). Modeling the heating and cooling energy demand of urban buildings at city scale. *Renewable and Sustainable Energy Reviews, 81*, 2318–2327.
89. Krebs, L. F., & Johansson, E. (2021). Influence of microclimate on the effect of green roofs in Southern Brazil–A study coupling outdoor and indoor thermal simulations. *Energy and Buildings, 241*, 110963.
90. Gobakis, K., & Kolokotsa, D. (2017). Coupling building energy simulation software with microclimatic simulation for the evaluation of the impact of urban outdoor conditions on the energy consumption and indoor environmental quality. *Energy and Buildings, 157*, 101–115.

91. Bouyer, J., Inard, C., & Musy, M. (2011). Microclimatic coupling as a solution to improve building energy simulation in an urban context. *Energy and Buildings, 43*(7), 1549–1559.
92. Malys, L., Musy, M., & Inard, C. (2015). Microclimate and building energy consumption: Study of different coupling methods. *Advances in Building Energy Research, 9*(2), 151–174.
93. Gros, A., et al. (2016). Simulation tools to assess microclimate and building energy–A case study on the design of a new district. *Energy and Buildings, 114*, 112–122.
94. Gros, A., Bozonnet, E., & Inard, C. (2011). Modelling the radiative exchanges in urban areas: A review. *Advances in Building Energy Research, 5*(1), 163–206.
95. Robinson, D., et al. (2009). CitySim: Comprehensive micro-simulation of resource flows for sustainable urban planning. In *Proceedings of Building Simulation*.
96. Robinson, D., et al. (2007). SUNtool–A new modelling paradigm for simulating and optimising urban sustainability. *Solar Energy, 81*(9), 1196–1211.
97. Tsoka, S., Tsikaloudaki, A., & Theodosiou, T. (2018). Analyzing the ENVI-met microclimate model's performance and assessing cool materials and urban vegetation applications-a review. *Sustainable Cities and Society, 43*, 55–76.
98. Sharmin, T., & Steemers, K. (2015). Exploring the effect of micro-climate data on building energy performance analysis. In *The 7th international conference of SuDBE2015*.
99. Morakinyo, T. E., et al. (2016). Modelling the effect of tree-shading on summer indoor and outdoor thermal condition of two similar buildings in a Nigerian university. *Energy and Buildings, 130*, 721–732.
100. Tsoka, S., Tsikaloudaki, K., & Theodosiou, T. (2019). Coupling a building energy simulation tool with a microclimate model to assess the impact of cool pavements on the building's energy performance application in a dense residential area. *Sustainability, 11*(9), 2519.
101. Santamouris, M., et al. (2018). On the energy impact of urban heat island in Sydney: Climate and energy potential of mitigation technologies. *Energy and Buildings, 166*, 154–164.
102. Djunaedy, E., Hensen, J., & Loomans, M. (2005). External coupling between CFD and energy simulation: Implementation and validation. *ASHRAE Transactions, 111*(1), 612–624.
103. Allegrini, J., et al. (2013). Modelling the urban microclimate and its influence on building energy demands of an urban neighbourhood. In *Proceedings of CISBAT 2013 Cleantech for Smart Cities and Buildings*. EPFL Solar Energy and Building Physics Laboratory (LESO-PB).
104. Barbason, M., & Reiter, S. (2014). Coupling building energy simulation and computational fluid dynamics: Application to a two-storey house in a temperate climate. *Building and Environment, 75*, 30–39.
105. Pandey, B., Banerjee, R., & Sharma, A. (2021). Coupled EnergyPlus and CFD analysis of PCM for thermal management of buildings. *Energy and Buildings, 231*, 110598.
106. Liu, J., et al. (2015). The impact of exterior surface convective heat transfer coefficients on the building energy consumption in urban neighborhoods with different plan area densities. *Energy and Buildings, 86*, 449–463.
107. Djunaedy, E., Hensen, J. L., & Loomans, M. (2003). Toward external coupling of building energy and airflow modeling programs. *ASHRAE Transactions, 109*(2), 771–787.
108. Robitu, M., et al. (2004). Energy balance study of water ponds and its influence on building energy consumption. *Building Services Engineering Research and Technology, 25*(3), 171–182.
109. Tsoka, S. (2019). *Urban microclimate analysis and its effect on the buildings energy performance* (PhD thesis). Aristotle University of Thessaloniki, Greece.
110. Nakicenovic, N., et al. (2000). *Special report on emissions scenarios*.
111. Moss, R. (2008). *Towards new scenarios for analysis of emissions, climate change, impacts, and response strategies*. IPCC expert meeting report. IPCC. http://www.ipcc.ch/pdf/supporting-material/expert-meeting-ts-scenarios.pdf
112. P Tootkaboni, M., et al. (2021). A comparative analysis of different future weather data for building energy performance simulation. *Climate, 9*(2), 37.
113. Jiang, A., et al. (2019). Hourly weather data projection due to climate change for impact assessment on building and infrastructure. *Sustainable Cities and Society, 50*, 101688.

114. Robert, A., & Kummert, M. (2012). Designing net-zero energy buildings for the future climate, not for the past. *Building and Environment, 55*, 150–158.
115. Pérez-Andreu, V., et al. (2018). Impact of climate change on heating and cooling energy demand in a residential building in a Mediterranean climate. *Energy, 165*, 63–74.
116. Shen, P. (2017). Impacts of climate change on US building energy use by using downscaled hourly future weather data. *Energy and Buildings, 134*, 61–70.
117. Chan, A. (2011). Developing future hourly weather files for studying the impact of climate change on building energy performance in Hong Kong. *Energy and Buildings, 43*(10), 2860–2868.
118. Remund, J., et al. (2010). The use of Meteonorm weather generator for climate change studies. In *10th EMS annual meeting*.
119. Jentsch, M. F., James, P., & Bahaj, A. (2012). *CCWorldWeatherGen software: Manual for CCWorldWeatherGen climate change world weather file generator*. Uppsala universitet, Elektricitetslära.
120. Aijazi, A., & Brager, G. (2018). Understanding climate change impacts on building energy use. *ASHRAE Journal, 60*(10).
121. Rey-Hernández, J. M., et al. (2018). Modelling the long-term effect of climate change on a zero energy and carbon dioxide building through energy efficiency and renewables. *Energy and Buildings, 174*, 85–96.
122. Tolika, C., Zanis, P., & Anagnostopoulou, C. (2012). Regional climate change scenarios for Greece: Future temperature and precipitation projections from ensembles of RCMs. *Global NEST Journal, 14*(4), 407–421.
123. Zhang, L., et al. (2020). Comparison of statistical and dynamic downscaling techniques in generating high-resolution temperatures in China from CMIP5 GCMs. *Journal of Applied Meteorology and Climatology, 59*(2), 207–235.
124. Tootkaboni, M. P., Ballarini, I., & Corrado, V. (2021). Analysing the future energy performance of residential buildings in the most populated Italian climatic zone: A study of climate change impacts. *Energy Reports, 7*, 8548–8560.
125. Tsoka, S., et al. (2021). Evaluating the combined effect of climate change and urban microclimate on buildings' heating and cooling energy demand in a Mediterranean City. *Energies, 14*(18), 5799.

Green Urbanism with Genuine Green Architecture: Toward Net Zero System in New York

Derya Oktay and James Garrison

Introduction

Changes that have taken place in the world over the past 30 years, including ecological disturbances and radical changes in traditional settlements, have produced cities that are not just chaotic and monotonous in appearance but have serious environmental problems threatening their inhabitants. In this context, environmentally sensitive design approaches at the building scale have been understood better comparing to those at the urban scale, and there have been significant developments in the field, although the contemporary architectural practice in the developing countries is still lacking many aspects of sustainable building design. On that ground, sustainable urbanism emerges as a sound framework that draws attention to the immense opportunity to redesign the built environment in a manner that supports a higher quality of life and human health. What is critical here is that a city cannot be green without genuine green buildings.

Today, most architects, unfortunately, continue to see architectural design as the design of an "object," although it is an undeniable reality that building design cannot be isolated from its environment, and an architect while designing a building affects the existing environment positively or negatively. As buildings are one of the principal users of energy and materials and the major causes for damaging nature, new approaches are needed in the conception, theorization, and implementation of architectural practices, which will generate ecologically responsive architectural

---

D. Oktay (✉)
Maltepe University, Istanbul, Turkey
e-mail: deryaoktay@maltepe.edu.tr

J. Garrison
Pratt Institute, New York, USA
e-mail: jgarrison@pratt.edu; garrison@garrisonarchitects.com

© The Author(s), under exclusive license to Springer Nature Switzerland AG 2023
A. Sayigh (ed.), *Towards Net Zero Carbon Emissions in the Building Industry*, Innovative Renewable Energy, https://doi.org/10.1007/978-3-031-15218-4_8

designs. In this context, "green architecture" emerges as the practice of creating buildings that are designed to reduce the overall impact of the built environment on human health and the natural environment by efficiently using energy, water, and other resources, safeguarding user health, increasing employee productivity, and reducing waste, pollution, and physical deterioration.

However, the concept of green architecture remains poorly defined and inadequately developed as a cultural and technical undertaking. Architecture produced in the context of industrialized, capitalist societies continues to be determined by expedience and profit rather than ecological principles. Even if the most sustainable design and technical solutions are utilized, their effectiveness is dramatically limited by supply chains, work patterns, and materials produced under capitalist priorities. These same priorities in the context of highly competitive and mediated cultures drive claims of sustainability without realizing complete, effective, or holistic solutions. Unsubstantiated or overstated "green washing" prevails across much of the product range and is liberally applied to architecture. Whether resulting from economic limits or partial commitment, we must move beyond such half measures if we are to successfully confront what has become the greatest existential threat of the modern world. It is very rare to see effective architectural examples that safeguard regional populations with locally appropriate sustainable design features that consider climate, health, and renewable energy [8]. The result is a lack of environmental sensitivity and the prevailing sameness in cities worldwide.

Early in the sustainability movement, many rejected the idea of "environmental balance" as an impossible goal that ignored the realities of the Anthropocene. Thirty years hence two questions remain: Can technological means save us from ecological disaster, or will we fail without disruptive cultural, economic, and behavioral change? It is now clear that both are necessary. We can take a lesson from US consumers' response to legislated vehicle mileage improvements; Americans now drive 40% more miles per person than they did 40 years ago.

Based on these shortcomings, this chapter focuses on the concepts of green urbanism and green architecture based on the ideas observed in the development of the ancient settlements and the traditional contexts, introducing the Seventy-Six, the second author's awarded project in Albany, New York, and interrogates the viability of the net zero concept through that exemplary project.

Looking Back for the Idea of Green Settlements

Although sustainability is considered new conceptually, it is not new as a worldview. The adaptation of building to the environment has been a continuous problem throughout the centuries. The use of local data, especially climatic features, in design has been a part of the rational approach of those who have been dealing with buildings since ancient times. In this context, the ancient builders learned to design houses that would benefit from solar energy on cool days and avoid the heat of the sun on hot summer days.

Furthermore, early evidence shows that solar energy and other climatic features were utilized not only on a single house scale but also when designing a group of houses in an urban context. Hippocrates, for example, suggested heading east in living spaces as the healthiest solution, emphasizing that the south direction is also acceptable. Vitruvius, on the other hand, drew attention to the importance of opening wide streets to the wind to clean the air of the city and to avoid the wind so that narrow streets could be used as a living environment [2]. As in Greek cities, streets in ancient Rome were created with walls and arcades that protected from the sun's rays and kept buildings cool, a principle that has become the norm for many urban settlements in the Mediterranean region as an excellent solution to protect pedestrians from climatic elements and enrich the urban space. Vitruvius's recognition of its importance is echoed even by Le Corbusier, who has not been enthusiastic about designing with the environment: "The symphony of climate along the curvature of the meridian, its intensity varies on the crust of the earth according to its incidence… In this play, many conditions are created which await adequate solutions. It is at this point that an authentic regionalism has its rightful place" [9].

In every region, a traditional building style or regional architecture has emerged due to the practical needs of the people living there, topography, and climatic conditions. These anonymous examples are important sources of information that should be considered in the design of new environments in that region, at the settlement and building scale. For instance, traditional Turkish (Ottoman) city is the perfect example of "design with nature." As analyzed by Oktay [6], the preexisting topographic character of the site is apparent at the urban scale even in intense built-up areas. Furthermore, gardens perforate an otherwise dense urban fabric, providing relief to streets and to public and private structures. With its trees, flowers, and small vegetable plot, the courtyard, *avlu*, is the closest relation the house has to nature; thus, it also provides the inhabitant with direct access to nature (Fig. 1). During the hot summer months, the well-defined, open-to-sky courtyard traps the dense, cool air in

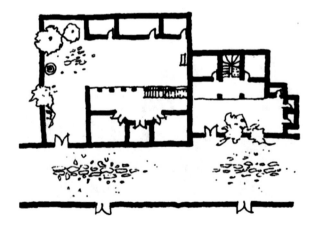

Fig. 1 Traditional courtyard house in the traditional Turkish (Ottoman) house [6]

the center of the house, helping air circulation and bringing down the general temperature inside. Trees, as nature's own evaporative coolers, are the important components of the courtyard; they also filter blowing dust from the air.

The Persian civilization is widely considered to have added windcatchers, the natural ventilation in buildings, to allow for better cooling – such as combining it with its existing irrigation system to help to cool the air down before releasing it throughout the home. As Roaf [10] highlights, in a hot and arid climate, windcatchers can be thought of as a zero-carbon cooling technology (Fig. 2).

Such indigenous environmental creativity was necessarily motivated and limited by available resources. It was, in a sense sustainable by circumstance, as its instigators had no choice but to work from what was at hand. Without glossing over the class and labor exploitation common to such endeavors, we can recognize that limitations are often the most effective way for cultures to work within ecological boundaries. With the arrival of the petroleum age, such boundaries were dissolved by a widely available and abundant source of energy. As is our nature, we have used this resource with abandon inventing means of production, movement, and materiality that sponsored unparalleled population growth and technical advancement without broad recognition of its effect on the world that sustains us.

The search for the cities of the future, against rapid and unbalanced urbanization, how to create the balance between the city and the rural environment, and how to shape ecologically sensitive environments were first brought to the agenda by Ebenezer Howard in 1898. Letchworth (England), designed by Howard as a "garden city" in line with these goals, was a city model with functional diversity and relatively self-sufficiency, surrounded by a production-oriented green belt that minimized dispersed development ([5]/1898). Here, the beauty of nature, easily accessible parks and fields, clean air and water, houses with gardens, high income, and social opportunities were combined with cooperative entrepreneurship (Fig. 3). With its features, Howard's model, unlike other garden city approaches, can be considered as a starting point for planning and design for sustainability.

Fig. 2 Traditional windcatchers in Yazd, Iran (https://gate-of-nations.org/wind-towers-or-windcatchers/ Retrieved: 08.04.2022)

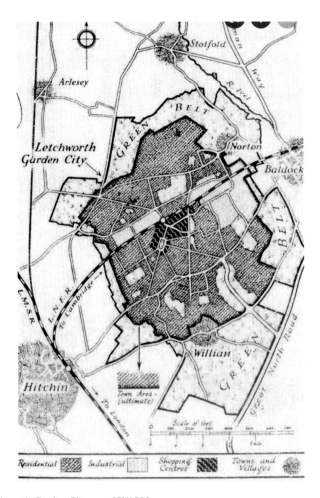

Fig. 3 Letchworth Garden City map [5]/1898

Towards the end of the twentieth century, by redefining urban and nonurban settlements in the United States in the context of sustainability, Peter Calthorpe started the urban and architectural movement that compelled the "New Urbanism" movement. Positioning all functions and services around an advanced and alternative public transportation system and considering pedestrian accessibility, including open spaces in the center that will support social life, creating a distinctive street pattern that does not give priority to vehicles, and designing the buildings in accordance with historical and climatic features constitute the main principles of the New Urbanism movement [3] (Figs. 4 and 5). Although the movement is open to criticism on several fronts, for being focused on better-designed suburban development, often for upper income groups, rather than the creation of truly "urban" places, together with the paradigm of "green architecture" provides the philosophical and practical framework of sustainable urbanism at the city level.

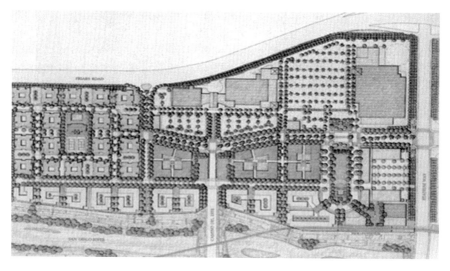

Fig. 4 Rio Vista West (San Diego, California) site plan [3]

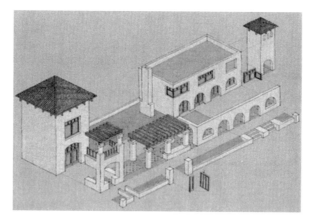

Fig. 5 Architectural language and components in Rio Vista West (San Diego, California) created in harmony with the tradition and climate [3]

Green Urbanism and the Need for "Genuine" Green Architecture in Contemporary Developments

When the theme of sustainability at the urban scale first came to the forefront, all those interested in the subject defined sustainability with great conviction, referring to the Brundlant Report (1987), as "development that meets the needs of the present without compromising the ability of future generations to meet their own needs and expectations." However, it was not very clear how it would be reflected in actual decisions and daily life. In this context, the general objectives were to develop energy efficiency and minimize air-polluting emissions, to develop mobility without the need for motor vehicles, to reduce the use of private vehicles by improving public transportation in this direction, and to develop new activity centers around the nodal points of public transportation. In this vein, the idea of "green" urbanism focuses on expanding our approach to the phenomenon of urbanism, which covers both urban and architectural scale to include ecological dimensions based on sustainability goals and is now indispensable on the agenda of the world of urban planning/design and architecture and nongovernmental organizations particularly in developed Western countries.

Since the failure of architectural practices to develop attitudes against globalization accelerates the loss of local resources, identities, and social values, what is needed to be emphasized is that the design approaches that take formal aesthetics as the "single and main goal" are extremely dangerous, but what it adds or how it affects the needs is important. In this context, the rapid proliferation of successful examples supporting the idea of "green urbanism," and the intense support and interest in it by being kept on the agenda both at the country and city administration level and through nongovernmental organizations are pleasing developments. One significant contemporary example of housing combining ecological and social principles is Ecolonia in Holland, master-planned by the Belgian architect Lucien Kroll. Ecolonia does demonstrate that significant savings in energy use and environmental impact can be achieved by optimizing the use of the existing methods and building materials. By paying regard to orientation (south, east, and west facing housing, not north), to differential window areas according to aspect, and to increased levels of insulation and efficient boiler systems, Ecolonia has met the target of a 25%

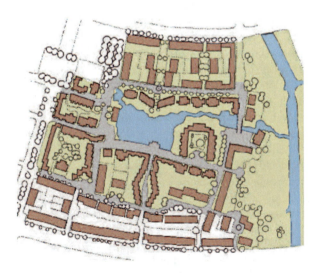

Fig. 6 The settlement layout of "Ecolonia" low-energy housing demonstration project in Alphen aan den Rijn (Netherlands) designed by Lucien Kroll (L. Kroll Archive)

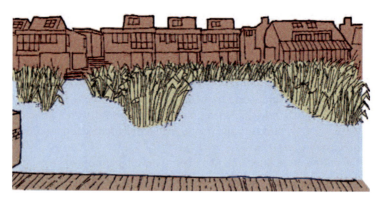

Fig. 7 The sketch view of "Ecolonia" drawn by Lucien Kroll (L. Kroll Archive)

reduction in household energy use in an area with a continental climate ([4], 195) (Figs. 6 and 7).

Our research and investigations reveal that green urbanism based on sustainability should be expected to include the following basic features beyond including "green" in the color palette: locally appropriate density, clear borders, integrated transportation, city and nature unity, climatic design, and energy. It is the "sustainable lifestyle" that we think is necessary for the conservation of these features and the sustainability of these features [7].

To put it simply, it is not possible to create a "green" city with low density and sparse textures dominated by unused spaces. What is important here is an integrated urban fabric with an appropriate density that is harmonious with the local climate and culture. In this way, agricultural areas outside the city will not be damaged, automobile fuel consumption and harmful substance (emissions) will decrease, and the local economy and social life will be strengthened due to the concentration of the people and their activities in the city.

Since ancient times, residential areas have been the basic unit of settlements. The *mahalle*, the traditional neighborhood unit, which determines the structure of the traditional Turkish (Ottoman) city, reflects the features to be learned in the context of sustainability as a social-spatial and recognizable module. It has high ecological efficiency, especially with its strong center and clear borders formed by a mostly production-oriented green belt [7]. Demarcated, recognizable settlements not only support ecological and social life but also encourage citizens to take more responsibility for their care and development.

Positioning the diverse functions and services around a developed and offering public transportation system and considering pedestrian accessibility, including public spaces in the center that will support social life, creating a street pattern with a strong spatial definition, emphasizing the comfortable circulation of people, and attractive pedestrian areas, the city will increase the "walkability" feature, which is important for its ecology, and will keep the city center alive at all hours of the day.

Integration with nature is the most important component of green urbanism. When the history of humanity is examined, it reflects the traces of the perfect harmony of the first examples of collective life with nature. In these examples, nature has been the main determinant of both the identity of the settlements and the physical boundaries of collective life. On the other hand, green spaces in a city contribute to human activities, climate, and ecological diversity without alienating people. Today, many European cities are trying to draw nature into the city and to establish physical and ecological connections between the urban structural areas and the surrounding natural and green spaces. Wetlands, forests, feeding points of underground waters, etc., in the city form the city's green infrastructure and are as important as structural elements because of the many benefits they provide (e.g., flood prevention); to strengthen the city ecologically, it is imperative to protect them.

The practices that will provide ecological benefits within the scope of planning or renewing the city are the use of productive landscape elements such as fruit trees in the arteries in the residential areas and the integration of the shared gardens, where the people of the city can grow their vegetables, with the settlement plan. Local and small-scale food production will not only reduce the environmental pressure caused by industrial agriculture but also form the main source for healthy nutrition.

Towards Net Zero Buildings: The Seventy-Six, New York

The Seventy-Six, the second author's awarded project in Albany, New York, is the brainchild of Corey Jones, a young African American developer, who is looking to create a new community from the ruins of the neighborhood he grew up in. Many environmental, economic, and cultural justice atrocities have confronted the people of Albany's South End where the Seventy-Six is located. Things many of us take for granted – safety, savings, healthy food, available parents, and friends are difficult to find there.

The Seventy-Six intends to be the first triple net zero (energy, water, and waste) multifamily/mixed-use project in the United States and seeks to create a complete transformation of the area, including the creation of economic and environmental equity by integrating scalable ownership models into businesses and homes. A radically sustainable infrastructure with high-quality, affordable, and flexible housing that meets universal design and accessibility requirements that can accommodate aging, changes in family size, and alternative living arrangements. The design team consisted of ME Engineering, The Levy Group (sustainability consultants), Go Energy Link (alternative energy specialists), Steve Ostrowski (an inventor), and the architectural team began with a Skunk Works[1] approach, a project developed by a

---

[1] A Skunk Works approach is the concept originally developed by Lockheed Martin, to quickly develop solutions by bypassing some of the time-consuming bureaucracy and allowing the team to make ad hoc decisions [1].

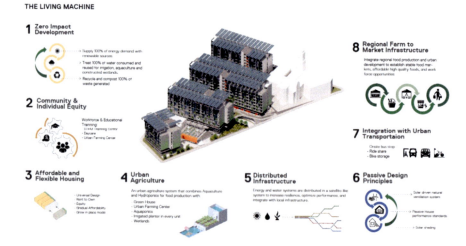

Fig. 8 The rationale of the Seventy-Six, NY. (Source: J. Garrison Archive)

Fig. 9 The general view of the Seventy-Six, NY. (Source: J. Garrison Archive)

relatively small and loosely structured group of people who research and develop a project primarily for the sake of radical innovation.

It always seems as though, by its simple physicality architecture cannot affect meaningful social and environmental change. And although it has real limits, it is a

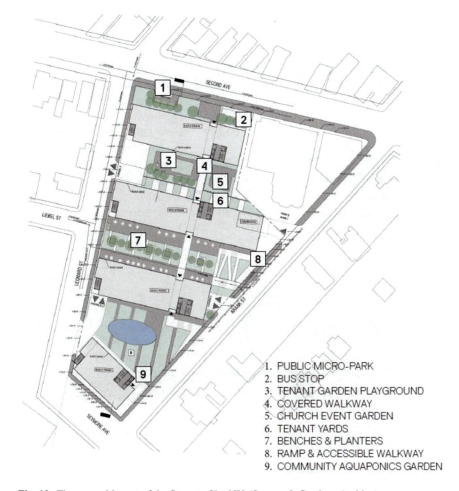

Fig. 10 The general layout of the Seventy-Six, NY. (Source: J. Garrison Archive)

great holistic enterprise that has the potential to create physical and cultural synergies that transcend much more than we may think. The Seventy-Six begins with a recipe for community restoration through effective programming including the creation of affordable housing, coworking, and commercial space, permanent jobs, STEM (science, technology, engineering, and mathematics) learning centers by local institutions, urban gardening resources, aquaculture, daycare, and a novel fresh foods cooperative with surrounding farmsteads. At the same time, these socially motivated programs reduce the need for daily travel while they introduce biodiversity and locally sourced foods. Taken together with the organizational and technical requirements of a triple net zero development, health, equity, affordability, and ecological balance are mutually reinforced (Figs. 8, 9, and 10).

Urban Context

The South End is in many ways typical of historically underprivileged, centrally located communities that exist in mid-sized cities throughout the United States. They are the remains of mid-twentieth-century suburbanization and are impoverished, low density, and bereft of available goods and services. In cities with robust twenty-first-century economies, these communities are often displaced by gentrification. However, this is far from universal in a country with dramatic inequities in wealth and access. Consequently, the Seventy-Six can leverage location, affordable land, and accessibility with the introduction of significant density and programming designed to serve its existing and future population. This opportunity allows us to imagine a new urban reality as we look to create an economically and culturally integrated community without the neighborhood homogeneity of cities like New York.

Equity

The South End will include both rental and ownership housing. Rental programs include market and subsidized rates. Homeownership is promoted and given its tax advantages and potential for personal and family equity generation and preservation. Purchase options include conventional co-op mortgages and limited equity

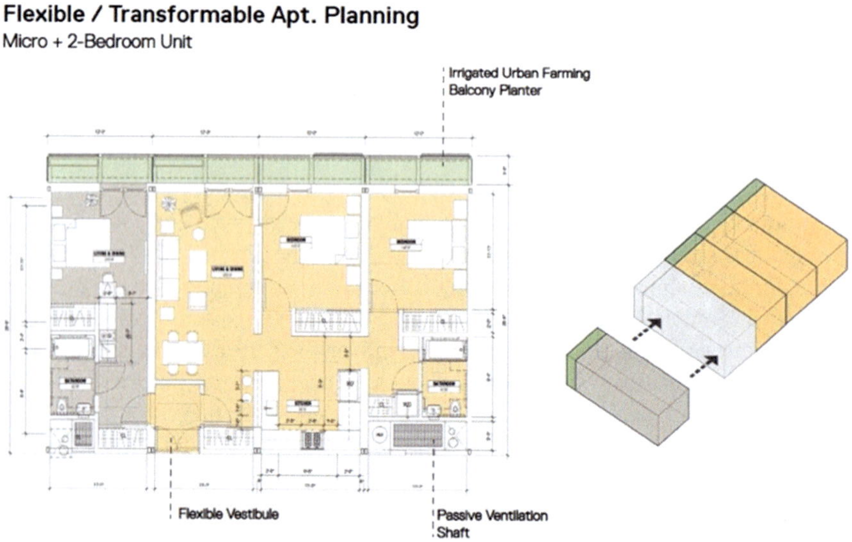

Fig. 11 The adaptability with flexible modules in Seventy-Six, NY. (Source: J. Garrison Archive)

programs for first-time purchasers. The limited equity model is inspired by London's Hackney Council programs[2] with multiple tiers and approaches to purchasing according to income and ability. At the Seventy-Six, affordability extends to retail and coworking spaces in recognition of the difficult economics of business startups.

Adaptability

The Seventy-Six includes a typical mix of urban apartment types with one notable exception; flats can be separated and combined to address growing and shrinking families without requiring residents to leave their neighborhood or building.

The introduction of a micro-studio that can be combined with or used separately from a two-bedroom flat allows several living arrangements; it can expand to serve a growing family, an elderly relative, or a child seeking independence, or it can be separated to provide rental income or a low-cost alternative for a pensioner whose family no longer needs a full-sized apartment (Fig. 11).

Permaculture

From the Permaculture Research Institute: "Permaculture integrates land, resources, people, and the environment through mutually beneficial synergies – imitating the no waste, closed-loop systems seen in diverse natural systems. Permaculture studies and applies holistic solutions that are applicable in rural and urban contexts at any scale. It is a multidisciplinary toolbox including agriculture, water harvesting and hydrology, energy, natural building, forestry, waste management, animal systems, aquaculture, appropriate technology, economics, and community development." Permaculture informs the holistic approach of the Seventy-Six as its more traditional meaning has led to urban forms of regenerative agriculture. As the proliferation of autoimmune diseases in the developed world has been linked to sanitized and hermetic living conditions, we now look to create a complete ecosystem in all architectural settings. To this end, every apartment in the Seventy-Six is outfitted with irrigated planters integrated into continuous terraces. Vertical farming support is provided to residents via an on-site agricultural resource center that advises regarding appropriate plant species and growing conditions as well as providing gardening tools, seeds, and stems.

---

[2] https://hackney.gov.uk/affordable-home-ownership

Recycling

Recycling programs are often stymied by the limited market for recycled materials, the lack of easy and available receptacles for multiple material streams, and the cooperation of residents. The Seventy-Six addresses these conditions by design, active management, and incentivizing residents to participate. Each flat includes four generous receptacles for composting, metals and plastics, paper, and incinerable waste. Adjacent to each elevator core is generous recycling rooms with multiple receptacles that weigh individual waste contributions. Immediate elevator access and basement collection routes bring recycling to a central hub where it is prepared for pickup. Residents receive a monthly quantity statement via the building operating system and are rewarded with reductions in common charges. Management negotiates contracts with neighborhood community gardens for composting and recycling companies that demonstrate effective sorting and placement.

Waste and Storm Water

While the northeast United States has, at this moment, an overabundance of rainfall that is changing its agricultural practices, its large urban centers require an extensive watershed and statutes to limit surrounding economic development to preserve water quality. This affects the economic growth of hundreds of communities and contributes to economic inequality. While this watershed is currently capable of providing high-quality water without filtration, it has seen steady deterioration due to land development. With filtration comes significant expense and an increasing tax burden for the communities that rely on this water. Subsequently, water, even when abundant, must be treated as the invaluable resource it is.

Like many older cities, Albany possesses a combined sewer system. This means that sewage and stormwater must share the same pipe. As storms and rainfall in the northeastern United States become increasingly intense, such systems are regularly overwhelmed and must release raw sewage into the surrounding water system. This has a direct impact on human health as pharmaceutical waste, viral matter, and microplastics are ingested by marine animals that contribute to the human food chain.

A net zero water program addresses such problems by treating each building and site as a closed loop. Such systems ideally recycle rainfall to supply all a building's water needs. This is, however, impossible for high-density urban development where rainfall can only satisfy a portion of water demand. While supplemental water from the watershed is necessary for the Seventy-Six, all discharged wastewater is retained, treated on-site, and returned to the original water source in a controlled manner.

The elements of this system are water-conserving fixtures, gray water recycling, on-site sewage filtration, stormwater retention, compactly constructed wetlands, and sufficient retention to eliminate storm surge contribution. Each of these elements requires significant engineering expertise as water treatment is highly regulated, and appropriate systems are chosen based on first and operating cost, flow consistency and quantity, and available technologies. The water system of the Seventy-Six was designed in collaboration with the landscape ecology firm Biohabitats and the engineering firm ME.

Energy Use

Solar radiation in the northeast United States typically generates adequate power through photovoltaic conversion to serve residential buildings up to three stories high when used in combination with significant conservation measures. As a high-density, mixed-use, urban development with seven floors of the residential area, ground floor commercial and community facility uses, and subgrade parking, the Seventy-Six presented what appeared to be an insurmountable challenge.

Recognizing that every available conservation means would be necessary to approach net zero, we began by employing aggressive-passive design strategies to control heat loss and heat gain and take maximum advantage of the northeast United States temperate spring and fall weather. By these means, we sought to increase the time during which apartments could be occupied without active heating or cooling to 4 months per year. Initial energy models revealed that more energy would be required to cool than heat the buildings. In response solar control, exterior envelope tuning and ventilation had to be optimized. Continuous south-facing balconies with integrated vertical farming were designed to completely shade south exposures while allowing low-angle winter sun to pass into apartments.

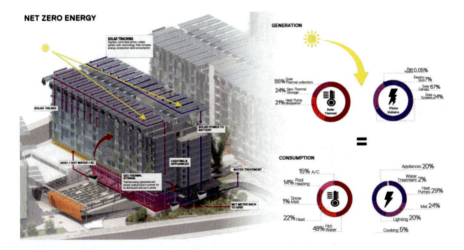

Fig. 12 Analysis of energy use in Seventy-Six, NY. (Source: J. Garrison Archive)

US affordable residential economic models require extreme efficiency and virtually dictate double-loaded corridors. This limits apartments to a single exposure and eliminates the potential for cross ventilation. To create warm-weather comfort without air conditioning, code-mandated ventilation air volume was made controllable and increased 20-fold. Inspired by Persian Wind Towers, large, vertical shafts are accessed from apartment corridors and extended above the roofs to generate air movement through buoyancy and prevailing winds. The shafts were also utilized for the module-to-module connections during erection and integrated with rooftop heat recovery ventilators with separate supply ducts to provide preheated ventilation air during winter months.

As energy consumption models were refined, multiple on-site energy production approaches were tested. Microturbines activated by water retained on roofs were modeled, as were façade-mounted piezoelectric devices activated by wind pressure and various wind turbine products and configurations. In each of these cases, energy output was insufficient to justify the cost of the devices. Single-axis rotating rooftop photovoltaic canopies were modeled and found to result in a 22% increase in electrical output. Façade-mounted panels were distributed where direct low-angle sunlight was available. Ultimately the optimized photovoltaic system was able to meet 50% of the energy needs of the Seventy-Six (Fig. 12).

An analysis of solar thermal energy potential indicated that it could theoretically supply the balance of the project's needs if rooftop space and energy storage were available. A thin, glycol-filled panel manufactured by the Sun Drum company (www.sundrumsolar.com) and designed to be placed directly below photovoltaic panels required no additional space. It also resulted in a 5% increase in photovoltaic efficiency due to the absorption of waste heat and subsequent cooling of the photovoltaics. Taken together the composite photovoltaic and solar thermal array can convert 80% of the sun's energy for direct use.

However, without storage, solar thermal energy is unavailable when it can be most efficiently used. In response, the team identified borehole technology given its ability to store energy in geothermal wells on a seasonal basis. When combined with water-to-water heat pumps, the resulting geothermal energy can provide efficient heating in the winter months and stay within the limits of a net zero system.

Conclusions

Early in the sustainability movement, many rejected the idea of "environmental balance" as an impossible goal that ignored the realities of the Anthropocene. Thirty years hence two questions remain: Can technological means save us from ecological disaster, or will we fail without disruptive cultural, economic, and behavioral change? It is now clear that both are necessary.

Early evidence shows that, in creating urban and architectural settings, the environment should not only be considered as supporting elements but also as essential elements of planning/ design, to reach a long-term and sustainable solution.

"Green urbanism" based on sustainability should be expected to include the following basic features beyond including "green" in the color palette: locally appropriate density, clear borders, integrated transportation, city and nature unity, climatic design, and energy efficiency.

The outcome of the green urbanism debate would be incomplete without people who are aware of the significance of adopting an environmentally responsive living. The ecological citizenship or the ecologically concerned citizens are therefore considered the new dimension of green urbanism, as they can be the civil power making pressure to their local and/or governmental institutions regarding the promotion of environmentally conscious everyday practices such as energy-saving, water conservation, waste management, recycling, green consumption, and sustainable transportation and movement in the city.

"Green architecture" targets to reduce the overall impact of the built environment on human health and the natural environment by efficiently using energy, water, and other resources, safeguarding user health, increasing employee productivity, and reducing waste, pollution, and physical deterioration. In this vein, the concept of the relationship between nature and the architecture as a design philosophy is restored, without resorting to superficial mimicry. The basic principles of green architecture include conformity of the building to its environs and to the climate, the use of renewable energy sources, the use of local and regional materials, the flexibility to adapt to changing conditions over time, and the rich variety of spaces extending from interior spaces to exterior spaces. Design with the climate and with a sense of place is an asset for ecological site design, as perfectly achieved in vernacular examples. In this context, the settlement plan and the block designs must form a cohesive and harmonious whole, in which the dwellers will feel at home in the literal sense of the world.

Over the last two decades, ecologically sensitive design approaches at the building scale have been understood better comparing to those at the urban scale, and there have been significant developments in the field, although the contemporary architectural practice in the developing countries is still lacking many aspects of sustainable building design. However, the absence of the urban or neighborhood scale in most of the environmental literature has been masked by the recent obsession with "green" buildings, most of which look green on their facades but lack energy saving ideas, climate-sensitive design, the use of locally appropriate materials, and so forth.

High-tech innovation and new sustainable technologies undoubtedly have an important role to play, but in an energy-depleted world, cities that can de-link from their dependence on these are likely to be more resilient. The emergence of the net zero buildings concept, taken holistically, anticipates the need to simultaneously disrupt and innovate even if, at this moment, our efforts remain largely technological. Energy and conservation technologies are advancing rapidly though we often cannot take full advantage of them as the necessary cultural and economic commitment remains wanting. Efforts to ascertain the real cost of human activity in environmental terms are also advancing as we develop the tools and concepts necessary to measure environmental impact before we act. Projects such as the Dutch nonprofit enterprise True Price are developing the means to assess the full environmental impact of goods. The US engineering firm, Thornton Tomasetti, has developed digital tools to analyze the embodied and operational carbon content of buildings. Constraints and opportunities will always coexist though with a committed client and extraordinary architectural and engineering effort we can make significant progress. We look forward to the day that we discuss the cost to build in human and environmental terms and create the buildings and communities we so desperately need.

While the Seventy-Six demonstrates that high-density operational net zero buildings are possible, they do not yet represent complete ecological balance. A true net zero building must also account for the energy required for its construction and physical maintenance. In the future, as we learn to accurately measure and predict the ecological consequences of our actions, we will be able to define just exactly what such a building requires. Human actions at their best are culturally intuitive. We proceed from a deep understanding of the values and priorities that inform our choices. Like the tennis player who overthinks their serve, it's impossible to break down every action and act effectively at the same time. Architecture and construction are inseparable arts, and to employ them in the regeneration of ecological balance, our acts must become second nature.

References

1. Bommer, M., DeLaPorte, R., & Higgins, J. (2002). Skunkworks approach to project management. *Journal of Management Engineering, 18*(1), 21–28.
2. Broadbent, G. (1990). *Emerging concepts in urban space design.* Van Nostrand Reinhold.
3. Calthorpe, P. (1993). *The next American metropolis: Ecology, community and the American dream.* Princeton Architectural Press.
4. Edwards, B. (1996). *Towards sustainable architecture.* Butterworth Architecture.
5. Howard, E. (1960, originally 1898). *Garden cities of tomorrow.* Faber.
6. Oktay, D. (2004). Urban design for sustainability: A study on the Turkish city. *Journal of Sustainable Development and World Ecology, 11*(2004), 24–35.
7. Oktay, D. (2017). A critical approach to sustainable urbanism: Lessons from traditional and contemporary paradigms. In E. Hepperle, R. Dixon-Gough, R. Mansberger, J. Paulsson, J. Hernik, & T. Kalbro (Eds.), *Land ownership and land use development: The integration of past, present, and future in spatial planning and land management policies* (pp. 295–306). vdf Hochschulverlag AG an der ETH.
8. Oktay, D. (2017). Lessons for Future Cities and Architecture: Ecology, Culture, Sustainability. In A. Sayigh & M. Sala (Eds.), *Green Buildings & Renewable Energy* (pp. 259–274). Springer, Cham.
9. Olgyay, V. (1963). *Design with climate: Bioclimatic approach to architectural regionalism.* Princeton University Press.
10. Roaf, S. (2008). The Traditional Technology Trap (2): More lessons from the Windcatchers of Yazd. In *PLEA 2008: 25th Conference on Passive and Low Energy Architecture,* Proceedings, Dublin, 22-24 October 2008.
11. Interview with Lucien Kroll by Derya Oktay. (2004). Brussels.

External Solar Shading Design for Low-Energy Buildings in Humid Temperate Climates

Seyedehmamak Salavatian

Introduction

Energy crisis issues and the attraction of international associations toward energy use optimization, particularly in the building sector, lead to the development of passive design strategies which are properly integrated in the building preliminary design stage and introduced them as the most practical and economical building energy management methods. Passive techniques in terms of natural ventilation improvement, maximizing natural lighting, optimization of building form and orientation, and applying the energy efficient materials are among them. Building envelopes are among the most crucial components; in the first step, thermal insulation and heat loss control were considered, and in the second place, applying the optimized shading devices over the transparent areas was surveyed to adjust the solar radiation inside.

Solar control can be simply controlled by introducing optimized shading to minimize solar transmission through glazing areas [1]. In the northern hemisphere, horizontal layouts can considerably reduce the unpleasant solar heat on the southern windows during the warm season while allowing solar incidence on windows in wintertime [2]. Indeed, the louver design has to take into account window specification and the geographical location and latitude which directly affect solar heat gain [3].

There is a complex system of classifying shading devices, but a simple grouping is based on their placement on the façade (external or internal) [4]. The literature review about external shading strategies is much extended. In the recent decades, several studies have been carried out to pinpoint the achievements regarding energy

S. Salavatian (✉)
Department of Architecture, Rasht Branch, Islamic Azad University, Rasht, Iran
e-mail: salavatian@iaurasht.ac.ir

saving by applying appropriate shading devices and showed shading systems results in energy saving and thermal comfort enhancement. Apart from the researches investigating daylight performance, some of studies focused on the building energy consumption affected by the shading design factors and declared that energy demand could be dramatically reduced depending on the shading design. In the majority of studies, some increases were observed in the lighting and heating energy load due to the shading application; nevertheless cooling energy and total energy consumption were reduced at the higher rate [5–7].

As one of the foremost studies, Datta illustrated the efficiency of external horizontal louvers in reduction of the building cooling demand by 70% through the optimal design of parameters. External louvers diminish direct radiation into inside ambient and disperse the solar heat to outside [8]. Another research in 2010 studied the louver performance comprehensively in different locations, and louvers layout was considered to the energy saving obtained [3]. In a subsequent research the potential of energy saving of fixed devices including horizontal louvers was calculated by Yassin [9].

A number of researches attempted to provide accurate guidelines for louver shading design. Hammad and Abu-Hijleh investigated the energy consumption of external dynamic louvers, integrated to office building facades, in Abu Dhabi [10]. However, movable shading systems are also widely developed. There are serious concerns on their cost and functional performance, and fixed louvers performed better in energy saving compared to dynamic ones [11].

Although numerous studies have been carried out to evaluate the advantages of external shading devices, most of researches are linked to the hot climates with a significant amount of solar gain; on the contrary, few studies are performed for moderate climates with medium amount of direct radiation. Therefore, there is a lack of specified guideline for designers to implement external devices for solar control in the aforesaid climates.

This chapter studies the benefits in annual energy consumption through installing appropriate louver shading devices in residential buildings in a representative humid temperate climate located on the north side of Iran. In order to assure that louver positioning is effective, the louver spacing and inclination should be investigated.

Climatic Region

In this chapter, a focused study is carried out to evaluate the effect of louver shading devices applied to the southern façade of a residential building in a city located in Csa zone in the northern area of Iran. The Guilan province is located in the northern part of the country. Guilan weather is generally mild, caused by the influences of both Alborz Mountains and Caspian Sea. This region has a humid temperate and Mediterranean climate with abundant annual rainfall and high relative humidity (between 40% and 100%), and its average temperature is 17 °C [12].

The weather data of Rasht with the Latitude of 37°2′N and 49°6′E was utilized for the simulation as the representative of a temperate humid climate. The weather data are obtained from the meteorological files of EnergyPlus. The corresponding sun path chart of the abovementioned latitude was derived from Autodesk Ecotect Analysis as shown in Fig. 1.

As found in the psychrometric chart (Fig. 2), both cooling in summer and heating in winter are needed. Therefore, and in consideration to reduce solar gain and consequently the cooling load, and probable heating load increase should be kept in mind. Thus, the optimization of the shading devices is done respecting annual primary energy loads.

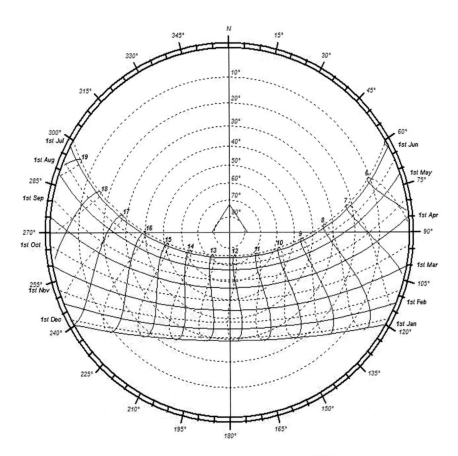

Fig. 1 Annual sun path chart in Rasht [Autodesk Ecotect Analysis 2011]

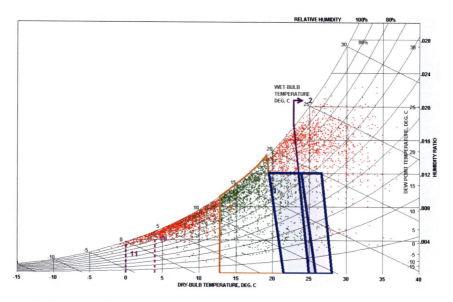

Fig. 2 "Givoni" bioclimatic chart of Rasht [Climate Consultant 6]

Table 1 Values of geometric features of louvers

Louver shading design parameters	Design value
w = width of slats	100, 200, 300, 400 (mm)
h = vertical distance between slats	300, 500 (mm)
α = slat tilt	90, 60, 45, 30 (°)
d = distance from window	100, 300 (mm)

Shading Device Variables

In countries of the north hemisphere, south, southeast, and southwest glazing facades require horizontal louvers. A model for a window with common proportional dimensions was applied. A shading control strategy based on horizontal louvers was proposed, and in this regard, a number of technical solutions were evaluated to obtain the most efficient louver design. It is considered that the optimal slat dimensions, placement, and angle are achieved as the shading design guideline in this climatic zone.

Table 1 shows different increments taken to evaluate various shading systems according to the existing specification sheets of a reliable national manufacturer [13]. The slats are made of extruded aluminum with opaque surface. Figure 3 schematically shows the initial louver section, made of aluminum. The abovementioned values taken for geometric parameters of louvers generated 32 different scenarios.

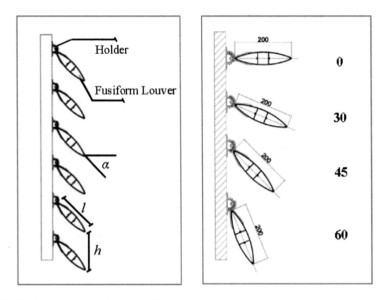

Fig. 3 Configuration of the louver shading system [13]

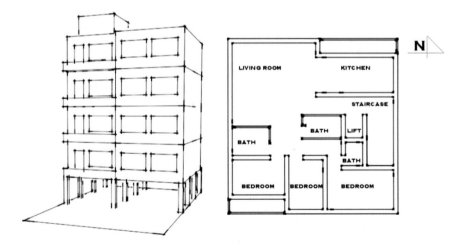

Fig. 4 View of the building model. Left: 3D view. Right: Plan view

Building Description

The case of study proposed in this paper is a four-story apartment building, as the most typical allowed building blocks in the urban zones of the selected city. It consists of four residential units (each in every floor) with the approximate net area of 120 m^2 and gross area of 145 m^2 (Fig. 4). Each unit includes three bedrooms with the floor-to-floor height of 3.40 m. Detailed building characteristics and external wall configuration are presented in Tables 2 and 3, respectively.

Table 2 Major characteristics of the building

Building details	Descriptions	Building details	Descriptions
Function	Residential	Net floor area	120 m²
No of story	4 stories over piloti	Gross floor area	145 m²
Floor area per capita	18.5 m²	Window type	Double glazed Clr pane,6 mm air gap
Floor height	3.40 m	Window-to-wall ratio	30%

External walls and window glazing type are according to the common practice in the construction sector of the region and also conform to the national building regulation requirements and the reference U-value attributed to the external wall of conditioned spaces, unconditioned spaces, and glazing [14]. The total wall thickness and the overall U-value are 0.27 m and 1.0 W/mK, respectively.

Table 3 shows the input setting based on ASHRAE standard 90.1 which is the energy standard for buildings except low-rise residential buildings that set the minimum requirements for the energy performance of building elements [15]. This building uses the electricity for cooling and lighting and gas for heating, and the total site energy consumption is taken for the building energy analysis. Site energy consumption can be useful to understand the performance of the building and building system [16].The building site energy is typically measured by the utility meters and is the total of the electrical, gas, and other energies delivered to the facility. Site energy consumption can be useful to understand the performance of the building and building system.

Simulation Tool

Simulation method is generally utilized to optimize the shading system design features. There are different simulation tools for thermal analysis, and Design Builder is one of the most utilized in the literature, and its reliability has been validated in previous studies. Design Builder as an integral and interface system provides reasonable accuracy in calculation of energy consumption [17], and designers are able to analyze energy performance of buildings by considering climatic data and other building characteristics in this software [18] (Fig. 5 and Table 4).

Results and Discussion

Cooling and heating energy consumption values of all louver scenarios were calculated with the Design Builder software for south-facing windows in order to determine the tendency of variations corresponding to slat width, tilt, and spacing as well as the most energy-efficient scenarios.

Table 3 Construction features of the tested building

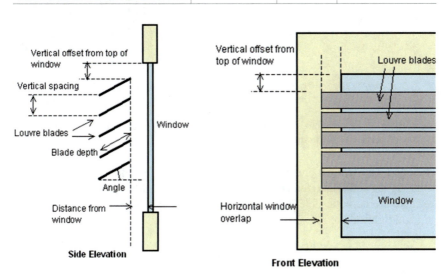

Wall layer	Thickness (m)	Conductivity (W/mK)
Brickwork (OUT)	0.10	0.84
Air gap	0.025	–
Mineral Wool	0.025	0.038
Brickwork (IN)	0.10	0.62
Plaster	0.015	0.16

Fig. 5 Configuration of Louver parameters in Design Builder

- Evaluation of scenarios according to slat width

For the louvers installed with 30 cm spacing and are perpendicular to the window ($\alpha = 0°$) (Fig. 6, left side), as the slat width increases from 100 cm to 400 cm, cooling energy reduced about 40%; although heating load had a rise of 13%, an amount of 14% reduction was observed in the total energy load. In the sloping angle of ($\alpha = 60°$), the cooling load reduction reached to 42%; likewise heating energy had a growth of 16%, meanwhile the total energy showed a reduction of 18%.

The same findings are valid for the right side figure (louver installation at 50 cm spacing), and the variations follow the same tendency.

- Evaluation of Scenarios According to Slat Tilt

Figure 7 presents information on the energy consumption values (KW/m²) according to the louver slat slope in the angles 0f (0°), (30°), (45°), and (60°). In the analysis of this test, it can be noted that at the installation of ($\alpha = 30°$), keeping the

Table 4 The input information applied for residential use in Design Builder

Parameters	Values/description	Parameters	Values/description
Heating fuel	Natural gas	Cooling fuel	Electric power
Heating set point temp	20 °C	Cooling set point temp	25 °C
Infiltration rate	0.7 ac/h	Window opening	50%

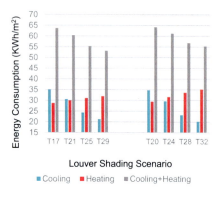
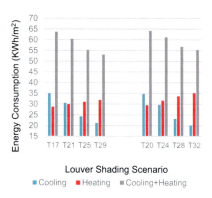

Fig. 6 Effect of slat width on the cooling (C), heating (H), and total (C + H) energy consumption; left: for scenarios with 30 cm spacing, right: for scenarios with 50 cm spacing

width as a constant parameter, a reasonable reduction in cooling energy and total energy is showed. The reference points are highlighted in the graphs.

There are no noticeable differences between the two states of louver distances (left and right side of Fig. 7).

- Evaluation of Scenarios According to Slat Spacing

The pair scenarios next to each other in the bar chart (Fig. 8) are identical in all design parameters except for the louver spacing (or the number of louver across the window height). Although cooling and total energy have decreased moving from left to the right of the chart – due to the scenarios with wider slats – there is no significant difference between two pair bars. Contrarily, the percentage of window covering achieved by slat tilt and width is the most effective determinant in energy load balance.

Among all the tested scenarios, T16 provided the highest saving (42%) in cooling energy consumption, while T17 provided the least saving and presented the largest cooling energy consumption (the louver type specifications are given in Table 5). T16 also ends in the greatest total energy saving (14%). It is found out that in each series of data with constant width, the rise in slat tilt ends in larger window coverage and obviously reduces the cooling load to a significant amount (Fig. 9).

As the main purpose of using shading device is to reduce cooling energy consumption, potential of the louver design in reducing cooling energy load is a principal decision factor. However, slats 200 and 300 cm wide can be preferred instead of the most optimized one (400 cm) since they are less obstructing the view.

External Solar Shading Design for Low-Energy Buildings in Humid Temperate Climates 191

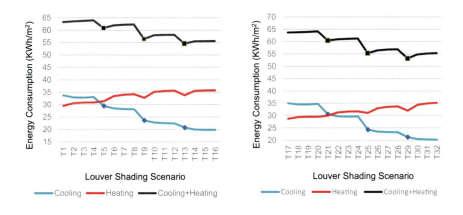

Fig. 7 Effect of slat tilt on the cooling (C), heating (H), and total (C + H) energy consumption; left: for scenarios with 30 cm spacing, right: for scenarios with 50 cm spacing

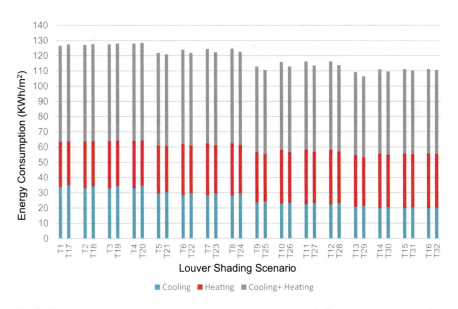

Fig. 8 Effect of slat spacing on the cooling (C), heating (H), and total (C + H) energy consumption

As observed in Fig. 10, the minimum and maximum data points were analyzed in two states of 10 cm and 30 cm distance from window in T_n and T_{n-1}, respectively. Louver distance from window has insignificant effects on the energy use.

Accordingly, it can be concluded that remarkable energy savings can be achieved by improving windows shield.

Table 5 Geometrical specifications of louvers

No of scenario	Slat spacing (cm)	Slat width (cm)	Slat tilt (°)	No of scenario	Slat spacing (cm)	Slat width (cm)	Slat tilt (°)
T1	30	100	0	T17	50	100	0
T2	30	100	30	T18	50	100	30
T3	30	100	45	T19	50	100	45
T4	30	100	60	T20	50	100	60
T5	30	200	0	T21	50	200	0
T6	30	200	30	T22	50	200	30
T7	30	200	45	T23	50	200	45
T8	30	200	60	T24	50	200	60
T9	30	300	0	T25	50	300	0
T10	30	300	30	T26	50	300	30
T11	30	300	45	T27	50	300	45
T12	30	300	60	T28	50	300	60
T13	30	400	0	T29	50	400	0
T14	30	400	30	T30	50	400	30
T15	30	400	45	T31	50	400	45
T16	30	400	60	T32	50	400	60

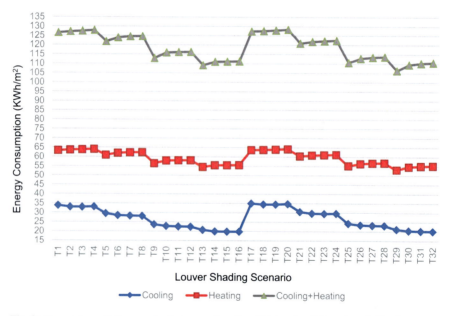

Fig. 9 Comparison of all scenarios in terms of cooling (C), heating (H), and total (C + H) energy consumption

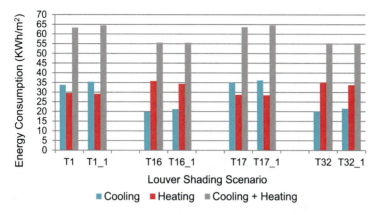

Fig. 10 Effects of distance from window on the cooling (C), heating (H), and total (C + H) energy consumption

Conclusion

The performance of the louver shading devices for the south façade of typical residential apartment buildings was evaluated in a city representative of humid temperate climates. Horizontal louver configurations, based on different values in the slat tilt, width, and spacing, were applied to the south-facing window and analyzed zone energy load affected by different louver installations.

Results show that louver devices are able to improve building energy balance and provide noticeable savings. Findings in the energy balance are directly influenced by the window coverage percentage obtained by higher sloping angles and wider slats. In all scenarios reduction in cooling loads is accompanied by an increase in heating needs. However, savings in cooling energy (40% in the upmost state) resulted in 20% higher heating load while showing acceptable decreases (up to 13%) in the sum of cooling and heating.

As found out by the results of this study, geometric parameters are determining factors in humid temperate climates and affect the building energy load which need to be considered simultaneously as a horizontal shading system is designed for south-facing facades.

Bibliography

1. Hashemi, A., & Khatami, N. (2016). Effects of solar shading on thermal comfort in low-income. *Energy Procedia, 111*, 235–244.
2. ASHRAE. (1997). *ASHRAE handbook, fundamentals*. American Society of Heating Refreigerating and cooling Engineers.

3. Palmero-Marrero, A. I., & Oliveira, A. C. (2010). Effect of louver shading devices on building energy requirements. *Applied Energy, 87*, 2040–2049.
4. Bellia, L., Marino, C., Minichierllo, F., & Pedace, A. (2014). An overview on solar shading systems for buildings. *Energy Procedia, 62*, 309–317.
5. Krstic-Frundzic, A., Vujosevic, M., & Petrovski, A. (2019). Energy and environmental performance of the office building facade scenarios. *Energy, 83*, 437–447.
6. Bellia, L., De Falco, F., & Minichiello, F. (2013). Effects of solar shading devices on energy requirements of standalone office buildings for Italian climates. *Applied Thermal Engineering, 54*, 190–201.
7. Lau, A., Salleh, E., Lim, C., & Sulaiman, M. (2016). Potential of shading devices and glazing configurations on cooling energy savings for high-rise office buildings in hot-humid climates : The case of Malaysia. *International Journal of Sustainable Built Environment, 5*, 387–399.
8. Datta, G. (2001). Effect of fixed horizontal louver shading devices on thermal performance of building by. *Renewable Energy, 23*, 497–507.
9. Yassine, F. (2013). *The effect of shading devices on the energy consumption of buildings : A study on an office building in Dubai*. The British University in Dubai.
10. Hammad, F., & Abu-Hijleh, B. (2010). The energy savings potential of using dynamic external louvers. *Energy and Buildings, 42*(10), 1888–1895.
11. Lee, B. (2019). Heating, cooling, and lighting energy demand simulation analysis of kinetic shading devices with automatic dimming control for asian countries. *Sustainability, 11*, 1253.
12. Iran Climaology Institiute. [Online]. Available: https://climatology.ir/. Accessed 20 Jan 2022.
13. I. C. Alumtechnic. (2022). [Online]. Available: https://alumtechnic.com/. Accessed 20 Jan 2022.
14. National Regulaion Affairs. (2012). *National building regulaion, energy saving*. Minisry of Roads and Urban Development.
15. ASHRAE. (2019). *Standard 90.1. energy standard for buildings except low-rise residential buildings*. American Society of Heating, Refrigerating and Air-Conditioning Engineers (ASHRAE).
16. Deru, M., & Torcellini, P. (2007). *Source energy and emission factors for energy use in buildings*. National Renewable Energy Laboratory.
17. Yu, J., Yang, C., & Tian, L. (2008). Low-energy envelope design of residential building in hot summer and cold winter zone in China. *Energy and Buildings, 40*(8), 1536–1546.
18. You, W., & Ding, W. (2015). Building façade opening evaluation using integrated energy simulation and automatic generation programs. *Architectural Science Review, 58*(3), 205–220.

What It Takes to Go Net Zero: Why Aren't We There Yet?

Carolina Ganem-Karlen, Gustavo Javier Barea-Paci, and Soledad Elisa Andreoni-Trentacoste

Introduction: Net Zero in the Building Industry: Why Aren't We There Yet?

Net zero is intrinsically a scientific concept. It is just a number, begging the question, 'net zero what?' For CO_2, the answer emerged in the late 2000s from understanding what it would take to halt the increase in global average surface temperature due to CO_2 emissions. If the objective is to keep the rise in global average temperatures within certain limits, physics implies that there is a finite budget of carbon dioxide that is allowed into the atmosphere, alongside other greenhouse gases. Beyond this budget, any further release must be balanced by removal into sinks [1].

The acceptable temperature rise is a societal choice but one informed by climate science. Under the Paris Agreement, 197 countries have agreed to limit global warming to well below 2 °C and make efforts to limit it to 1.5 °C. Meeting the 1.5 °C goal with 50% probability translates into a remaining carbon budget of 400–800 GtCO_2. Staying within this carbon budget requires CO_2 emissions to peak before 2030 and fall to net zero by around 2050 [2].

To limit climate change to 1.5 °C, global carbon emissions from energy supply and demand need to be reduced, and any remaining carbon emissions may need to be offset to prevent further warming. Carbon neutrality in the building industry can be achieved by lowering energy consumption with energy efficiency measures in the entire life cycle of a building and, at the same time, by generating energy from renewable sources to cover the baseline energy demand.

C. Ganem-Karlen (✉) · G. J. Barea-Paci · S. E. Andreoni-Trentacoste
CONICET – UNCUYO, Mendoza, Argentina
e-mail: cganem@mendoza-conicet.gov.ar; carolinaganem@gm.fad.uncu.edu.ar; gbarea@mendoza-conicet.gov.ar; sandreoni@mendoza-conicet.gov.ar

Notably, net zero energy systems imply using less energy overall and/or using energy more efficiently, as well as shifting to low-carbon energy sources (e.g. renewable energy) and technologies (e.g. solar water heater/electric vehicle). It is therefore crucial to find the sweet spot in the balance between lowering energy use on the one side and generating renewable energy on the other side.

However, net zero is much more than a scientific concept or a technically determined target. It is also a frame of reference through which global action against climate change can be (and is increasingly) structured and understood.

Achieving net zero requires implementation in varied social, political and economic spheres. There are numerous ethical judgements, social concerns, political interests, fairness dimensions, economic considerations and technology transitions that need to be navigated and several political, economic, legal and behavioural pitfalls that could derail a successful implementation of net zero [1].

The financial, socio-economic and environmental costs of switching from fossil fuels to solar electricity, or other renewable energy sources, eventually need to be offset by savings or new 'benefit streams'. Some scenarios carry the risk that for one or more of the indicators the balance never moves into the positive.

Moreover, net zero energy systems are not limited to one type of behaviour, technology or region. Instead, it is a system-level transition of a global scale that requires multiple solutions tailored to different cultural, economic, geographic, historical, political and social structures across different countries and regions [3].

It is imperative to understand what motivates people to change their behaviours to reduce carbon emissions and what influences public acceptability and adoption of low-carbon technologies and energy system changes. Therefore, human dimensions are at the heart of net zero energy systems, and the conjunction of aspects must contribute setting the agenda for future research.

This chapter discusses four key challenges to achieve net zero in the building industry: (1) the global challenge referred to climate change and future climates according to the 6th IPCC report, (2) the technical challenge related mainly to the achievement of net zero energy parameters in new and existing buildings and the difficulties to certify energy efficiency in a massive attempt, (3) the social challenge contending with the human angle of net zero perception and commitment (different net zero configurations imply different system and lifestyle changes, and strongly depends on people supporting and adopting these changes) (4) the future challenge in the next transition from the building itself to a low carbon community and energy-positive buildings.

The Global Challenge: Climate Change and Future Climates

About 100 years ago, only 14% of the population lived in cities, and in 1950, less than 30% of the world population was urban [4]. Nowadays, around 3.5 billion people live in urban areas around the world, and by 2050 more than two-thirds of the urban population will live in cities [5].

Today, at least 170 cities support more than one million inhabitants each. The situation is even more dramatic in developing countries. Already, 23 of the 34 cities with more than five million inhabitants are in developing countries, and 11 of those cities have populations of between 20 and 30 million inhabitants. Estimations show that urban populations will occupy 80% of the total world population in 2100 [6].

Since the 1950s many of the changes observed are unprecedented. Numerous studies by the Intergovernmental Panel on Climate Change (IPCC) confirm that climate change is caused by human activity and warn that the associated risks are significant. Its latest report [7] states that global surface temperatures will continue to rise until at least mid-century under all emissions scenarios studied. Global warming of 1.5 °C and 2 °C above pre-industrial levels will be exceeded during the twenty-first century unless there are deep reductions in CO_2 and other greenhouse gas emissions in the coming decades.

The Intergovernmental Panel on Climate Change has developed climate scenarios to improve our understanding of how future climate might change [8]. In the 5th Assessment Report (AR5), four emissions scenarios called Representative Concentration Pathways (RCPs) were published. These emissions scenarios are incorporated into general circulation models (GCMs) and regional circulation models (RCMs), which are models used to understand climate behaviour and to forecast likely changes [9]. GCMs/RCMs simulate changes in climate over time and illustrate how the components of climate (surface, atmospheric and oceanic) interact with each other to develop an understanding of its variability. They are called projections because future GHG emissions are unknown. Projections present a snapshot of the possibilities that may occur in the future based on the current state of emissions and assumptions about socio-economic factors such as population, economic and technological developments [10].

Recently, a group of experts from the IPCC [11] set out the results of their assessment of the impact of climate change (1 °C increase above pre-industrial levels in 2017) and the associated risks of exposure of natural and human systems to climate hazards. The same study also shows how these risks would increase significantly under the scenario of a 1.5 °C increase between 2030 and 2052.

While it is theoretically possible that we can keep global warming under 1.5 °C above pre-industrial temperatures, that task is monumental. In fact, models struggle to generate future scenarios that 'keep' warming below 1.5 °C but can pull temperatures back under 1.5 °C after they have overshot that target. Even then, some of these models assume that there has been a global carbon tax in place since 2010 (and there has not) [12].

Robbie Andrews' statements are based on IPCC [7] and GCP [13] projections, in which constant emissions for 8 years will use up the remaining carbon budget. In his projections of CO_2 mitigation curves, assuming a 1.5 °C increment in temperatures by 2100 (Fig. 1) shows a rate of about 4%/year if mitigation would have started in 2000.

Keeping global surface temperatures no more than 1.5 °C above pre-industrial levels entails primarily achieving near zero greenhouse gas (GHG) emissions by 2050. Fossil fuels became the main source of energy after the industrial revolution.

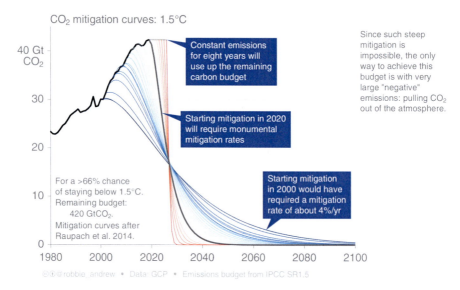

Fig. 1 CO$_2$ mitigation curves: 1.5 °C [12]

Today, the burning of fossil fuels such as coal, oil or gas is the cause of about three quarters of global greenhouse gas emissions. They are also the largest source of air pollution [14].

If we suppose a scenario in which mitigation starts in 2020, monumental mitigation rates will be required. Since such steep mitigation is impossible, the only way to achieve this budget is with very large 'negative' emissions. That is, pulling CO$_2$ out of the atmosphere. Frankhauser et al. [1] state that carbon dioxide removal will probably be constrained by cost considerations and geopolitical factors, as well as by biological, geological, technological and institutional limitations on our ability to remove carbon from the atmosphere and store it durably and safely.

Moreover, Dyke et al. [15] emphasise that there are also concerns about moral hazard risks arising from an over-reliance on carbon removal strategies, which may enable business as usual rather than the drastic scaling back of fossil fuel use.

Figures 1 and 2 present historical emissions to 2017 from CDIAC/Global Carbon Project, projection to 2018 from Global Carbon Project [13].

In Fig. 2 are shown mitigation curves assuming a 2 °C increment in temperatures by 2100. In this scenario, constant emissions for 10 years lead to a required mitigation rate of 10%/year. If we suppose mitigation has started in 2020, the required mitigation rate is about 6%/year. And if mitigation would have started in 2000, it would have required mitigation rates of about 2%/year.

This second temporal scenario presents better chances of achievement, but also depending on the models used to estimate global greenhouse gas emissions, there are even greater risks if the increase in global warming reaches or exceeds 2 °C. Heat waves, which already affect agricultural systems and human health, are projected to become more intense and prolonged. Some regions will suffer an increased risk of

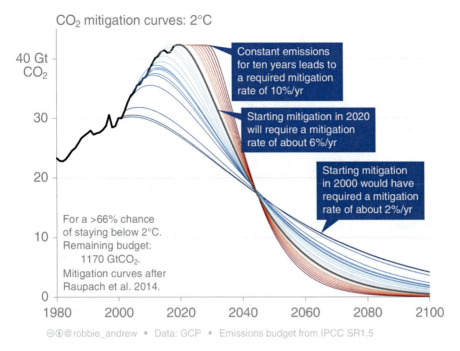

Fig. 2 CO₂ mitigation curves: 2 °C [12]

drought, while others will be severely affected by severe flooding due to increased rainfall intensity, frequency and volume.

In order to slow the advance of climate change and thus preserve people's health, it is essential to aim for a reduction in energy consumption and a transition to renewable energy sources. It is vital to implement measures in the short term, given that under current consumption patterns, global energy consumption is expected to increase by 27% by 2040.

Among other aspects, a range of possibilities associated to the building industry should be considered: the integration into the architectural envelope of small-scale renewable energy producing systems (photovoltaic and solar thermal) and passive conditioning strategies (thermal insulating materials, materials with thermal inertia, absorptive, reflective and selective materials, thermo-economic balances, use of thermal energy in industry, etc.).

Therefore, the identification and application of methodologies to decrease the energy demands of buildings is very important given the global energy challenges as well as the imminent consequences of climate change. Technologies that are applied for efficient energy use in buildings with payback periods for the consumer of less than 5 years have the economic potential to reduce carbon emissions by 25% by 2020 and up to 40% by 2050.

In the face of the global challenge scenarios related to climate change and future climates, the first measures should be the adjustment of bioclimatic strategies for buildings based on in situ measurements and computational simulations and the quantification of the impact towards climate change and future climates of new and retrofitted buildings, using IPCC climate models.

The Technical Challenge: Net Zero in the Building Industry

The construction sector is one of the main causes of pollution, due to the excessive emissions released into the environment as a result of the processes of heating and cooling systems in buildings. Therefore, a turnaround is required, entrusted to the design of new buildings with a reduced annual energy requirement, supported by the inclusion of systems powered by renewable energy sources (RES) [16].

According to the International Energy Agency [17], the energy intensity per square metre of the building sector needs to improve by 30% by 2030 to meet the climate targets of the Paris Agreement. This will require almost doubling the current energy performance of buildings and means that nearly zero energy buildings (nZEB) need to become the global standard in the next decade. At the same time, there is a clear need to invest in the energy renovation of existing buildings.

Most contemporary authors agree that the building stock is inexorably aging, composed of 75% buildings built before 1990, and the replacement of existing building stock by new (environmentally sound) construction is in the order of 1.2% per year [18–22].

This situation leads to the assumption that, in the case that the current regulations consider adequate standards for new building, the negative impacts due to energy consumption and polluting emissions in cities due to the permanence of a large amount of existing building stock would be little reduced. It is important to keep in mind that in a decades-long perspective, the remaining buildings of the twentieth century will cause most of the environmental impacts in cities.

For this reason, improvements to the existing building stock will be the only way in which the benefits of energy efficiency will be made available to the majority of the population, and absolute reductions in domestic energy use and carbon emissions can be achieved. Increasing energy efficiency is the fastest and least costly way to meet the challenges of energy security, the environment and the economy. This fact leaves the main responsibility to the refurbishment of existing buildings, and this is a very complex task.

Another difficulty to overcome is related to passive design as we know it. As a response to the last century energy crisis, passive strategies have been an interesting approach. With the aim of displacing fossil fuels for space conditioning and lighting, the design of construction and shape of the building itself play major roles in capturing, storing and distributing wind and solar energy.

But there are limits to passive design today, and these limits will be greater constrains to the maintenance of interior temperatures within acceptable parameters in the future's more extreme climatic conditions, even with the complementary use of auxiliary energy. A key question is whether a particular energy-optimised design under the present climate and use conditions would remain energy optimised in the future emission scenarios.

Extreme hot climates will become even hotter, and the possibilities of using passive strategies will be further reduced. The use of renewable energy as a primary response to extreme weather conditions will be a constant in many regions of the world, and passive design criteria in these cases will be an accompanying factor in reducing energy demand.

However, there will be nuances, and in complex climates with significant daily and seasonal temperature variations, passive design still presents itself as a viable alternative in the future. However, consideration of dynamic forward projections will allow for flexibility and adaptation to future changes, as opposed to the static simplification of current passive design recommendations.

In the case of temperate continental climates, throughout the twentieth century and specifically since the energy crisis of the 1970s, research into passive or low-energy strategies that contribute to the thermal conditioning of buildings in winter has been a priority. Well into the twenty-first century, the urban heat island is a well-known and sufficiently studied phenomenon, which, together with the prospective climate scenarios proposed by the IPCC for the next 50–80 years, redirects the disciplinary field to give priority to research on the thermal and energy situation of buildings in summer.

These changes not only imply a sustained increase in temperature values but also a change in the relationship between daily maximum and minimum temperatures, which were the methodological basis for the application of passive cooling strategies such as night-time ventilation. In other words, an effective and massive assessment of the current condition of existing buildings and their likely future behaviour is necessary in order to reorient energy efficiency and passive cooling strategies. Only in this way will it be possible to make accurate recommendations from academia to decision-making bodies for the implementation of public policies that will ensure the resilience of our cities throughout the twenty-first century (coinciding with the estimated useful life of buildings).

As an example, it is presented a study performed for the city of Mendoza, Argentina (south latitude −33°9′, west longitude 69°15′, elevation 1.950 m above sea level), a continental temperate cold desert climate (Bwk, according to Koeppen) [24] in which is shown how the different climate change projected scenarios could impact on passive strategies effectiveness.

Notice in Fig. 3 that, in a sun, shading of windows and cooling with dehumidification will be crucial in a years' perspective in summer. And in winter passive strategies such as passive solar direct gain will lose importance as climate will be warmer [23].

Based on the obtained results for the Metropolitan Area of Mendoza, the impact of climate change on urban microclimate and expected changes in energy consumption of buildings during the next century will present new challenges. The heat waves with respect to the present will be more extensive and are estimated to be incremented in 115 more days by 2050 [25].

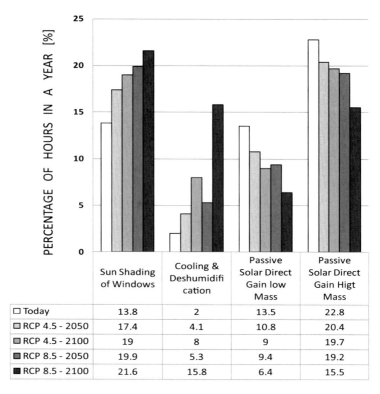

Fig. 3 Percentage of hours in a year where the different passive strategies are recommended for the city of Mendoza in RCP 4.5 and RCP 8.5 (2050 and 2100) [23]

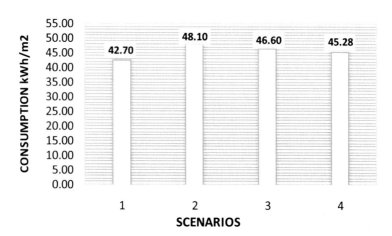

Fig. 4 Total energy consumption results for each scenario [23]

The previous analyses on passive strategies show a clear tendency to the increment of cooling loads and a reduction of the heating requirement. Moreover, when heating and cooling needs are added in a single figure, it is interesting to analyse the whole energy consumption trends. In the case under study, as well as similar constructions, total energy consumption presents maximum variation of less than 10% when compared with the worst-case scenario 4: CRP 8.5 (2100). See Fig. 4.

It is important to take into account that even though consumptions seem similar and variations are perceived as low between scenarios, the main aspect to discuss is in the use of the energy required. In the current situation prevails energy consumption for heating with 57.5% of the total, mainly covered by using natural gas as auxiliary energy. While in the worst scenario, predominates energy consumption for cooling with 77.8% of the total, mainly covered by using electricity.

A third aspect to consider is that the application of energy savings measures in related directly with the people that manage them, and how the risks related to climate change are underestimated. There is a tendency to be unduly optimistic about the likelihood of adverse events occurring, and 'follow the herd', such that our choices are often influenced by other people's behaviour, especially under conditions of uncertainty. It is important to take responsibility for the impact associated with our way of living; this involves a commitment to reduce the negative effects that our actions can cause in the future.

The Social Challenge: Net Zero Perception and Commitment

Change – including climate change – is inherent to planet Earth, which, over its billions of years of history, has undergone much more intense changes [26]. However, there are two characteristics of current climate change that make the associated global biophysical and social impacts unique in the history of the planet: the speed and intensity with which this change is taking place, in such a short space of time for the evolution of the planet as decades, and human activity as a driver of all these changes.

It is almost a self-evident fact to remember that societies are constantly changing, although sometimes more evolutionary (slow) and sometimes more revolutionary (fast). The speed of change in societies is a key factor for the analysis of social impact, especially in terms of its interrelation with the biogeophysical environment, since a large part of the problem of so-called climate change is being produced by the high speed of social change in contemporary societies (e.g., the increase in the demand for energy and basic resources), which produces pressures on the biogeophysical environment, whose possibilities for renewal of resources and, above all, for 'integration' of toxic and hazardous waste require a much longer time and a slower speed of pressure.

The social impact or consequences of global change is ultimately what will result from the interactions between changes in the biophysical environment and changes in the specific social environment. However, these interactions are almost never direct, as they are also – and mainly – mediated by the various spheres of social activity, including social organisation (economy, social relations, norms, values, etc.) and technology.

Human activities on earth impact globally in such a way that a virus initially spread in the city of Wuhan in China in December 2019 caused the death of 4,962,000 people worldwide by September 2021. In March 2020, the WHO declared a COVID-19 pandemic, with 437,923,303 people infected with the virus worldwide. Governments in all countries immediately took steps to contain the spread of the virus by implementing collective actions that would drastically change the way we relate to each other. The closure of schools and public buildings, the cancellation of flights, the wearing of face masks, and the closing of restaurants and entertainment venues, among many others.

What has happened shows that global challenges can be met with massive behavioural responses by governments and citizens, especially if they are implemented immediately. At the same time, there is an opportunity for people to recognise that our individual actions cause global impacts, both positive and negative.

Numerous studies now draw parallels between the pandemic and another global externality, climate change. Similarities have been found between the two issues, mainly in the negative effects on developing countries. Both the impacts of the current pandemic and many consequences of climate change, such as more frequent and intense natural disasters, can be characterised as low probability-high consequences (LP-HC) risks. Recent studies [27] show that people underestimate LP-HC risks, such as climate change and COVID-19, until they experience the consequences themselves or see the life of a close friend or family member affected. People are likely to make decisions focusing on the low probability of a disaster happening to them or its possible consequences, rather than making a rational assessment of the overall risk. Many climate change-related risks, such as natural disasters, have a low probability that individuals simplify to being zero or falling below their threshold level of concern. The same case occurs with pandemics.

However, there are striking similarities between the climate and COVID-19 crises; they also differ in many fundamental ways, including the speed at which they develop [28]. The COVID-19 crisis can both occur and be controlled rapidly, in comparison to the more slowly looming climate crisis, whose impacts may be even greater. There are no safe climate change vaccines or potential treatments to be developed that could 'solve' the climate crisis, and any activities aiming for the reversal of climate change would likely take decades or more before coming to fruition. The COVID-19 crisis shows us the importance of prevention and early action, and this may be even more important in averting the worst outcomes of climate crisis.

Even though it is important to take immediate action on the occurrence of an event that involves an imminent risk, it is even more important to be prepared before the effects are devastating. Many factors lead to delay response, related to emotional reactions and cognitive biases such as the failure of the human mind to grasp the concept of exponential growth and the misperception of risk [29]. Several other biases lead people to ignore the potential consequences of a looming event. The 'simplification bias' implies that individuals view the likelihood of LP-HC events as

What It Takes to Go Net Zero: Why Aren't We There Yet?

falling below their threshold level of concern and fail to take risk reduction measures, unless they experience the impacts of a disaster according to the 'availability bias'.

As with COVID-19, it will be important for political leaders at the national, state and local levels in every country in the entire world to recognise cognitive biases and turn to experts for advice on how to deal with the impacts of climate change. Addressing the 'simplification and availability biases' necessitates the development of communication strategies that stress the consequences of risks associated with climate change and COVID-19 to ensure that individuals start paying attention. The need to accelerate climate action can be managed by linking policies and measures that are currently adopted to limit the risks from pandemics to actions that also reduce the risks from climate change.

In an image that has now gone viral (Fig. 5), cartoonist Graeme Mackay reflects how society takes action in the face of imminent risks, currently the pandemic, paying less attention to those they consider less urgent, such as climate change. The associated impacts would also be devastating for human life and ecosystems. However, many of the actions that can be implemented in the future to prevent a new pandemic can also help combat climate change and vice versa.

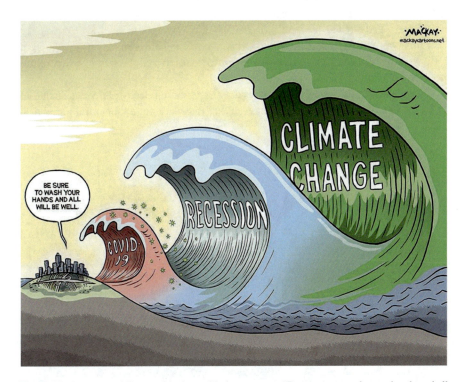

Fig. 5 Viral cartoon of Graeme Mackay with the message: "Be sure you wash your hands and all will be well" [30]

Society's response to the COVID-19 pandemic crisis has demonstrated convincingly that people can adapt quickly and change habits and incorporate new ones if there is an imminent threat. It is therefore possible to believe that achieving behavioural changes in people that aim to mitigate climate change is a feasible reality.

The Future Challenge: From the Building Itself to a Low-Carbon Community – Energy-Positive Buildings

Our remaining carbon budget to keep warming under 1.5 °C is tiny. Modelling shows the 'best' way to achieve this is by actively removing CO_2 from the atmosphere: 'negative emissions'. Doing this allows more room for actual, positive emissions. Various technologies are being explored, as well as good, old-fashioned tree planting, but all have significant limits. It is partly this hope in future technologies – technological optimism – that delays action [12].

Figure 6 presents projections based in the functional form from Raupach et al. [31] for positive emissions, adding residual, hard-to-mitigate emissions of 5% of the current level, and negative emissions using the ramp of a cosine function. Σ indicates the cumulative emissions in 2019–2100. Global cumulative CO_2 emissions budgets are from the IPCC Special Report on 1.5 °C: 420 GtCO2 for a 66% of 1.5 °C and 1170 GtCO2 for a 66% of 2 °C [32].

If this was actually possible, and the positive and negative emissions on Fig. 6 in 2100 were to continue beyond 2100, we would begin to pull the global temperature increase back under 1.5 °C.

Mitigation curves describe approximately exponential decay pathways such that the quota is never exceeded [31]. But note that these are not exponential pathways:

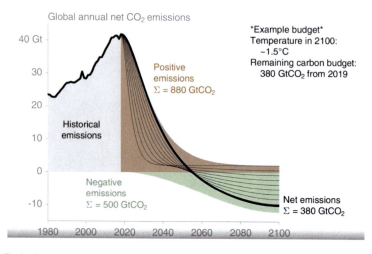

Fig. 6 Projections based in the functional form from for positive emissions [31]

the rate of mitigation is not the same every year. This comes from a recognition that an oil tanker cannot turn on a dime: we have enormous infrastructure (social, political, physical) that cannot be changed overnight. So these curves allow for some inertia in the early years of mitigation.

A further step is represented by the possibility of connecting the individual NZEBs to an intelligent energy distribution network, also called smart grids. Positive energy buildings (PEBs) could be a new target, through their contribution to the energy support of other buildings connected to them, producing more energy than necessary to their needs. A system of units connected together at the neighbourhood level, aiming of achieving neutrality or, in extreme cases, energy positivity is the next challenge. The buildings will thus become collectors and energy storage structures [16].

According to Carlisle et al. [33], the net zero energy community (NZEC) is to be considered by four following perspectives:

- The central energy system should consist of renewable energy sources and will accommodate the energy for the whole community.
- The percentage of energy loss should be kept in mind to deliver the energy.
- The financial aspects of the community should be considered.
- The environmental impacts including GHGs emissions are to be considered.

Ulla et al. [34] reviewed 23 case studies of net zero energy communities around the globe revealing that they mainly focus on the onsite energy generation. Solar energy is the widely used renewable energy resources for NZEC, as 22 settlements have considered utilising it. Beside the solar energy, wind energy is the second most popular renewable energy source, while geothermal energy, biogas plants and other sources are not considered significantly.

NZEC components are highlighted in the Fig. 7. It is also clear that energy storage, management, and control systems are also the vital parts of the NZES to increase the self-consumption and reduce the cost and size of energy generation and supply systems [35].

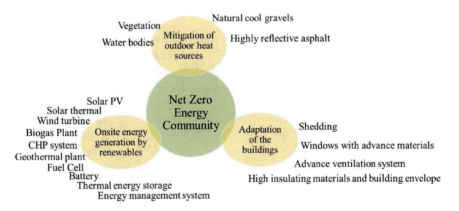

Fig. 7 The various components of a NZEC. Ulla et al. [34]

The impacts of climate zones and building types have influenced the selection methods of these building adaptation tools and techniques along with the renewable. Among them, 17 settlements additionally consider various adaptation techniques that have significantly lessened the energy demand of the buildings. However, some case studies did not thoroughly consider the adaptation tools and new technologies for the adaptation of buildings. A few studies (four settlements only) incorporate some mitigation strategies to mitigate the outdoor heat sources that enable to lower the ambient temperature along with diminishing the GHGs emissions as well as minimising the energy demand of the community.

Therefore, to establish the NZEC, further efforts can be given to the following areas:

– Mitigation strategies to reduce the ambient temperature, especially for the settlements at hot climate zones, such as vegetation, water bodies, natural cool graves and highly reflective asphalt.
– Adaptation of the buildings with high-performance materials would be a promising option to reduce the energy demand and improve the indoor air quality as well as control the indoor temperature of the buildings.
– Diversified use of renewables: The solar energy (PV and thermal collector) and wind energy are mainly considered as renewable energy resources.

Discussion: Our Role in the Transition to Net Zero: We Must Be There as Soon as Possible

Despite being a widely used term, sustainability is a concept that acquires different connotations according to its interpretation. In its most basic meaning it is understood as 'maintaining the status quo and not disappearing' [36]. Maintaining the status quo, however, translates into a notion that avoids or mitigates change [37]. Thinking of sustainability as 'longevity' is another way of approaching the concept, meaning 'the longer the system can be maintained, the more sustainable it is' [38, 39].

Leach et al. [40] define sustainability as the "capacity to maintain over an indefinite period of time specific qualities of well-being, social equity and environmental integrity" and argue that, while the concept of sustainability focuses on reducing negative impacts on the environment to avoid change, the concept of resilience is about adapting to change.

Holling [41] introduced the term resilience as a measure of the persistence of a system and the ability to absorb change and still maintain the same relationships between variables. The twenty-first century requires buildings to be not only sustainable but also resilient, that is, not contributing to the deterioration of the environment but also flexible to change as here are changes in climate that will be inevitable in the future.

Even if emissions of carbon dioxide and other pollutants were to be drastically reduced, these climate scenarios would not change much until 2040 because gas concentrations remain in the atmosphere for a long time and thermal adjustment is slow. In other words, over the next 30 years the outlook is inexorable, and we must prepare to adapt finding means to achieve net zero in the building industry.

Therefore, net zero commitments are not an alternative to urgent and comprehensive emissions cuts. Indeed, net zero demands greater focus on eliminating difficult emissions sources than has so far been the case. The 'net' in net zero is essential, but the need for social and environmental integrity imposes firm constraints on the scope, timing and governance of both carbon dioxide removal and carbon offsets. It is possible to align net zero with sustainable development objectives, allow for different stages of development and secure zero-carbon prosperity [1].

From an economic perspective, while NZEBs require higher investment costs, there are much greater long-term benefits, both in micro- and macro-perspectives. For NZEBs generating power on site, the speed of reaching the payback period may be influenced by the local feed-in tariff structure.

Preferential government policies, such as tax subsidies, will help make NZEBs more economically attractive. Financing opportunities can be found in various investment schemes, such as green bonds and green investments. Making these more accessible to the general population can help accelerate the adoption of the NZEB concept in each country.

Also, different net zero configurations imply different system and lifestyle changes and strongly depend on people supporting and adopting these changes. Behaviour change by individuals, commercial entities, and policy makers is critical to achieving net zero in all domains [42]. People's knowledge of which behaviours generate the most greenhouse gases is generally poor. It is important to develop new communication strategies to ensure that individuals start paying attention.

While legislative mechanisms, geographical and climatic conditions are essential factors, it is the technological applications and innovations that are at the heart of

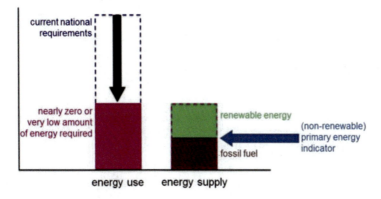

Fig. 8 The NZEB model according to EPBD indications [16]

the design of an NZEB. These apply to both the energy generation and energy saving characteristics of a building [43, 44].

The introduction of the NZEB target in the building design will encourage a decrease in the amount of energy required, thus abandoning fossil fuels [45]. A graphical interpretation of the NZEB energy balance is indicated in the scheme of Fig. 8. With fairly limited options for on-site renewable energy generation systems, it is crucial that NZEBs diversify in design and energy efficiency technologies.

It is important to take into account that no single technology alone, negative emission technologies included, can reduce global warming to 1.5 °C. There is agreement among climate scientists that reducing overall energy demand, using energy more efficiently and shifting to cleaner energy sources and technologies, is critical for reducing global carbon emissions. Changes in everyday energy behaviours of individuals and households, in particular, behaviours related to mobility, housing and food, have a substantial potential to reduce carbon emissions.

Real progress is frozen until a solution is found that works for (almost) everyone at (almost) the same time. Our role in the transition to net zero is that energy-positive buildings and low-carbon communities will become the new normal, and at the same time, we will grow more connected on a human scale.

Acknowledgments Authors would like to thank the National Research Council of Argentina (CONICET) and the Fund for Scientific and Technological Research of Argentina (FONCYT) for funding this research.

References

1. Fankhauser, S., Smith, S. M., Allen, M., Axelsson, K., Hale, T., Hepburn, C., Kendall, J. M., Khosla, R., Lezaun, J., Mitchell-Larson, E., Obersteiner, M., Rajamani, L., Rickaby, R., Seddon, N., & Wetzer, T. (2022). The meaning of net zero and how to get it right. *Nature Climate Change, 12*, 15–21.
2. IPCC. (2018). Global warming of 1.5°C. An IPCC Special Report on the impacts of global warming of 1.5°C above pre-industrial levels and related global greenhouse gas emission pathways, in the context of strengthening the global response to the threat of climate change. Retrieved from https://www.ipcc.ch/sr15/. Last consulted: May, 2022.
3. Rogelj, J., Geden, O., Cowie, A., & Reisinger, A. (2021). Three ways to improve net-zero emissions targets. *Nature, 591*, 365–368.
4. Bitan, A. (1992). The high climatic quality city of the future. *Atmospheric Environment., 26*(3), 313–329.
5. DOE – Department of Energy. (2019). World population prospects: The 2019 revision, key findings and advance tables. https://population.un.org/wpp/. Last consulted: May, 2022.
6. Santamouris, M., Papanikolaou, N., Livad, A. I., Koronakis, I., Georgakis, C., Argiriou, A., & Assimakopoulos, D. N. (2001). On the impact of urban climate on the energy consumption of buildings. *Solar Energy, 70*(3), 201–216.
7. IPCC. (In press). *Climate change 2021: The physical science basis. Contribution of working group I to the sixth assessment report of the Intergovernmental Panel on Climate Change.* Cambridge University Press. https://www.ipcc.ch/report/ar6/wg1/downloads/report/IPCC_AR6_WGI_Full_Report.pdf. Last consulted: May, 2022.

8. Pachauri, R. K., Allen, M. R., Barros, V. R., Broome, J., Cramer, W., Christ, R., Church, J. A., Clarke, L., Dahe, Q., Dasgupta, P., et al. *Climate change 2014: Synthesis report. Contribution of working groups I, II and III to the fifth assessment report of the Intergovernmental Panel on Climate Change*. IPCC: Geneva, Switzerland.
9. Herrera, M., Natarajan, S., Coley, D. A., Kershaw, T., Ramallo-González, A. P., Eames, M., Fosas, D., & Wood, M. (2017). A review of current and future weather data for building simulation. *Building Services Engineering Research and Technology, 38*, 602–627.
10. Nik, K. (2013). Impact study of the climate change on the energy performance of the building stock in Stockholm considering four climate uncertainties. *Building and Environment, 60*, 291–304.
11. Hoegh-Guldberg, O., Jacob, D., Taylor, M., GuillénBolaños, T., Bindi, M., Brown, S., Camilloni, I. A., Diedhiou, A., Ddjalante, R., Ebi, K., Engelbrecht, F., Gguiot, J., Hijioka, Y., Mehrotra, S., Hope, C. W., Payne, A. J., Pörtner, H. O., Seneviratne, S. I., Thomas, A., & Zhou, G. (2019). The human imperative of stabilizing global climate change at 1.5°C. *Science, 365*, 6459.
12. Andrews, R. (2020). *It's getting harder and harder to limit ourselves to 2 °C*. CICERO Center for International Climate Research. University of Oslo, Norway. https://folk.universitetetioslo.no/roberan/t/global_mitigation_curves.shtml. Last consulted: May, 2022.
13. Le Quéré, C., Andrew, R. M., & 71 other authors. Global carbon budget 2018. Earth system. *Science Data, 10*(2018), 2141–2194.
14. Ritchie, H., & Roser, M. (2020). Argentina: Energy country profile. *Our World in Data*. https://ourworldindata.org/energy. Last consulted: May, 2022.
15. Dyke, J., Watson, R., & Knorr, W. (2021, April 22). Climate scientists: concept of net zero is a dangerous trap. *The Conversation*. https://theconversation.com/climate-scientists-concept-of-net-zero-is-a-dangerous-trap-157368. Last consulted: May, 2022.
16. Magrini, A., Lentini, G., Cuman, S., Bodrato, A., & Marenco, L. (2020). From nearly zero energy buildings (NZEB) to positive energy buildings (PEB): The next challenge – The most recent European trends with some notes on the energy analysis of a forerunner PEB example. *Developments in the Built Environment, 3*, 1–12.
17. International Energy Agency IEA. (2018). *Energy efficiency 2018, analysis and outlooks to 2040*. International Energy Agency.
18. Lowe, R., Bell, M., & Johnston, D. (1996). *Directory of energy efficient housing*. Chartered Institute of Housing Edition.
19. Boonstra, C., & Thijssen, I. (1997). *Solar energy in building renovation*. International Energy Agency – IEA. Solar Heating and Cooling Programme. Task 20. James & James.
20. Voss, K. (2000). Solar energy in building renovation – results and experience of international demonstration buildings. *Building and Environment, 32*, 291–302.
21. Vilches, A., Garcia-Martinez, A., & Sanchez-Montanes, B. (2017). Life cycle assessment (LCA) of building refurbishment: A literature review. *Energy and Buildings, 135*, 286–301.
22. European Commission, European Construction Sector Observatory. (2018, November). *Analytical report on improving resource efficiency*. https://ec.europa.eu/growth/sectors/construction/observatory_en. Last consulted: May, 2022.
23. Ganem-Karlen, C., & Barea-Paci, G. (2022, November). The complex challenge of sustainable architectural design. Assessing climate change impact on passive strategies and buildings' opportunities for adaptation. In *Proceedings of PLEA 2022: Will cities survive?* Santiago de Chile.
24. Kottek, M., Grieser, J., Beck, C., Rudolf, B., & Rubel, F. (2006). World Map of the Köppen-Geiger climate classification updated. *Meteorologische Zeitschrift, 15*(3), 259–263.
25. Crawley, D. B., Lawrie, L. K., Pedersen, C. O., & Winkelman, F. C. (2000). EnergyPlus: Energy simulation program. *ASHRAE Journal, 42*(4), 49–56.
26. Duarte, C. M. (Coord.). (2006). Cambio Global. Impacto de la Actividad Humana sobre el Sistema *Tierra. Colección divulgación, 3*, 167.

27. Botzen, W., Duijndam, S., & Van Beukering, P. (2021). Lessons for climate policy from behavioural biases towards COVID-19 and climate change risks. *World Development, 137*, 105214.
28. Manzanedo, R., & Manning, P. (2020). COVID-19: Lessons for the climate change emergency. *Science of the Total Environment, 742*, 140563.
29. Kunreuther, H. C., & Slovic, P. (2021). Learning from the COVID-19 pandemic to address climate change. *Management and Business Review, 1*(1), 92–99.
30. Mackay, G. *Cartoon originally drawn for March 11, 2020*. Revised May 23, 2020. Source: https://mackaycartoons.net. Last consulted: May, 2022.
31. Raupach, M. R., Davis, S. J., Peters, G. P., Andrew, R. M., Canadell, J. G., Friedlingstein, P., Jotzo, F., & Quéré, C. L. (2014). Sharing a quota on cumulative carbon emissions. *Nature Climate Change, 4*, 873–879.
32. Rogelj, J., Shindell, D., Jiang, K., Fifita, S., Forster, P., Ginzburg, V., Handa, C., Kheshgi, H., Kobayashi, S., Kriegler, E., Mundaca, L., Séférian, R., & Vilariño, M. V. (2018). Mitigation pathways compatible with 1.5°C in the context of sustainable development. In V. Masson-Delmotte, P. Zhai, H.-O. Pörtner, D. Roberts, J. Skea, P. R. Shukla, A. Pirani, W. Moufouma-Okia, C. Péan, R. Pidcock, S. Connors, J. B. R. Matthews, Y. Chen, X. Zhou, M. I. Gomis, E. Lonnoy, T. Maycock, M. Tignor, & T. Waterfield (Eds.), *Global Warming of 1.5°C. An IPCC Special Report on the impacts of global warming of 1.5°C above pre-industrial levels and related global greenhouse gas emission pathways, in the context of strengthening the global response to the threat of climate change, sustainable development, and efforts to eradicate poverty*. Cambridge University Press.
33. Carlisle, N., Van Geet, O., & Pless, S. (2009). *Definition of a 'Zero Net Energy' Community*. National Renewable Energy Lab. (NREL).
34. Ulla, K. R., Prodanovic, V., Pignatta, G., Deletic, A., & Santamouris, M. (2021). Technological advancements towards the net-zero energy communities: A review on 23 case studies around the globe. *Solar Energy, 224*, 1107–1126.
35. Sokolnikova, P., Lombardi, P., Arendarski, B., Suslov, K., Pantaleo, A. M., Kranhold, M., & Komarnicki, P. (2020). Net-zero multi-energy systems for Siberian rural communities: a methodology to size thermal and electric storage units. *Renewable Energy, 155*, 979–989.
36. Sayer, C. (2004). *The science of sustainable development: Local livelihoods and the global environment*. Cambridge University Press.
37. Lizarralde, C., & Bosher, D. (2015). Sustainability and resilience in the built environment: The challenges of establishing a turquoise agenda in the UK. *Sustainable Cities and Society, 15*, 96–104.
38. Lew, A. A., Ng, P. T., Ni, C. C., & Wu, T. C. (2016). Community sustainability and resilience: Similarities, differences and indicators. *Tourism Geographies, 18*(1), 18–27.
39. Marchese, D., Reynolds, E., Bates, M. E., Morgan, H., Clark, S. S., & Linkov, I. (2018). Resilience and sustainability: Similarities and differences in environmental management applications. *Science Total Environment, 613*, 1275–1283.
40. Leach, M., Stirling, A. C., & Scoones, I. (2010). *Dynamic sustainabilities: Technology, environment, social justice*. Taylos and Francis.
41. Holling, C. S. (1973). Resilience and stability of ecological systems. *Annual review of ecology and systematics, 4*(1), 1–23.
42. Marteau, T. M., Chater, N., & Garnett, E. E. (2021). Changing behaviour for net zero 2050. *BMJ (Clinical research ed.), 375*, 2293.
43. Liu, Z., Liu, Y., He, B.-J., Xu, W., Jin, G., & Zhang, X. (2018). Application and suitability analysis of the key technologies in nearly zero energy buildings in China. *Renewable Sustainable Energy Reviews, 101*, 329–345.
44. Cao, X., Dai, X., & Liu, J. (2016). Building energy-consumption status worldwide and the state-of-the-art technologies for zero-energy buildings during the past decade. *Energy and Buildings, 128*, 198–213.
45. Erhorn, H., & Erhorn-Kluttig, H. (2016, November). *New buildings & NZEBs. Concerted action – energy performance of building*.

The Integrated Design Studio as a Means to Achieve Zero Net Energy Buildings

Khaled A. Al-Sallal, Ariel Gomez, and Ghulam Qadir

Introduction

This chapter discusses the authors' philosophy about the notion of the integrated design studio (IDS) and explains their methodologies that support the operation of the IDS. Achieving a zero net energy (ZNE) design has always been a main consideration when teaching the course "ARCH-430: Integrated Building Design Studio" at the Architectural Engineering Department, UAE University. The course was developed by the chapter's author and others throughout many years. It is described in the syllabus as follows: "it develops a comprehensive design process with focus on systems design and integration of a mixed-use building, issues of technology, ecology and energy. Exercises focus on the design of building systems and components, building structural design, building codes, design for safety in buildings, architectural expression, integration strategies and applications involving the mechanical, electrical, energy, and building management systems."

The chapter highlights a specific experience that took place in 2018, when the students of the ARCH-430 participated in the reputable MultiComfort House Student Contest, Edition 2018, by Saint Gobain [3]. Although the contest did not stipulate ZNE as an obligation to win, the instructors (authors of this chapter)

K. A. Al-Sallal (✉)
Concreto, Coquitlam, BC, Canada
e-mail: khaled@concreto.ca

A. Gomez
UAE University, Al Ain, UAE
e-mail: a.gomez@uaeu.ac.ae

G. Qadir
The University of Sydney, School of Civil Engineering, Darlington, NSW, Australia
e-mail: ghulam.qadir@sydney.edu.au

© The Author(s), under exclusive license to Springer Nature Switzerland AG 2023
A. Sayigh (ed.), *Towards Net Zero Carbon Emissions in the Building Industry*, Innovative Renewable Energy, https://doi.org/10.1007/978-3-031-15218-4_11

encouraged their students to consider it in their participation. They believed it would represent a great incentive for the participating students to generate their best designs under support from the course's instructors and direction from the studio's professor (i.e., first author). Achieving a ZNE target in the extreme hot climate of Dubai was a real challenge. Whether the students achieve it or come very close to it, the real merit was not really in reaching to the exact zero number as much as it was in the great learning experience the students attained and the enjoyment they felt when they found themselves capable of challenging such a tough design problem. The great efforts made by everybody were highly rewarded when the students won the third prize in the contest, and their faces were glowing with success smiles.

Background

The Global Energy Problem

Fast urbanization and higher need of habitat for more people in urban areas in addition to enhanced lifestyle of the people demands increasing levels of energy every year. Most of the world energy is still produced by nonrenewable sources of power generation that are major contributors of greenhouse gas (GHG) emission, out of which carbon dioxide (CO_2) is crucial. Buildings consume more energy than any other sector (nearly half of all energy produced in the USA and 20–40% of the total energy consumption in Europe) [1]. Around 75% of all the electricity is used just to operate buildings in the USA. Moreover, the International Energy Agency (EIA) indicated that the energy consumption of the building sector is expected to grow faster than that of industry or transportation [2]. Reducing energy demand through building efficiency is significantly cheaper than producing the same amount of energy by coal or nuclear power. The top four end-uses in the commercial sector are space lighting (20%), space heating (16%), space cooling (14.5%), and ventilation (9%). All together they represent 60% of commercial primary energy consumption [2].

In hot-climate countries, energy needs for cooling can amount to two or three times those for heating on an annual basis. The UAE was ranked among the highest energy consumers per capita in the entire world due to its high electrical consumption that was almost 127,000-gigawatt-hours (GWh) in 2017 [4]. Most electrical energy in the UAE in 2018 was generated from natural gas-fired [5]. The buildings' sector is responsible for 80% of the total consumed energy in the UAE [6]. In very hot climates such as in the UAE, 40% of the total cooling energy can be utilized to offset heat gains from walls and roofs, and it could reach 75% when combined with the glazing effect [7].

The Efficient World Scenario (EWS), developed for the World Energy Outlook 2012 (WEO-2012), enabled quantifying the implications on the economy, environment, and energy security when taking major steps toward energy efficiency (WEO-2012: [8]). It was developed to help in putting policies in place to allow the

The Integrated Design Studio as a Means to Achieve Zero Net Energy Buildings 215

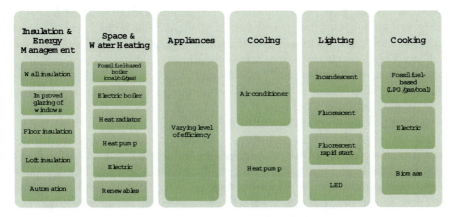

Fig. 1 Technical potentials identifying key technologies to improve energy efficiency in the building sector considered in the Efficient World Scenario. (WEO-2012; [8])

market to realize the potential of energy efficiency measures that are economically viable. For the building sector, the potential viable measures were divided into six categories (Fig. 1) based on consultation with a large number of companies, experts, and research institutions at national and international levels. It also involved the conduction of an extensive literature search to catalogue the technologies that are now in use in different parts of the world, as well as judgment of their probable evolution. These measures represented technical potentials identifying key technologies to improve energy efficiency in the building sector.

Low Carbon Strategies

Designing a new green building requires estimating CO_2 equivalent emission footprint. The calculations of the CO_2 emissions are usually simple and based on the use of emissions factors. Computing the reduction in emissions associated with energy conservation measures for existing buildings has been proposed [9]. Calculation of the greenhouse gas emissions should be based on source energy and not on energy consumed onsite alone. A good reference source for emission factors is contained in a National Renewable Energy Laboratory (NREL) report released in 2007 [10].

When searching for design solutions to improve sustainability, it is important to give priority to tackle problems that contribute the most to building carbon emissions. The common use of excessively glazed envelopes especially in commercial buildings is a major problem due to the cause of substantial thermal loads. To make considerable reductions of building carbon emissions, it is necessary to minimize building energy by improving the thermal performance of the envelope. A saving of carbon emission of 31–36% can be achieved via retrofitting and the selection of appropriate U-factors for building envelope materials [11]. Reuse, recycling, and regeneration of building energy together can save up to 10% of total energy; which subsequently mitigates further emissions. The architectural design and how it is integrated with lighting design are other common problems leading to high carbon emissions and high electrical power demands, especially in office buildings and schools. The problems are caused mostly by the inappropriate designs for the architectural form, the floor-plate, and other design factors, which necessitate the use of electrical lighting, rather than natural lighting that eventually leads to increased internal heat gains and cooling loads.

Building systems are usually categorized as passive systems, active systems, integrated (passive and active) systems, water conservation strategies, wastewater and sewage recycling systems, and power generation systems. Available strategies that would substantially reduce building GHG emissions can be divided into three general categories (Table 1): (1) the planning and design strategies, (2) the building envelope and material and equipment selection, and (3) the added technologies category [12]. The factors affecting the buildings' energy performance were discussed in numerous studies [13–17]. The notion of creating zero net energy (ZNE) building can only be realized by maximizing reliance on passive and low-energy systems while minimizing reliance on active systems. Technically, this is done through an appropriate design of the built-form configuration, site layout, façade design, solar-control devices, daylight strategies, envelope materials, vertical landscaping, passive cooling and heating strategies, water-efficient systems, and onsite renewable

Table 1 Available strategies that can substantially reduce building sector GHG emissions

Planning and design strategies	Building envelope and material and equipment selection	Added technology
Building shape, orientation, and color	Adequate insulation values	Solar hot water heating
Spatial layout	Radiant barriers	Photovoltaic systems
Window shape and orientation	Low-e coatings and argon gas-filled glazing	Micro-wind electric generation
Daylighting	Thermal break windows and systems and movable insulation	Community-scale solar thermal, wind, and biomass electric generation
Natural ventilation		
Exterior shading		Combined heat and power systems
Vegetation and microclimate control	Sunlight and daylight fixtures and systems	
Passive solar heating systems	Cool roofs	
Night-vent and night-sky radiation cooling systems	Green roofs	
Double-envelope systems	Occupancy and CO_2 sensors	
Common wall design strategies	Daylighting controls and photo sensors	
Building and unit density	Energy management systems	
Mixed-use development	High-efficiency equipment, lighting and appliances	
Pedestrian- and transit-oriented development (reduced miles traveled)	Geothermal heat pump	
	Air-to-air heat exchangers and heat recovery systems	
	Building commissioning	

Source: Kharecha et al. [12]

energy systems [18]. The following sections include discussions on the strategies and technologies that help achieve the ZNE building.

Passive and Low-Energy Cooling Systems

Passive and low-energy cooling systems help save cooling energy and operation costs, but it might add costs for construction. These costs can be minimized if the architect succeeds first in minimizing building heat gains so that the remaining cooling requirements can be limited and hence covered by using passive systems. Minimizing heat gains through architectural design can be done by choosing and designing a site that helps improve the microclimate around the building, developing a building form that helps shade itself and reduce exposure to the sun, arranging the building spaces to receive only the desirable environmental factors (e.g., desirable breezes) while blocking undesirable ones (e.g., intense afternoon west sun), shading the building envelope and openings using porches and shading devices, painting or cladding the building envelope with light colors, using high-resistance thermal insulation to minimize heat transfer by conduction, and maximizing reliance on daylighting and using energy-efficient lights and equipment to control internal heat gains. After determining the sources of heat and applying methods for minimizing heat gains through architectural design, passive and low-energy systems can be applied. The process starts by identifying the sources of coolness (or heat sinks) available in the natural environment and then choosing the passive systems that have greater potential to utilize these sources. The passive cooling systems are usually divided into three main categories: (1) systems that are more effective in hot and arid regions, (2) systems that depend on airflow systems for comfort cooling, and (3) systems that have potential to mitigate discomfort problems due to high humidity levels in warm and humid climates [19].

HVAC Equipment

Climate, building function, and design decisions determine whether heating, cooling, or both are needed for a given building. The interactions between climate and function, which is extremely affected by building layout and enclosure, are varied and complicated [20]. The major divisions of the HVAC equipment include source, distribution, delivery, and control components. These, respectively, produce heating and/or cooling effect, move this effect around a building, introduce it into the spaces being conditioned, and regulate the magnitude and timing of such effects [21]. The source components in an HVAC system are intended to provide heat (heating effect), coolth (cooling effect), and/or both. There are several fundamental options for cooling effect sources [22]. The three basic (nonpassive) means of producing a cooling effect for a building are vapor compression refrigeration, absorption refrigeration, and evaporative cooling [23]. The most thermally efficient (potentially approaching infinity) and almost-carbon-neutral approach is evaporative cooling. Yet, it is substantially constrained by climate conditions (being most effective in hot dry climates), although its applicability can be extended by indirect equipment configurations. Evaporative cooling can also consume a fair amount of water in an arid climate, which is of concern in desert locations.

There are a number of HVAC systems and subsystems that are getting current attention by incorporation into high-performance building projects [23, 24]. These are as follows:

1. Ground-source heat pumps
2. Dedicated outdoor air systems (DOAS)
3. Chilled beams
4. Underfloor air distribution (UFAD)
5. Ventilation air heat exchanger

Lighting and Daylighting

Electric power consumption is largely attributed to lighting, around 34% of tertiary-sector electricity consumption, and 14% of residential consumption in OECD (Organization for Economic Co-operation and Development) countries according to a report by the International Energy Agency (IEA) in 2006 [25]. Energy-related greenhouse gas emissions are largely attributed to electric lighting in buildings due to buildings' high dependence on electric lighting that is mainly generated from fossil fuels. Thus, shifting to daylight in buildings is one of the top priority solutions that need to be considered. A good daylight design can assist in the overall cutback of energy consumption of a building. This has been proven in numerous studies that showed potential energy savings ranging between 20 and 60% in office and retail buildings [26]. When combined with electric lighting controls in largely daylit spaces, daylighting can reduce electric lighting energy consumption by up to 30–80%. Diminishing lighting energy can also be beneficial in saving cooling load energies, particularly in climates that require heavy cooling. For instance, reducing 100 W of lighting energy translates to 30 W of cooling energy being saved.

Teaching the IDS

IDS Challenges

In the Integrated Design Studio (IDS), students learn how to develop a comprehensive design approach in which the design and performance evaluation are practiced in one integrated process. The design goal is to resolve a challenging problem with multifaceted issues (social, cultural, functional, environmental, energy, human comfort, etc.). High consideration is given to the integration of sustainable technologies (passive and active) into the building construction components. To tackle the design challenges, the students must complete several design/analysis assignments that focus on the site/weather, case studies, cultural and heritage background, architectural expression, construction systems and components, building codes, structural design, design for safety, and integration strategies involving the mechanical, electrical, energy, and building management systems. The instructors train them how to use and integrate various tools such as 2D/3D CAD, digital fabrication, energy simulation, artificial sky, heliodons, and advanced photography.

The main challenges for the instructors are:

- How to create a learning environment that inspires students to gain new knowledge through their own choice and encounter broad experiences similar to those that could be faced in real life after they graduate.
- How to develop a comprehensive process for a mixed-use complex building in the design studio that stimulates creative architectural thinking and in the meantime helps students tackle the highly technical problems of the construction and engineering systems.

To manage the studio operation and progress of the students' work effectively, the IDS is taught by a teaching team:

1. A senior professor in architectural design with deep knowledge of building performance simulation who acts as the main advisor of the studio.
2. One or two highly experienced engineering instructors in environmental control issues including building energy, lighting, and carbon emissions.
3. A junior teaching assistant usually a master's or Ph.D. student who helps in following up the progress of the projects' development.
4. Other faculty members who possess expertise on other areas such as structural design. This is done with prior arrangement between the studio professor and faculty who show interest in sharing the experience with the studio team.

Teaching Philosophy

The student is the center of the learning process; therefore, it is always more fitting to describe the process as a "learning" rather than a "teaching" process. From the student's perspective "to learn" means to explore new unknown facts, i.e., the discovery of new knowledge through acts motivated by the teachers, who basically act here as learning facilitators. Learning is similar to research, except that research is about exploration of new facts, or unveiling of new unknown knowledge, while learning is about re-exploration of a previously revealed knowledge – often by someone else – to a new person who seeks to learn it, i.e., the student. In school, research and learning are applied in one integrated process in which the student is the center of this process. Therefore, the goal of the IDS can be seen as to support this procedure through well-planned activities that use efficient delivery methods, with the support of IT systems, which all take place in a highly inspiring studio environment.

Teaching Objectives

1. Develop students' sense of responsibility toward saving the environment, energy, natural resources, and appreciation for cultural values and heritage.
2. Broaden the students' scope of knowledge to fully realize that design is an integrated teamwork process done by professionals/experts in several areas, including architecture, construction, and technology, and leads to achieving the goal of sustainable development.
3. Raise the level of students' technical knowledge and improve their self-confidence in the technical aspects of architectural design.
4. Help students become proficient in discovering several alternatives of design solutions to a technical problem through analysis of building systems' performance and integration and equip students with advanced, rigorous simulation tools to evaluate design impacts on human needs and building performance.

How Objectives Are Achieved

To achieve objective 1, the instructors include mini lectures on international and regional protocols and policies that respond to environmental issues and their relation to the building and construction industry. Policies that are established locally, which are more fitting to the culture and environment, are highlighted to the students, and their consideration is emphasized. Other mini lectures are designed to make students aware of different kinds of real case studies, some of which are selected to show problems caused by irresponsible practices as opposed to other ones that show innovative design solutions.

To achieve objective 2, the instructors include mini lectures on the meaning and goals of sustainable development and sustainable design and the integrated design process layout, components, and stakeholders. They design assignments and term projects that give students opportunities to work in teams, learn from their experiences, and resolve any teamwork conflicts. They also train them on how to phase in design solutions using planning and project management techniques. Depending on the problem, they are sometimes advised to divide a design problem into segments and apply the induction discovery method (IDM). The IDM helps in improving the designer's focus of attention by seeking alternative solutions for each segment of the problem separately and evaluate them and then, in a later stage, integrate the most promising solutions and resolve any conflicts between them.

To achieve objective 3, the instructors explain through presentations and giving examples how design and performance analysis should go hand in hand until design goals are achieved. Depending on the academic level of the students and the prerequisite or corequisite of the course, sometimes the instructors are obliged to give mini lectures and application assignments to the students on some or all of the following subjects:

- Theories and principles of building science physics and foundations of heat transfer
- Thermal and lighting behavior in buildings and the effect of different materials and components
- How to calculate design loads (e.g., heating and cooling loads, lighting load) based on specifications for the designed spaces and activities/tasks
- How to size and optimize systems based on cost and energy cutdowns

To achieve objective 4, the instructors organize for the students training workshops and hands-on training on the following subjects:

- How to set design criteria and targets for high-performance buildings based on codes and standards and how to explore and verify solutions based on engineering criteria.
- How to operate and get reliable results from a range of simulation and analysis tools (from simplified to most rigorous) that help during the conceptual design and up to the final design development.
- How to infer and discuss results, establish correlations among different variables, and show and present significant findings.

Teaching Methodologies

Table 2 outlines the teaching methods that are applied in the IDS. Each methodology is accomplished by a number of planned actions by the instructors and assigned students' activities as described in the second column.

Building Systems and Integration

In the prerequisite courses prior to ARCH-430, the students learn the building mechanical and lighting systems. They also learn the HVAC fundamental options and how they present a wide range of thermal efficiencies and carbon emissions. This can be outlined as follows:

- The major divisions of the HVAC equipment that include source, distribution, delivery, and control components.

Table 2 Teaching methodologies with corresponding instructors' actions and students activities

Applied method	Instructors' actions & students' activities
Self-learning	Distribute list of reading assignments, request student's expanded readings
	Assignments on case studies, site and context, and functional analyses
	Assignments on students' thoughts and exchanging of views in pre-planned seminar-like discussions
Knowledge inquiry	Develop preplanned sessions with stimulus material for opening inquiry sessions
	Depend on a number of technologies, such as images, presentations, animations, videos, data shows, and multimedia systems in class
Creative thinking	Identify obstacles that limit the student's creativity, which could be due to the complex nature of the project, inappropriate design studio environment, limitations of the attempted design method, lack of information, or a combination of any of these.
	Teach nontypical methods of discovery:
	The deduction discovery method (for more complex problems) helps improve the student's focus of attention toward the most general/important design issues that have greater impact on the final results while delaying the tiny details to later stages after a cohesive concept has been developed
	The induction discovery method (for simpler problems) helps in solving functional relationships; one might start with designing the components internally (e.g., design a classroom based on design standards for functional space size, proportion, and furniture layout) then proceed to the more general issues such as space zoning that achieve higher efficiency among relationships of spaces and types of circulations.
Cooperative learning	Divide students into a number of design teams
	Design assignments for teamwork
	Teamwork involves collection of data, literature review, analysis, synthesis of ideas, evaluation, and presentation of reports and final drawings
Performative design	Train students to integrate simulation in design using a number of tools: Revit, Radiance, DIVA for Rhino, eQuest, DesignBuilder, Enerwin, Ecotect, Climate Consultant, and Weather tool
	Use simulation to demonstrate complex concepts/systems to students that can make a considerable difference in enhancing students' skills in design and performance verification
	Develop assignments to engage students in laboratory/field experimental work involving physical simulation and data measurement activities using artificial sky, heliodons, microscopic enhanced digital cameras suitable for architectural models, luminance and illuminance meters with photometric sensors and data loggers, thermal cameras, etc.
Model-making	Train students to create physical models to verify the design performance (such as daylight) of a major space in their projects using physical simulators based on some predefined design targets (such as Lux level, or $DA_{300lux-50\%}$)
Projects' exhibitions	Create a wider learning environment by organizing exhibitions that show students' projects
	Encourage students to participate in the events and engage in discussions to improve their self-confidence and capabilities of socializing with the public

(continued)

Table 2 (continued)

Applied method	Instructors' actions & students' activities
Outdoor activities	Organize field trips for students to learn from vernacular architecture
	Organize field trips for students to learn from real projects that have achieved green building ratings or international design awards
	Develop assignments that engage students in site visits to collect project data, take photographs, record information related to sites or buildings, test physical models in the outdoor environment for solar access/shading, experience/measure daylighting, and/or sketch existing buildings/landscape

- How heating and/or cooling effect is produced, moved around a building, and introduced into the spaces being conditioned and how the magnitude and timing of such effects are regulated.
- The source components in an HVAC system that provide heat (heating effect), coolth (cooling effect), and/or both.
- The three basic means of producing a cooling effect for a building: vapor compression refrigeration, absorption refrigeration, and evaporative cooling.

During the ARCH-430, the students learn how the end results of the interactions between climate and building function will lead to the required capacity and load factors for heating and cooling equipment. They employ computer simulations to determine both capacity and projected energy consumption. A demand for a high-performance ZNE building suggests the wisdom of using such simulations iteratively to optimize building and HVAC system coordination rather than simply using a simulation as a one-pass system sizing tool.

In any green building, especially the ones aiming to achieve ZNE, systems integration is vitally critical. Many aspects of integration need to be considered by the students of the IDS. The first one is to examine the owner's project requirements (OPR) for possibilities of HVAC integration. The student team needs to compare the characteristics of the many possible options for HVAC systems with the criteria contained in the OPR document as this will ensure that the system can provide the outcomes anticipated by the project client [27]. They also should check the capabilities of a potential HVAC system against the demands and opportunities presented by the project site climate. They study in their projects many possibilities, such as looking at solar resources relative to heat source options, soil temperatures relative to ground-source heat pump options, and relative humidity relative to the potential for evaporative cooling and/or the applicability of cooling towers versus air-cooled condensers.

It is exceedingly important to integrate the HVAC system with the many and diverse aspects of the architecture of a ZNE building. The HVAC system should mitigate the effects of the exterior environment that are allowed to enter the building through the building envelope. The students learn through computer simulation how the less successful the envelope design lead to greater load on the HVAC system and how the high loads result in high energy consumption and high carbon emission. The computer simulation helps students optimize the architecture of the envelope design that lead to minimum load on the HVAC. They also comprehend through simulation how the architectural design decisions will affect other interior loads such as electric lighting and, to a lesser extent, plug loads. Conceptually speaking, the loads that are handled by an HVAC system can be categorized as heating, cooling, lighting, and plug load related. In very hot climates, the heating load is almost zero, and the cooling load is extremely high while the lighting load can also be very high, as that depends on the building size and function though. At this point, the students realize the likely distribution of these loads, which assist them in designing appropriate and efficient HVAC systems.

Another major area of integration of HVAC systems that the students deal with is with other building support systems, such as the electrical, structural, and fire protection systems. They learn here how to coordinate spatially in the drawings the structural system components (e.g., columns and beams) with the HVAC ducts and other distribution elements. It can also involve design philosophies regarding exposure of support system components. The fire protection systems require also integration at several levels; the most important and complex one is the conceptual coordination of system interactions during emergency events.

MultiComfort House Student Contest 2018

Aim

The MultiComfort House is an annual contest open for students of architecture, design, and construction engineering or other disciplines from universities in countries where the contest is organized [3]. Participation is open for all students from 1st to 6th year of study as an individual or as teams. Upon request of the local SAINT-GOBAIN organization and depending on the specificity of the countries, up to 3 members per team can be accepted. The competition is structured in two (2) stages: national stages followed by an international stage, where the best projects from each country are invited.

The aim of the contest is to spread the message for the need of integration of energy efficiency, sustainable design, and MultiComfort concept in the contemporary architecture planning and building process and to underline the importance of the different dimensions of comfort (thermal, acoustic, indoor air quality, visual). The task for the 14th International Edition of the contest as developed by ISOVER in close collaboration with Dubai Municipality and Dubai Properties Group is the development of a vision for a transcultural vibrant community located in the perimeter of Cultural Village of Dubai.

Project Site in the Cultural Village

The traditional ship in the local UAE dialect is called Dhow. The dhows (or Al Boom) are traditional timber ships used to sail between the shores of the Arabian Gulf to East Africa, India, and China since the eleventh century A.C. The area of the site is called Al-Jaddaf, which literally means The Rower. It has been known as the place for constructing the dhows. Unfortunately, this beautiful tradition of dhow building has been increasingly diminishing in the last two decades because the need for such old-fashioned ships has become so limited. The modern urbanization plans of Dubai gives priority to allocate the most attractive sites, especially those with historical significance at the waterfronts to projects that could achieve high momentum to boost the tourism economy; indeed this site is one of them.

The Culture Village at Al Jaddaf encompasses four phases. The Saint-Gobain Contest Site is part of Phase-2 – a new thriving destination within Dubai, with a land area of 19 ha approximately and a total Gross Floor Area of 400,189 m^2 [3]. The goal is to create a dynamic and vibrant development offering attraction to the residents and visitors while maximizing the benefits of the strategic public transport, proximity to the waterfront, and history of the site. Figure 2 shows the Cultural Village with the contest location.

Task

The students are required to develop a vision for a transcultural vibrant community development located in the perimeter of Cultural Village 2 of Dubai [3]. The design will have to propose a viable combination of residential and public spaces (cultural, commercial, others) while respecting the plot characteristic and its history. Special focus will be given on the development of the sustainability dimensions as well as the comfort ones. The site allocated to the students is the E-001 parcel – orange part of the plan bellow (Fig. 3).

The main characteristic of the project parcel are as follows:

- Land mixed use: residential [88%] and retail [12%]

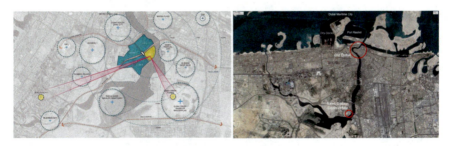

Fig. 2 The contest location and Cultural Village in Dubai [3]

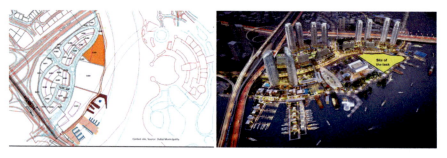

Fig. 3 The allocated site to the students [3]

- Plot area parcel: 26,936 sqm
- Maximum plot coverage: 62%
- Maximum GFA: 40,000 sqm
- Maximum building height: 45 m
- Residential apartment type: studio, 1–3 bedroom
- Number of units: 242
- Average surface per person is about 50–70 sqm

The Students' Project

Project Brief/Design Goals

Create a sustainable architecture integrated into the urban space while responding to the Saint-Gobain MultiComfort criteria and taking into account the climatic conditions and regional context of the site and Dubai City. The design should be sustainable and innovative (original and creative) and should drive the city further into sustainable development. It should create a dynamic and vibrant development offering attraction to the residents and visitors while maximizing the benefits of the strategic public transport, proximity to the waterfront, and history of the site. The design goals and expected outcomes of the project as outlined by the students design team are as follows:

- Sustainable transcultural vibrant community development
- Innovative green design
- Consideration of social, economic, and environmental aspects
- Considerable reduction of energy consumption (ZNE building)
- Comfort design development
- Architecture has to fit in the surrounding of the site
- Design proposes a viable combination of residential and public spaces (cultural, commercial, etc.) while respecting the plot characteristics and its history

Integrated Design Paradigm/Design Process

When designing a building, the students must deal with several specialty domains such as architectural, construction, structural, electrical, or mechanical design (see Fig. 4). In the design process, the design team collaborate together to achieve the project's goals, which requires many sessions of planning, developing ideas, discussions, and making decisions. At the end of the process, the efforts lead to assimilation of the best ideas/results made in all the specialty modules to create one integrated design. The diagram in the left side of Fig. 4 shows seven design modules that each team of students in the IDS deal with and the tasks they have to accomplish in each module. The flow chart in the right side of Fig. 4 shows the design process from start to end. The blue nodes represent the kinds of effort or activities made by the design team, such as project definition (problem statement and objectives), research and analysis, synthesis, evaluation, and design development. The white spinning cycle (i.e., the white circle with the arrows) is the core of the whole design process. It is basically a loop mechanism representing how students exert their most intensive efforts between the main design activities (i.e., synthesis, analysis, evaluation, and development) to search for the most promising final design concept. The red nodes represent the design specialty modules discussed previously. The green nodes are subareas within each design specialty module.

Architectural Design Concept

The form was inspired by the shape of the traditional dhow. The site shape and geometry and the need to offer high visual access to the waterfront was the reason to divide the project into three major masses that included the residential apartments (Fig. 5).

The heights of the masses were stepped to maximize the views to the waterfront. They look together as three traditional dhows at the final stage of construction,

Fig. 4 Design modules of the IDS and the design process

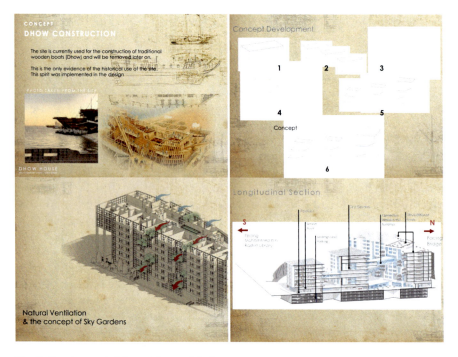

Fig. 5 Development of the architectural design concept

almost ready to plunge into the gulf water. Like a real dhow hull, the building was designed to be thinner near the ground and gradually increase in width as it goes higher. It was also cladded on the west facade by brown wood logs, which gave the impression of a traditional dhow. The dhow form design along with the wood cladding helped maximize solar shading and reduce heat. The dhow-like masses rest on a podium; which housed the functions that serve the public services such as an exhibition, a gym, shops, restaurants, and coffee shops. To ease circulation of the people, three bridges were designed to connect between the residential masses. To promote effective natural ventilation, several measures were taken into consideration. One of them was the sparse massing and the form orientation that was made to infiltrate airflow in the site and building to increase opportunities for cross ventilation. Another one was the idea of placing several sky gardens at different floors that generate passive cooling.

Improving Building's Energy Performance

The students improved the building's energy performance by using several passive and active systems. They managed to reduce the energy requirements up to 60% from the baseline case using the following strategies (Fig. 6):

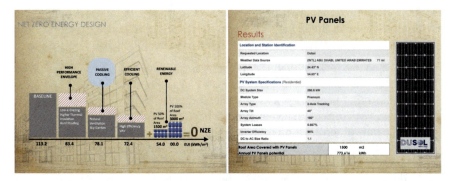

Fig. 6 The process of achieving the ZNE building through optimized building design and improvement of energy performance and use of renewable energy (PV system)

- Shading the roof and building skin
- High-performance envelope
- Natural ventilation assisted by sky gardens
- Optimizing form design for daylighting
- High-efficiency variable air volume (VAV) HVAC system

Renewable Energy Using PV System

After considerable reduction of the building loads and energy, the problem became easier to the students to consider renewable energy systems and turn their design into an NZE building. They tested two options to cover the top surfaces of the building with PV panels; option 1 had 50% (1500 m^2), while option 2 had 100% (3000 m^2) of the total top surfaces of the building. The idea in option 1 was to install the PV panels on the roof of the top floor only (area = 1500 m^2) while leaving the stepped masses (i.e., the terraces) open to the sky (i.e., no PV panels). The idea in option 2 was to expand the PV panels' coverage to include not only the roof of the top floor (area = 1500 m^2) but also the stepped terraces (installed on sheds over the terraces). Option 1 was believed to give architecturally a more appealing design; furthermore, it would help lower the construction cost. On the other hand, option 2 would help achieve the target for an NZE building yet with higher construction costs. The images of the final design that appear in Appendix A show only option 1.

Conclusion

Designing a ZNE building is a very challenging task even to experts. It requires many talents; most importantly are a creative architect's mind, a great deal of engineering knowledge and specialized skills, and a highly coordinated teamwork with enormous determination and patience. The implemented IDS model is believed to have high potential in stimulating architectural students to acquire these talents. The success of the implemented model can be referred to the following:

- First, the course in which the IDS model was implemented was supported by a number of prerequisite/corequisite courses that taught the students the foundation engineering knowledge and skills needed to deal with the technical aspects of design. The curriculum layout of the pedagogical program took into consideration the flow of information from the prerequisites to the design studio.
- Second, the course was taught by a well-balanced team from architecture and two areas of engineering backgrounds.
- Third, the teaching team identified themselves as learning facilitators (rather than as typical teachers) whose major role was to encourage students by supporting the idea of providing a highly inspiring studio environment to them through well-planned activities.
- Fourth, the teaching team applied a variety of efficient knowledge delivery methods in the studio and through interactive educational IT systems that included self-learning, knowledge inquiry through stimulus material, creative thinking, cooperative learning through teamwork, performative design, model-making, projects' exhibitions, and outdoor activities. They also supported the students with mini lectures and hands-on training to cover areas that were not very familiar to them and improve their skills, especially in the advanced building performance simulation area.

Acknowledgment The authors would like to thank UAE University for providing all the needed means to support the architectural students who won the third prize of the reputable MultiComfort House Student Contest, Edition 2018: Shamma Al Kaabi, Maha Al Luwaimi, and Fatima Ba-tuq. Special thanks go also to the contest's sponsors Saint-Gobain and Dubai Municipality and to the organizing team, particularly Dr. Nada Chami, Head of Product & Innovation chez Saint-Gobain UAE, who provided unlimited support to the participating students.

Appendices

Appendix A: Slides of the Final Design

These are some of the slides the students produced to present their project in the contest.

The Integrated Design Studio as a Means to Achieve Zero Net Energy Buildings

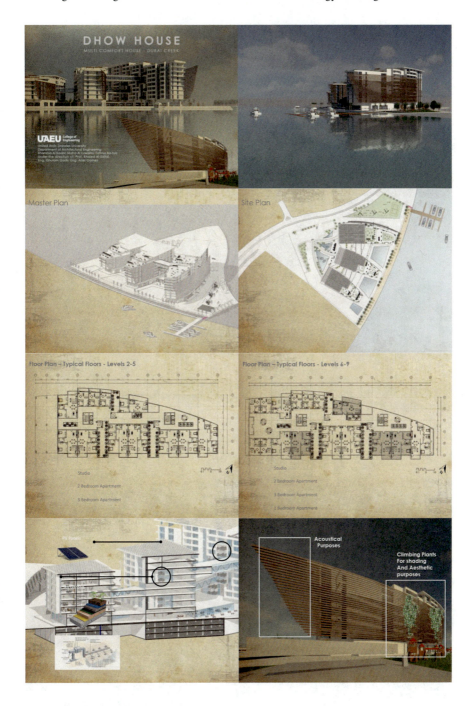

Appendix B: Celebration Photos

References

1. Architecture 2030: 2011, http://architecture2030.org/ (accessed 1/11/2013).
2. EIA:Annual Energy Outlook 2012; Early Release, US Energy Information Administration (EIA)/Department of Energy (DoE), January 2012.
3. MultiComfort. (2018). *MultiComfort House Student Contest Edition 2018, invitation for competition submissions, MultiComfort by Saint Gobain*. https://architecture-student-contest.saint-gobain.com/last-editions/edition-dubai-2018. Accessed 15 Jan 2018.
4. IEA. (2017). *World energy statistics*. OECD Publishing 2017, International Energy Agency (IEA). https://doi.org/10.1787/world_energy_stats-2017-en
5. BP. (2019). *Statistical review of energy 2019, 68th edition*. https://www.bp.com/content/dam/bp/business-sites/en/global/corporate/pdfs/energy-economics/statistical-review/bp-stats-review-2019-full-report.pdf
6. Clarke, K. (2016). 80% energy consumed by buildings. *Khaleej Times*. https://www.khaleejtimes.com/nation/abu-dhabi/80-energy-consumed-by-buildings
7. Aboul-Naga, M., Al-Sallal, K. A., & El Diasty, R. (2000). Impact of city urban patterns on building energy use: Al-Ain city as a case study for hot-arid climates. *Architectural Science Review, 43*, 147–158.
8. IEA. (2012). *Efficient world scenario: Policy framework, world energy outlook 2012 (WEO-2012)*. International Energy Agency, Paris, France. http://www.worldenergyoutlook.org/media/weowebsite/energymodel/documentation/Methodology_EfficientWorldScenario.pdf
9. Lawrence, T., Darwich, A. K., Boyle, S., & Means, J. (2013). *ASHRAE GreenGuide: design, construction, and operation of sustainable buildings* (4th ed.). ASHRAE.
10. Deru, M., & Torcellini, P. (2007). *Source energy and emissions factors for energy use in buildings*. Technical report NREL/TP_550-38617. National Renewable Energy Laboratory, Golden, CO.
11. Rijksen, D. O., Wisse, C. J., & van Schijndel, A. W. M. (2010). Reducing peak requirements for cooling by using thermally activated buildings. *Journal of Energy and buildings, 42*, 298–304.
12. Kharecha, P., Kutscher, C., Hansen, J., & Mazria, E. (2010). Options for near-term phase-out of CO_2 emissions from coal use in the United States. *Environmental Science & Technology, 44-11*, 4050–4062.
13. Gratia, E., & De Herde, A. (2003). Design of low energy office buildings. *Energy and Buildings, 35*(5), 473–491. https://doi.org/10.1016/S0378-7788(02)00160-3
14. Hemsath, T. L., & Alagheband Bandhosseini, K. (2015). Sensitivity analysis evaluating basic building geometry's effect on energy use. *Renewable Energy, 76*, 526–538. https://doi.org/10.1016/j.renene.2014.11.044
15. Olgyay, O. (1963). *Design with climate: Bioclimatic approach to architectural regionalism* (1st ed.). Princeton University Press.
16. Pacheco, R., Ordóñez, J., & Martínez, G. (2012). Energy efficient design of building: A review. *Renewable and Sustainable Energy Reviews, 16*(6), 3559–3573.
17. Roos, A., & Karlsson, B. (1994). Optical and thermal characterization of multiple glazed windows with low U-values. *Solar Energy, 52*(4), 315–325. https://doi.org/10.1016/0038-092X(94)90138-4
18. Yeang, K. (1999). *The Green Skyscraper: The basis for designing sustainable intensive buildings*. Prestel Verlag.
19. Al-Sallal, K. A. (2016). Passive and low energy cooling. In K. Al-Sallal (Ed.), *Low energy low carbon architecture, Vol. 12 – Sustainable energy developments* (pp. 237–248). CRC Press.
20. ASHRAE. (2013). *ASHRAE handbook – Fundamentals*. American Society of Heating, Refrigerating and Air-Conditioning Engineers.
21. ASHRAE. (2012). *ASHRAE handbook – HVAC systems and equipment*. American Society of Heating, Refrigerating and Air-Conditioning Engineers.
22. ASHRAE. (2014). *ASHRAE Handbook – Refrigeration*. American Society of Heating, Refrigerating and Air-Conditioning Engineers.

23. Grondzik, W. (2016). Energy-efficient HVAC systems and systems integration. In K. Al-Sallal (Ed.), *Low energy low carbon architecture, Vol. 12 – Sustainable energy developments* (pp. 237–248). CRC Press.
24. Grondzik, W., & Kwok, A. (2015). *Mechanical and electrical equipment for buildings* (12th ed.). Wiley.
25. IEA. (2006). *Light's labour's lost: Policies for energy-efficient lighting.* IEA Publications.
26. Galasiu, A. D., Newsham, G. R., Suvagau, C., & Sander, D. M. (2007). Energy saving lighting control systems for open-plan offices: A field study. *Leukos, 4*(1), 7–29.
27. ASHRAE. (2011). *ASHRAE handbook – HVAC applications.* American Society of Heating, Refrigerating and Air-Conditioning Engineers.

Indicators Toward Zero-Energy Houses for the Mediterranean Region

The Average Buildings as Means of Optimized Refurbishment Strategies for the Residential Buildings in Cyprus

Despina Serghides, Martha Katafygiotou, Ioanna Kyprianou, and Stella Dimitriou

Introduction

The increasing energy demand already threatens the future of the planet, as researches show that 10% of the world population exploits 90% of energy resources [1]. Under this threat, the European Union aims to achieve a more sustainable future and therefore focuses on the existing building stock by identifying the potential energy conservation of the building sector.

In 2016, it was estimated that 45% of the energy produced in Europe for heating and cooling was used in the building sector, and 50% of air pollution was caused by the same sector [2]. The existing residential building stock exceeds by a large amount the number of newly built dwellings in most developed countries. Moreover, residential buildings are responsible for approximately 2/3 of the energy consumption of the building sector [3]. While new buildings add at most 1% per year to the existing stock, the other 99% of the buildings are already built and produce about 26% of the energy use-induced carbon emissions [4]. Thus, greater potential of energy savings can be achieved in the existing residential buildings than the newly built structures.

The improvement of the energy performance of existing residential stock in every country is essential because the operational cost, energy consumption, and carbon dioxide emissions are major issues worldwide. The paper is based on

---

D. Serghides (✉) · I. Kyprianou · S. Dimitriou
The Energy, Environment & Water Research Centre, The Cyprus Institute, Nicosia, Cyprus
e-mail: d.serghides@cyi.ac.cy; i.kyprianou@cyi.ac.cy; s.dimitriou@cyi.ac.cy

M. Katafygiotou
Real Estate Department, Neapolis University Pafos, Paphos, Cyprus
e-mail: m.katafygiotou@nup.ac.cy

© The Author(s), under exclusive license to Springer Nature Switzerland AG 2023
A. Sayigh (ed.), *Towards Net Zero Carbon Emissions in the Building Industry*, Innovative Renewable Energy, https://doi.org/10.1007/978-3-031-15218-4_12

information laid out by the European, IEE Episcope Project [5], and focuses on the possibilities of effective energy refurbishment studies through the determination of a set of average buildings. Although average dwellings and buildings are an advanced manifestation of the reference or typical building concept, because they encompass all the characteristics of the subset which they represent and thus more adequate for strategic feasibility studies, they are not used for investigation purposes. Instead, common practice is the utilization of reference buildings, those sharing most of their characteristics with the other subgroup buildings.

Various studies have used reference or typical buildings to quantify the energy-saving potentials of the existent building stock. Stefano Paolo Corgnati et al. [6] have developed three general methodologies for the definition of reference buildings to be used for cost-optimal analysis. Through a case study of an office building, used as a reference building for the Italian existing building stock, the definition process was demonstrated, and the modeling was carried out by using EnergyPlus, a dynamic energy simulation program.

For the housing sector, Tommerup and Svendsen from the Technical University of Denmark investigated the energy savings in Danish residential building stock based on two typical buildings [7].

In order to estimate the total savings potential, detailed calculations were performed in a case study with two typical buildings representing the residential building stock. The calculations served for the assessment of the energy-saving potential, resulting in possible profitable savings of energy used for space heating of about 80% from 2006 to 2050, within the residential building stock, if the energy performance is upgraded when buildings are renovated.

Building typologies were explored as well by Dascalaki et al. [8], as useful instruments toward facilitating the energy performance assessment of housing stock in Greece and by Ballarini et al. [9] for the evaluation of the energy-saving potentials of the Italian housing stock. Both papers were based on the harmonized TABULA structure for European housing typologies, where national typologies were developed as sets of model buildings with characteristic energy-related properties representative for each county. The model buildings were used in both cases as a showcase for demonstrating the energy performance, the potential energy savings, and CO_2 reduction occurring after applying basic/typical and advanced energy conservation measures (ECMs) on the thermal envelope and the heat supply system. In particular, the Italian approach modeled the energy balance of a subset of the national building stock, with the results being indicative of the enormous potential of energy savings even with basic energy retrofit actions.

In Cyprus, it was observed that with the current trends, taking into consideration the depth and rate of the performed energy refurbishments and the energy performance of the new buildings, the national climate protection energy targets are unattainable [10]. Reference buildings as means of energy performance assessment and energy profile development of the housing stock of Cyprus, as well as for the evaluation of the energy savings potential from energy refurbishments, were introduced during the EPISCOPE project. Serghides has authored several papers where reference dwellings are used in order to determine possible energy savings, from all the typologies and construction periods [11–15]. The summarized results of the investigation can be found in the Cyprus National Typology Brochures [16]. The potential percentage reductions of annual primary energy need for heating and DHW, obtained after deep refurbishment, range from 40% to 71% depending on the type of the dwelling and its construction period.

A report by Hermelink et al. [17] discussed the implementation of the Energy Performance of Buildings Directive (2013) regarding new and publicly owned buildings, which should conform to nearly zero-energy requirements by the end of 2020, offering a definition for a "nearly zero-energy building" (nZEB). A 2017 status report suggests that nZEBs are indeed a priority in the improvement of energy intensity in construction sector, with many European countries already adopting this approach and increasing the share of nZEBs in new construction [18]. In Cyprus, the transition to nZEBs poses a challenge, which is met by capacity building and training of the workforce of the construction industry [19]. Taking nZEBs a step further, Dimoudi et al. [20] and Serghides et al. [21] introduced ZenHs (zero-energy hospitals), investigating the specificities of healthcare facilities in the Balkan and indicating that hospitals of the region, whether publicly or privately owned, are not in line with the EPBD requirements.

Focal to this type of investigations is the use of indicators as proxies of the energy performance of a building. Factors such as carbon footprints (measured as CO_2 emissions), thermal properties of buildings (such as the u-value), design and fenestration elements, orientation of buildings, and the depth and level of energy refurbishments, among others, are being used as indicators of energy performance [22]. For instance, Dimitriou et al. [23] investigated scenarios of refurbishment aiming to transform standard office buildings into nZEBs in Cyprus, employing design solutions such as greenery, cool, and insulating materials, to assess changes in thermal properties and make energy saving achievable. In a Romanian case study, Prada et al. [24] showed that transforming energy-inefficient healthcare facilities into intelligent buildings, integrated with renewable technologies, has the potential for significant energy savings and related reductions in CO_2 emissions.

This chapter is the synopsis of an extensive research that was co-funded by the Intelligent Energy Europe Programme of the European Union. It is based on reliable primary data on the observed building stock, and the indicators are presented to describe the refurbishment optimization studies of residential building typologies to achieve nearly zero-energy houses in Cyprus.

Methodology

For the analysis, several residential units constructed from 1980 to date have been examined. The focus is on pilot houses in Cyprus, and the housing units selected were representative of the main building typologies in Cyprus (multi-family, single-family houses, terrace family houses, and apartments), and based on their construction period, they were divided into two chronological periods, before and after

2007, due to the launch of the legislation concerning energy efficiency in 2007. The relevant data for all the residential buildings were collected, including architectural drawings, electromechanical equipment, actual energy consumption, and energy refurbishments.

When the collection of data was completed, the average buildings were determined. Consequently, these were inserted in the Tabula.xls software, which was developed during the TABULA project and further enhanced in the EPISCOPE project [25]. The Tabula.xls calculated the annual energy balance for heating and DHW and showcased the energy loses from the building envelope for each average building and the energy need for the whole stock by average building category. This is an advanced excel workbook using algorithmic equations. It encompasses the climatic data for each country along with the constructional data of the typical buildings and the performance of the supply systems, providing results regarding the primary energy consumption, the CO_2 emissions, and the operational costs of each building.

The information about the most common constructional characteristics of the envelope elements and installed electromechanical equipment for heating, currently available in the Tabula.xls database, was collected on a national level [26]. The constructional characteristics of the roof, wall, floor, building structure, and openings were recorded for all the building typologies, and their U-values, along with their thermal capacity, were calculated. The remaining information required for the simulation of the average buildings, such as the square meters of each envelope element, the orientation and percentage of fenestration, and the depth and level of energy-refurbished elements per typology, were manually inserted in the software. Envelope refurbishment scenarios were applied to the average dwellings based on the impact of each typology energy need on the overall need and the possibilities of energy improvements of the envelope elements, with the objective of achieving at least a total of 30% energy need reduction [27]. The effectiveness of the utilization of the average dwellings as means of strategic refurbishment planning was concluded from the results, and the potential of national energy savings per housing typology was highlighted.

The buildings were also modeled using the software interface of the official Simplified Building Energy Model tool (iSBEM_Cy), the official government software for issuing Energy Performance Certificates (EPC). Energy refurbishments of the envelope elements were performed focusing on the monitoring, which includes a set of energy-related aspects of the building envelope, such as insulation levels of walls, windows, roofs, and floors as well as their respective refurbishment rates.

The efficiency of each strategy and technique employed toward zero-energy houses and the greenhouse gas emissions was evaluated, and its efficacy toward achieving the 2030 nearly zero-energy target was examined.

Primarily, the goal is to make a reliable prediction of the future energy situation and to fill in the gaps of the database in order to create a clear map that will lead to the right decision making to achieve the 2030 targets.

The Average Residential Dwellings

For the determination of the average dwellings, information was collected and extracted from the study of architectural plans and onsite surveys. The numbers of dwellings under study for each chronological period are presented in Table 1.

A total of six (6) categories were developed and represented through 2 Single-Family Houses (SFH), 2 Terrace Houses (TH), and 2 Apartments in Multi-Family Houses (MFH). The dwellings constructed prior to 2007 are denoted with I, whereas those after 2007 are denoted with II. From the study of the architectural drawings, the average square meters, volume of heated space, and square meters of each envelope element were calculated. As can be seen in Table 2, the geometrical characteristics of the old and new building stock differ among the typologies in terms of the square meter prevalence and between the chronological periods in their thermal performance (U-value). The difference in the thermal performance of the elements is essential, because it is improved by over 50%, from the old to the new building stock, due to the addition of insulation, which was made obligatory after 2006 [28].

Table 1 Monitoring samples from the CLDC stock

Total No. of buildings	I – Old building stock (1981–2006)	II – New building stock (2007 – up to date)
2484	2006	478
Architectural drawings (m² of heated space and envelope elements)		
84	37	47
Onsite surveys (heat supply systems and refurbishment trends)		
104	76	28

Table 2 Geometrical and thermal performance characteristics of average buildings

Thermal envelope average building						
Basic data	SFHI	SFHII	THI	THII	MFHI	MFHII
Floor area (in m²)	132.0	132.0	127.4	127.4	105.7	105.7
Number of dwellings represented by the average one	16	15	899	119	908	301
Thermal envelope areas (external dimensions in m²)						
Roof	60.3	60.3	60.5	60.5	29.9	29.9
Wall	174.8	174.8	114.4	105.3	83.5	83.5
Window	24.1	24.1	23.3	23.3	13.6	13.6
Floor	59.6	59.6	60.5	60.5	30.7	30.7
Original state/unrefurbished fraction of the envelope area U-values						
Roof	3.42	0.77	3.42	0.77	3.42	0.77
Wall	1.39	0.82	1.39	0.82	1.39	0.82
Window	5.78	3.16	5.77	3.16	6.10	3.20
Floor	0.91	0.91	1.01	1.01	2.34	2.34

Table 3 Envelope refurbishment

Refurbishments (averages)	SFH I	TH I	MFH I
Refurbished fraction of envelope areas (%)			
Roof	–	41%	–
Window	41%	41%	91%
Total (indicative)	*3%*	*13%*	*8%*
U-values of the refurbished fraction (averages)			
Roof	–	0.79	–
Window	3.20	3.20	3.20

In Cyprus it is usual practice to raise the Multi-Family Houses in columns and create a free, open space in the ground floor, referred to as "Pilotis," to be used mainly as a parking space. A 50% of the Multi-Family buildings have Pilotis. In the building stock under study, floors raised in Pilotis were not insulated after 2006; thus the floor U-value of the Multi-Family Houses is significantly higher than those of the Single-Family Houses and Terrace Houses in both chronological periods, reaching 2.12 W/(m^{2}K).

From the 104 dwellings surveyed, a total of 7 had refurbished their roof by adding insulation, and 19 had changed their windows, replacing the single-glazed windows with double-glazed ones. Another finding of the survey was that the only energy-efficient refurbishments documented refer to the old building stock and concern only roof insulation and windows replacement with ones of better thermal performance [29] (Table 3).

Energy Profile

Heat Losses

After inserting the data in Tabula.xls, heat losses from the envelope elements are provided and demonstrate that the old buildings, as expected, present higher losses than the new ones. The highest losses are attributed to SFH I, followed by TH I and MFH I. The least-energy demanding dwelling is TH II, as it has the highest percentage of refurbished envelope (Fig. 1). The relatively high losses from the Multi-Family Houses are due to their floor, which is partly in Pilotis and thus uninsulated.

Overall the most heat flow occurs from the walls, in all the average buildings, because walls have the highest exposed area. In the old average dwellings, roof is the second envelope element with the highest heat losses, whereas in the new dwellings are windows for the SFH and TH and floors for the MFH.

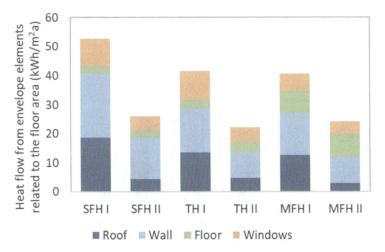

Fig. 1 Envelope heat losses – current state

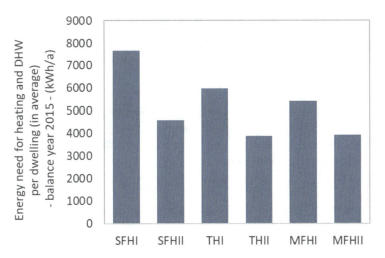

Fig. 2 Energy needs for heating and DHW – current state

Energy Need

Regarding the energy need for heating and DHW, per average dwelling, the highest gap is observed between the old and the new SFHs, which exceeds the 3000 kWh/a, whereas the lowest between the MFHI and MFHII is less than 1500 kWh/a. This difference is due to the total envelope area of the SFHs, which is more than double that of the MFHs, and therefore the insulation is more impactful in the SFHs rather

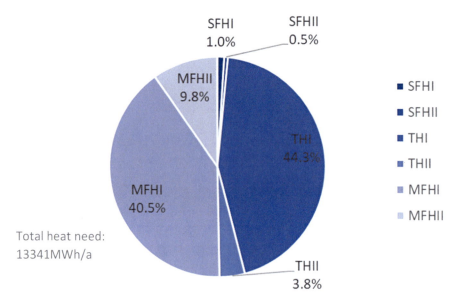

Fig. 3 Percentage distribution of energy need for heating and DHW among the average building categories – current state

than in the MFHs. Overall, the highest energy need is that of the average SFHI at 7658 kWh/a, whereas the lowest is that of the THII, at 3883kWh/a (Fig. 2).

The percentage distribution of heat needs between the different average dwelling categories in the total examined stock can be seen in Fig. 3. Almost 95% of the total energy need comes from the THI, MFHI, and MFHII, whereas the remaining three categories (SFHI, SFHII, and THII) are responsible for near one twentieth of the heat need. As a result, any strategic energy refurbishment of this building stock should primarily address the energy needed for heating the dwellings of the THI, MFHI, and MFHII categories and then those of the less energy-demanding ones.

Energy Refurbishment Scenarios

Based on the prior analysis, targeted improvements were performed on the envelope of the dwellings, with the objective of 30% decrease in the energy need for heating and DHW, through the energy refurbishment of two elements for the MFHII and two for the THI and MFHI, in order to comply with the target of the 2030 energy policy. The selection of the envelope elements to be thermally upgraded was made

Table 4 Requirements for houses in Cyprus according to Directive 119/2016

Directive 119/2016 requirements for refurbished houses	
Technical specifications – construction element	U-Value W/(m²K)
Horizontal elements (roof, floor in Pilotis)	0.40
Vertical elements (walls, columns)	0.40
Double-glazed windows	2.90

Table 5 Thermal performance characteristics of average buildings–refurbishment state

Original state/unrefurbished fraction of the envelope area						
U-values of the refurbished state	SFHI	SFHII	THI	THII	MFHI	MFHII
Roof	3.42	0.77	0.40	0.77	0.40	0.77
Wall	1.39	0.82	0.40	0.82	0.40	0.40
Window	5.78	3.16	5.77	3.16	6.10	3.20
Floor	0.91	0.91	1.01	1.01	2.34	0.49

based on the level of heat losses originating from the envelope. The upgrade follows the requirement of the Directive 119/2016, as seen in Table 4.

For the old average dwellings, roof and walls were upgraded, whereas for the average new apartments (MFHII), walls and floor were in Pilotis. The resulting U-values of the refurbished elements used in the simulations are presented in Table 5.

Table 6 presents the mean U-values of the building stock envelope elements. It is observed that the "new building stock" has improved the U-values compared with those of the old building stock. The mean percentage improvement reaches 78% for the opaque elements and 39% for the windows, which is indicative of the effectiveness of the insulation and the impact of the EPB Directives on the construction practices and the improvement of the thermal performance of the building envelope.

From the simulation results, it is observed that the energy need for heating has decreased in all the average dwellings under refurbishment, and the energy need gap between the old and the new THs and MFHs is reduced to less than 200 kWh/a and 800 kWh/a, respectively, resulting in a more uniform distribution of heating need among the average dwellings (Fig. 4).

The percentage distribution of the energy need among the categories is the same as in the current state, with the THI corresponding to the most demanding category and the SFHII to the least one, as depicted in Fig. 5. Nevertheless, the combined heating need of the THI, MFHI, and MFHII is reduced from the previous 95% to around 91%, whereas the impact of the remaining three categories (SFHI, SFHII, and THII) has increased from 5% to 9%. The total energy need is 8861MHh/a, reduced by over 33% from the current energy need for heating.

Table 6 The dwelling envelope mean U-values

	Old building stock, 1981–2007		New building stock, after 2007	
	Typical construction (from exterior to interior)	Mean U-value (W/m²K)	Typical construction (from exterior to interior)	Mean U-value (W/m²K)
Walls	2.5 cm plaster – 20 cm brick – 2.5 cm plaster	1.39	2.5 cm plaster – 20 cm brick – 2.5 cm plaster	1.39
Columns/Beamsbeams	2.5 cm plaster – 25 cm reinforced concrete – 2.5 plaster	3.13	2.5 cm plaster – 3 cm expanded polystyrene – 25 cm reinforced concrete – 2.5 cm plaster	0.76
Flat roof	Asphalt membrane – 7 cm tapered screed – 15 cm reinforced concrete	2.88	Asphalt membrane – 7 cm tapered screed – 4 cm expanded polystyrene – 15 cm reinforced concrete	0.60
Upper ceiling	1 cm tiles – 7 cm screed – 15 cm reinforced concrete	3.13	1 cm tiles – 7 cm screed – 15 cm reinforced concrete	3.13
Apartment floor/floor in contact with non-heated space	1 cm tiles – 7 cm screed – 15 cm reinforced concrete	2.18	1 cm tiles – 7 cm screed – 15 cm reinforced concrete	2.18
Floor in contact with the ground	15 cm reinforced concrete – 7 cm screed – 1 cm tiles	1.05	15 reinforced concrete – 7 cm screed – 1 cm tiles	1.05
Exposed floor (Pilotis)	15 cm reinforced concrete – 7 cm screed – 1 cm tiles	3.03	15 cm reinforced concrete – 4 cm expanded polystyrene – 7 cm screed – 1 cm tiles	0.60
	Old building stock, 1981–2003/2014		New building stock, 2003–2014/2014	
	Typical construction (from exterior to interior)	Mean U-value (W/m²K)	Typical construction (from exterior to interior)	Mean U-value (W/m²K)
Windows	Single-glazed aluminum frame	6.1	Double-glazed aluminum frame	3.7

Refurbishment Indicators of the Building Envelope

The retrofitted houses correspond to 13.5% of the building stock for the insulation installment on the roof and 18.27% for the window replacement. The percentages are higher for the old building stocks, reaching 18% and 50%, respectively. There was no indication of any energy upgrade of the walls and the floor.

None of the surveyed households had previous energy refurbishment of both roof and windows. Thus, the foremost energy-related improvement of the households is the upgrading of their fenestration, with the replacement of the single-glazed windows with double-glazed ones.

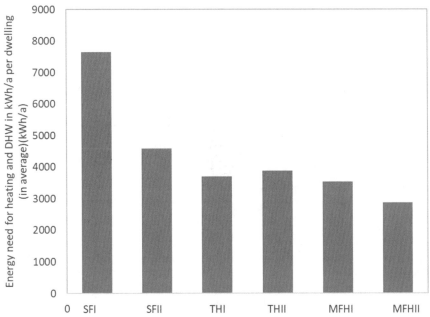

Fig. 4 Energy needs for heating and DHW – refurbishment state

The annual refurbishment rate for the roof insulation installment is 0.41%/a for the complete building stock and 1.14%/a for the windows replacement. The old building stocks present a 0.45% annual refurbishment rate for the roof insulation improvement and 2.27% for the window replacement.

In the refurbished dwellings, the average thickness of the installed insulation on the roof is found to be 4 cm, and the average U-value of the improved windows is 3.2 W/m^{2}K. These values result from the common construction practices and the average values of the double-glazed windows with aluminum frame, which comply with the Directive 466/2009 [30] and the *Guide for Building Insulation* published by the Ministry of Energy [31] (Table 7).

When comparing the external envelope elements of the new building stock with those of the old building stock, it is observed that U-value improvements of 78% for the opaque elements and 39% for the glazed elements are implemented.

From the survey, it is furthermore observed that the initial refurbishment carried is the replacement of the windows. This tendency might be attributed to the aesthetics associated with the fenestration of a dwelling, as well as the perception of the inhabitants that most of the energy is lost through the windows.

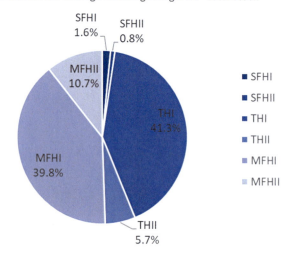

Fig. 5 Percentage distribution of energy need for heating among the average building categories – refurbished state

Table 7 Building insulation: trends of modernization

Roofs/upper floor ceilings	Complete building stock, after 2007	Old building stock, 1981–2007
Insulation improved (from original state)	6.52%	8.93%
Average thickness of improved insulation	–	4 cm
Annual rate of insulation improvement	0.41%/a	0.45%/a
Windows	Complete building stock, after 2007	Old building stock, 1981–2007
Insulation improved (from original state)	18.27%	50.00%
Average U-value (W/m^{2}K) of improved windows	–	3.2 W/m^{2}K
Annual rate of insulation improvement	1.14%/a	2.27%/a

Future Projection of the Energy Renovations towards Toward Zero-Energy Dwellings

Based on the average residential dwelling and the energy performance indicators discussed above, a projection is made of the future refurbishments of the building envelope elements of the building stock. In the projection, it is assumed that the number of dwellings comprising the old building stock is not changed and that the annual refurbishment rate per element remains the same. The energy upgrading of

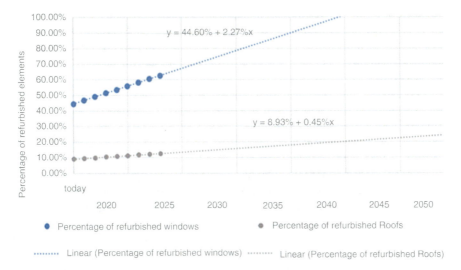

Fig. 6 Future projection of current energy refurbishments rates for the building stock

the fenestration and the roof were considered. The tendency, as shown in Fig. 6, is indicative of the current low annual refurbishment rate.

Conclusions

In accordance to the EU energy strategy, there are specific targets to be met by each country, both mid-term and long term, for which policy making, promoting targeted energy renovations, is deemed necessary. Average buildings are proven to be useful tool toward developing an effective energy conservation policy, because they provide a compact set of information to be employed for strategic planning.

In this chapter one example of the possible decision-making alternatives is showcased through the creation of six average categories of buildings, representative of the main building typologies used for developing the energy profile of the Cyprus dwellings. From the targeted refurbishments performed after evaluating the current energy performance of the dwellings, a 33% reduction in the energy need was achieved, surpassing the 30% goal of 2030. The decision-making approach used in this study was focused on the heat losses from the envelope elements and the share of heating need per category.

The results validated the effectiveness and the expediency of the introduction of the average buildings, as well as the usefulness of monitoring energy performance indicators as tools toward determining the most energy-efficient refurbishment measures per typology and for subsequently obtaining the highest possible overall energy reduction. Their categorization is significant because they are filling the current knowledge gap, they highlight the lurking obstacles, and they portray that building typologies and the monitoring indicators can be useful instruments to facilitate the energy performance assessment of the building stock. Nevertheless, this specific approach is not intended to be utilized as a model to follow but to serve as one of the multiple options to be developed and choose from when formulating energy planning.

If the current annual refurbishment rate is to be projected in the future, all the single-glazed windows will be replaced with double-glazed ones by 2044, and the roofs of the corresponding old building stock will be insulated in approximately 200 years. From the current thermal insulation levels of the different envelope elements, it can be deduced that the recent European Directives referring to the minimum energy requirements of the building envelope have contributed significantly in the upgrading of the insulation levels of the building elements. Nevertheless, they are not adequate if the low rates of energy refurbishment of the old building stock are not improved.

In future studies the notion of average buildings and renovation energy performance indicators could be expanded to include other building typologies, besides housing units. The extension of the scope of energy performance and energy refurbishment from the private to the public domain is also possible, placing into perspective the urban scale and developing average urban blocks.

References

1. NATO Science Series. (2005). *Thermal energy storage for sustainable energy fundamentals, case studies and design* (Vol. 234). Springer Science & Business Media. Preface ix.
2. EUROPEAN COMMISSION, Brussels, 16.2.2016, COM(2016) 51 final.
3. Economidou, M., Atanasiu, B., Despret, C., Maio, J., Nolte, I., Rapf, O., ... & Zinetti, S. (2011). Europe's buildings under the microscope. A country-by-country review of the energy performance of buildings. Chicago.
4. Konstantinou, T., & Knaack, U. (2013). An approach to integrate energy efficiency upgrade into refurbishment design process, applied in two case study buildings in Northern European Climate. *Energy and Buildings, 59*, 301–309.
5. IEE EPISCOPE Cyprus. http://episcope.eu/building-typology/country/cy/. Accessed 31 Aug 2016.
6. Corgnati, S. P., Fabrizio, E., Filippi, M., & Monetti, V. (2013). Reference buildings for cost optimal analysis: Method of definition and application. *Applied Energy, 102*, 983–993.
7. Tommerup, H., & Svendsen, S. (2006). Energy savings in Danish residential building stock. *Energy and Buildings, 38*(6), 618–626.
8. Dascalaki, E. G., et al. (2011). Building typologies as a tool for assessing the energy performance of residential buildings – A case study for the Hellenic building stock. *Energy and Buildings, 43*(12), 3400–3409.
9. Ballarini, I., Corgnati, S. P., & Corrado, V. (2014). Use of reference buildings to assess the energy saving potentials of the residential building stock: The experience of TABULA project. *Energy Policy, 68*, 273–284.
10. Serghides, D. K., Dimitriou, S., & Katafygiotou, M. C. (2016). Towards European targets by monitoring the energy profile of the Cyprus housing stock. *Energy and Buildings, 132*, 130–140.

11. Serghides, D. K., Markides, M., & Katafygiotou, M. C. (2015). Energy retrofitting of the mediterranean terrace dwellings. *Renewable Energy and Sustainable Development, 1*(1), 138–145.
12. Serghides, D. K., et al. (2016). Energy refurbishment towards nearly zero energy multi-family houses, for cyprus, proceedings. In *5th international conference on renewable energy sources & energy – new challenges, Nicosia, Cyprus* (pp. 481–492).
13. Serghides, D. K., et al. (2015). Energy efficient refurbishment towards nearly zero energy houses, for the mediterranean region. *Energy Procedia, 83*, 533–543.
14. Serghides, D. K., Markides, M., & Katafygiotou, M. C.. (2014). Nearly zero energy terrace houses (in Greek), 10th national conference for renewable energy sources, Thessaloniki, Greece.
15. Serghides, D. K., et al. (2015). Energy-efficient refurbishment of existing buildings: A multiple case study of terraced family housing. In *Renewable energy in the service of mankind* (Vol. I, pp. 551–560). Springer.
16. IEE EPISCOPE Typology Brochure. http://episcope.eu/fileadmin/tabula/public/docs/brochure/CY_TABULA_TypologyBrochure_CUT.pdf. Accessed 29 Aug 2016.
17. Hermelink, A., Schimschar, S., Boermans, T., Pagliano, L., Zangheri, P., Armani, R., Voss, K., & Musall, E. (2013). Towards nearly zero-energy buildings – Definition of common principles under the EPBD, Ecofys 2012 by order of: European Commission. www.ecofys.com
18. UN Environment and International Energy Agency. (2017). Towards a zero-emission, efficient, and resilient buildings and construction sector, Global status report. www.globalabc.org
19. D'agostino, D., & Zangheri, P. (2017). Final report transition towards NZEBs in Cyprus (D2.5), JRC Technical Reports. https://ec.europa.eu/jrc
20. Dimoudi, A., Kantzioura, A., Toumpoulides, P., Zoras, S., Serghides, D., Dimitriou, S., Thravalou, S., Metaj, M., Mara, E., & Dorri, A. (2022). The energy performance of hospital buildings in the South Balkan Region: The prospects for zero-energy hospitals. In *Sustainable energy development and innovation* (pp. 757–763). Springer. https://doi.org/10.1007/978-3-030-76221-6_83
21. Serghides, D., Dimitriou, S., & Kyprianou, I. (2022). Paving the way towards zero energy hospitals in the mediterranean region. In *Green buildings and renewable energy* (pp. 159–167). Springer. https://doi.org/10.1007/978-3-030-30841-4_11
22. GlobalABC/IEA/UNEP. (2020). GlobalABC roadmap for buildings and construction towards a zero-emission, efficient, and resilient buildings and construction sector. www.iea.org
23. Dimitriou, S., Kyprianou, I., Papanicolas, C., & Serghides, D. (2020). A new approach in the refurbishment of the office buildings–from standard to alternative nearly zero energy buildings. *International Journal of Sustainable Energy, 39*(8), 761–778. https://doi.org/10.1080/14786451.2020.1749629
24. Prada, M., Prada, I. F., Cristea, M., Popescu, D. E., Bungău, C., Aleya, L., & Bungău, C. C. (2020). New solutions to reduce greenhouse gas emissions through energy efficiency of buildings of special importance – Hospitals. *Science of the Total Environment, 718*, 137446. https://doi.org/10.1016/j.scitotenv.2020.137446
25. IEE EPISCOPE Project: http://episcope.eu/index.php?id=97. Accessed 25 Aug 2016.
26. Loga, T. (2016). *Introduction to the excel workbook TABULA.xlsm*. Tabula Project Team. Available from: http://www.building-typology.eu/downloads/intern/development/TABULA-XLS_Introduction.pdf. Accessed 20 Aug 2016.
27. EU, Europe 2030 official webpage, http://ec.europa.eu/energy/en/topics/energy-strategy/2030-energy-strategy. Accessed 28 Aug 2016.
28. Directive 568/2007. Website of the ministry of energy, commerce, industry and tourism. http://www.mcit.gov.cy/mcit/mcit.nsf/0/FBFBEE85D45A6CD5C22575D30034F1A1/$file/KDP568_2007%20%20peri%20Apaithseon%20Elaxistis%20Energeiakis%20Apodosis%20Diatagma.pdf. Accessed 28 Aug 2016.
29. Serghides, D. K., et al. (2016). Monitoring indicators of the building envelope for the optimisation of the refurbishment processes. *International Journal of Contemporary Architecture "The New ARCH", 3*(1), 1–10.

30. Directive Κ.Δ.Π.446/2009, Website of the ministry of energy, commerce, industry and tourism. http://www.mcit.gov.cy/mcit/mcit.nsf/All/DF8E187B6AF21A89C22575AD00 2C6160/$file/KDP446_2009%20peri%20Rythmisis%20Energeiakis%20Apodosis%20 Ktirion(Apaitiseis%20Elaxistis%20Energeiakis%20Apodosis%20Ktiriou)%20Diatagma.pdf
31. «Guide for Building insulation» published by the ministry of energy, website of the ministry of energy, commerce, industry and tourism. http://www.mcit.gov.cy/mcit/mcit.nsf/0/ E074577C58AD9EFCC22575B60047BEA8/$file/ODIGOS%20THERMOMONOSIS%20 KTIRIWN%202H%20EKDOSI_%20PINAKAS%20DIORTHOSEWN.pdf

Toward NZEB in Public Buildings: Integrated Energy Management Systems of Thermal and Power Networks

Ana Beatriz Soares Mendes, Carlos Santos Silva, and Manuel Correia Guedes

Introduction

In recent years, climate change and environmental degradation have grown to be major threats to the well-being of the global population.

The European Commission has required each EU member state to establish a 10-year integrated national energy and climate plan (NECP) for the period of 2021–2030 [1].

Buildings are one of the largest energy consumer sectors in Europe, being responsible for approximately 40% of EU energy consumption and 36% of its greenhouse gas (GHG) emissions [2].

Nearly zero energy buildings (nZEB) have a very high energy performance, which means that they have low energy needs being largely covered by energy from renewable sources, produced onsite or nearby. Therefore, the renovation and rehabilitation of the existing buildings turning them into nZEB are also a priority in the Portuguese NECP, making it possible to achieve other objectives such as a reduction in energy bills and emissions and an improvement in the levels of health and comfort of these buildings [3]. In Portugal, a commerce and services building is considered nZEB when the maximum value of the energy efficiency indicator is equal or smaller than 75% of the reference value and the energy class ratio is not greater than 0.50 [4].

The first step for the conception of nZEB buildings is the integration of bioclimatic/passive design strategies in their architectural project. This step can allow for a reduction in energy needs of up to 60% (thermal and lighting) compared with conventional buildings. Renewable energy systems will provide the remaining

---

A. B. S. Mendes · C. S. Silva · M. C. Guedes (✉)
Instituto Superior Técnico, Universidade de Lisboa, Lisbon, Portugal
e-mail: manuel.guedes@tecnico.ulisboa.pt

energy needs, being the complement (rather than the remedy) to a good building design. This articulation between bioclimatic design and renewable energy systems should be considered in the first stages of the project – as was the case of the LNEG building.

The local small-scale power supply technologies and storage systems for energy consumption in buildings, such as microgrids, will play a key role in the future power system. They offer environmental benefits, by using locally produced renewable energy; social benefits, due to their reliability, affordability, and resilience; and economic benefits, by increasing self-sufficiency.

The main objective of this work is to develop integrated energy management strategies for thermal and power networks in a building. This work will focus on the study of a microgrid implemented on a pilot bioclimatic office building at *Laboratório Nacional de Energia e Geologia* (LNEG), part of the IMPROVEMENT research project that aims to transform existing public buildings into nZEB, integrating renewable energy microgrids with combined heat, cold, and electricity generation as well as storage systems [5].

Literature Review

Microgrids

According to the US Department of Energy, a microgrid (MG) is a "group of interconnected loads and distributed energy resources (DERs) within clearly defined electrical boundaries that acts as a single controllable entity with respect to the grid" and which has the capability to "connect and disconnect from the grid to enable it to operate in both grid-connected or islanded-mode" [6].

A MG consists of loads, DERs, a control system, and a point of common coupling (PCC). DERs are composed of distributed generation (DG) units, which may include, for example, solar photovoltaic (PV) panels, wind turbines, combined heat and power generators, and energy storage systems (ESS). In grid-connected mode, ESS establishes optimal periods to interchange power with the utility grid in a more convenient way. Thus, ESSs improve the reliability of the system and support the DGs when they cannot supply the full power required by the consumers [7].

Microgrid Energy Management Systems

An energy management system (EMS) is an information and control system that ensures that generation, transmission, and distribution supply energy at a minimum cost [8]. MG involves a software that optimizes the operation of the system by considering the two MG operation modes (isolated and interconnected) and the minimal required cost of operation [9].

Control Structure

Two approaches to the control structure were identified: centralization and decentralization. In a centralized control system, the data from all components is gathered into a central controller (CC) that performs the required calculations and determines the control actions for all the units at a single point. The CC uses the input data to solve the optimization problem and then transmits the optimal control decisions, through local controllers (LCs), to the correspondent DERs, where they are implemented. In a decentralized (or distributed) control system, there is no CC, and several LCs are set up to measure signals of the different DERs [10].

A compromise between fully centralized and fully decentralized control schemes is achieved by means of a hierarchical control scheme [11] with three control levels: primary, secondary, and tertiary. They differ in (i) their speed of response and the time frame in which they operate as well as (ii) infrastructure requirements [12].

Control Strategies

According to [11], most studies use model predictive control (MPC) strategies and optimal control, followed by multiagent systems (MAS) and rule-based control (RBC).

RBC is a static control strategy that relies on "IF-THEN" commands. It is popular in commercial building automation systems, because it is simple to implement [13]. It does not require any future data profile to make a decision and is thus suitable for real-time applications [14]. Homeostatic control (HC) is a form of RBC that employs an adaptive strategy that balances positive and negative feedback mechanisms, inspired by the biological concept of homeostasis.

Optimal control aims to optimize an objective function, subject to a set of constraints [15]. Such methods are often divided into three categories: classical, metaheuristic, and stochastic. In MG applications, the objective is systematically to minimize the total operational cost but may also have a second objective, e.g., minimizing GHG emissions or occupant thermal discomfort or considering undesirable outcomes by introducing a corresponding penalty. Common decision variables are: the amount of power that is taken from each dispatchable source; requested from the grid at each time; injected into the grid; charged to or discharged from a battery; reactive power support from renewable energy systems and/or batteries; state of charge (SOC) of ESS; controllable loads; and temperature setpoints.

A MAS is a collection of intelligent agents that interact with each other in such a way that the entire system learns and evolves toward a better solution. MGs allow coordination and control in a decentralized way [16].

MPC encompasses a set of control methods that rely on the dynamic model of the studied system [17]. Here, the current state of the system and its model and outside disturbances are inputs of the controller, which in turn outputs the future state of the system. It has the advantage of considering the future state of the system and disturbances, anticipating future events, and acting on that foreknowledge.

State of the Art
The literature covering the situation of EMSs that make use of thermal and power systems simultaneously was found to be sparse. In the following, an overview of the relevant case studies is given.

In [18], HC strategies were used for power management and EMS while considering the thermal behavior of the building. The MG under study contained solar panels, a wind turbine, an inverter, storage based on batteries, an HVAC, and a smart meter to measure the amount of energy that was consumed from the grid or injected to it. It also included multiple temperature and humidity sensors. The control block received as input the power consumption limit from the utility grid, power consumption, SOC of batteries, availability of the utility grid, and temperatures (walls, external and internal). The outputs were on-off for the grid selector, the battery, and the HVAC selector. There were two parts to the control strategy, a reactive part that ensured that the batteries had enough charge to maintain the MG running and a predictive one that used the thermal model of the room to maintain the temperatures within the comfortable interval.

In [19], a grid-connected MG located in a sport center facility was investigated through dynamic simulations, considering thermal and electrical loads. This MG was composed of a solar PV installation, a building energy storage system, and a heat pump. The goal was to balance self-sufficiency with electricity cost. To this end, two RBC strategies were employed, taking into account the impact of an HVAC system and of the heat pump. These strategies involved (1) peak shaving of the MG consumption with off-peak grid power and (2) pricing-based operation of the building energy storage system, according to the main grid electricity price. To benchmark the strategies, the resulting power flows and electricity costs were assessed. This analysis revealed that strategy 2 yielded the best results.

In [20], systems consisting of HVAC, battery energy storage and renewable generation in buildings were approached through an optimal control framework. In this case, the goal was to reduce peak load demand and electricity costs while maintaining thermal comfort levels within an acceptable range. The control strategy took into account the thermal dynamics of the building, battery SOC, renewable generation status, and actual operational data and constraints. For the thermal dynamics, a simple model was developed and trained with actual thermal and electrical data. The controller was tested using data from a real building. Preliminary results suggested that it yields a significant reduction in peak electrical power demand.

In [21], an experimental room was considered. It featured temperature, relative humidity, and lighting sensors, as well as an HVAC unit and an electricity microgeneration system with PV panels and an energy storage system. Here, the main objective was to ensure users' comfort and minimize cost, considering electricity

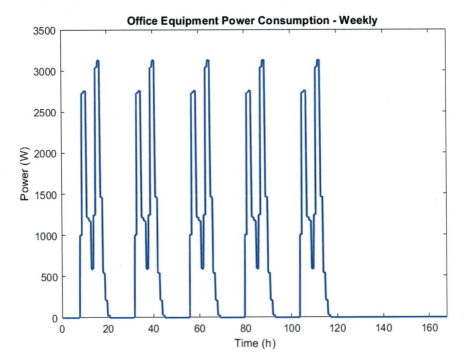

Fig. 1 Office equipment power consumption for a week

price and available energy from renewable sources. To this end, three algorithms were developed: (i) dynamic programming with simplified thermal model, (ii) genetic algorithm with simplified thermal model, and (iii) genetic algorithm with EnergyPlus. They were tested and validated in real conditions and benchmarked against each other. This study found that (iii) generally achieved higher convergence to the optimal value, with more energy being used from the PV system to operate the HVAC. It was noted that the simplified thermal model is less accurate in simulating the indoor temperature (Figs. 1 and 2).

Bioclimatic Building Design

Research on Bioclimatic, or Passive, building design began essentially in the 1970s, during the first oil crisis, and then expanded in the 1990s, with the global warming awareness. The objective of passive design is to reduce the weight of (fossil fuel) energy-consuming mechanical systems in the building, such as HVAC and artificial lighting, through natural means – by taking advantage of the sun's energy for heating and lighting, the local winds for natural ventilation, etc. Today, there are numerous publications on bioclimatic design strategies concerning a variety of climates [22]. These strategies involve design considerations such as solar orientation,

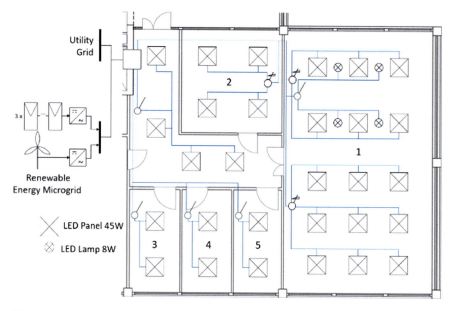

Fig. 2 Building's lighting single-line diagram

building form, shading, insulation, adequate glazing ratio in the facades, thermal inertia, and opening design for ventilation and natural lighting. The LNEG Building followed bioclimatic strategies since the first stages of its design. It is an atrium building (for natural lighting and ventilation), with adequate glazing ratio on the different facades, adequate insulation, and thermal inertia, and uses special passive systems for heating and cooling, such as buried pipes. PVs are embedded in the main (South) façade, doubling as Trombe walls for heating in winter. This bioclimatic approach accounted for a reduction of over 50% of its energy needs (thermal and lighting) from the onset, compared with a conventional office building. It is a fully passive, naturally ventilated building, with no (need for) air-conditioning.

Case Study

The work developed in this study is integrated in the IMPROVEMENT project that has the main goal of converting existing public buildings, which have high energy consumption of electricity, heating, and air conditioning, into nZEB. With this goal in mind, MG pilot plants in LNEG, in Portugal – which, as previously referred, has a bioclimatic building design conception which significantly reduces its energy needs from the onset, i.e., this paper focuses on its complementary renewable energy system design.

The building, represented in Fig. 2, is composed of five rooms and an unconditioned area. Room 1 is a multiuse room, can accommodate 8 people, and has 15

Table 1 Equipment power consumption

Equipment		Power consumption (W)
Desktop		150
Laptop		100
Projector		365
Coffee machine		1560
Printer	Printing/copying	750
	Stand-by	30

45W LED panels and 4 8W LED lamps. Room 2 is a meeting room with a capacity for five people and has four LED panels. Rooms 3, 4, and 5 are individual offices with two LED panels each. The unconditioned space is a small corridor with four LED panels.

The consumption was estimated considering the purpose of each room. It was assumed that the building has one printer and one coffee machine in the corridor. Room 1 has one projector and one desktop. The individual offices have one desktop each and Room 2 two desktops. During the working hours, a variable number of laptops were assumed to be used in the meeting room. The power consumption of each equipment is shown in Table 1. In Fig. 1, the weekly equipment power consumption is displayed considering that the week starts on Monday.

The LNEG MG can be divided into two separate systems: a thermal system and an electrical system. The thermal system is composed of two solar collectors that are 2 m^2 each and have two tanks, one air/water heat pump, a storage tank, and fan coils. The electrical system is made up of five subsystems: four energy generation systems that consist of two PV systems, one with 4050 W and the other with a 560 W rated power; a photovoltaic-thermal (PVT) system with an electricity rated power of 690 W and a 2500 W rated wind power system; and an energy storage system, composed of a 48 V lithium-ion battery with 660-Ah energy capacity and a depth of discharge (DOD) of 85%. The battery has a maximum charging and discharging power of 4200 W.

Thermal and Power Microgrid Integration

Model Integration

As the systems were model in Simulink, the integration itself is done using the same software. Henceforth, the time unit associated with the simulations will be 1 h.

The main difference in the two models was in the solvers. For the electrical model, a fixed time-step (Ts) solver (with Ts necessarily between 1e−5 and 5e−5) with discrete states was used, whereas the thermal model relied on the variable Ts solver ode23s. It was found that the solver ode23t, with a maximum time step of

5e−5, could accommodate the requirements of both models, allowing them to run simultaneously.

Both systems receive weather files as input. They both rely on irradiance and temperature data, and additionally, the electrical model requires wind velocity data. The weather data used was obtained using Meteonorm, with Lisbon as the chosen location and the irradiation values referent to a surface with a tilt angle of 30°.

The second step of the integration was to introduce the thermal model as a subsystem of the electrical model in Simulink. To ensure that the systems were working as intended, a simulation in these conditions was run. It yielded the same results as the separated models, thus confirming the correctness of the implementation so far.

Afterwards, the electrical power of the heat pump was set as a load in the electrical system, together with the office equipment consumption loads.

The thermal model developed in [23] features an on-off control for the heat pump, where the difference between the reference temperature of the storage tank and the observed temperature was used as the control variable. The pump switches on when the temperature of the storage tank is 5 °C lower than the reference temperature, taken to be 50 °C. The pump only switches on if the hot water tank has the capacity to heat the storage tank. To prevent frequent switching of the pump, a minimum operation time is set. This value is equal to 0.1 h (6 min). In the summer, the chosen reference temperature for the heat pump is 10 °C, and a new condition is set to prevent too low temperatures in the hot water tank. The heat pump is always switched off when the tank reaches 7 °C.

The electrical model developed in [24] has a battery charging/discharging control unit responsible for charging the battery when there is simultaneously a surplus in energy generation and the battery SOC is lower than 100% and limited by the charging rate. When the demand surpasses the energy production, this control unit is responsible for discharging the battery until a SOC of 15% is reached, respecting the DOD of 85%. From that point on, if the demand is still higher than the production, the control unit stops discharging the battery to respect the DOD. It is also important to note that when there is a surplus in energy generation and the battery SOC is equal to 100%, the left-over energy is injected into the grid. Similarly, when the demand surpasses the production and the battery SOC equals 15%, the required energy is extracted from the grid. The new model also receives as input the energy tariff. The values were obtained by consulting the website of a Portuguese retailer [25] and correspond to an energy supply tariff scheme with four different periods: super empty, normal empty, floods, and peak. The schedule was consulted in the Entidade Reguladora dos Serviços Energéticos website [26]. The daily term of the price corresponds to 0.7476 €/day, and the power term is divided in two parcels, the peak time power, 0.4874 €/kW.day, and the contracted power, 0.0256 €/kWh.day (Fig. 3). The prices corresponding to each period and the schedule are presented in Table 2.

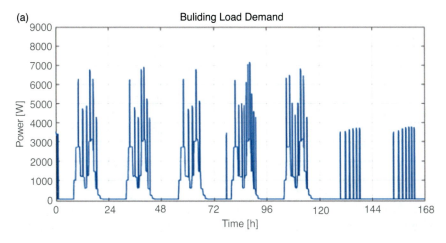

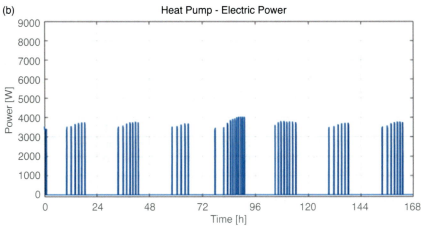

Fig. 3 Building load demand for a week in July: (**a**) total, (**b**) contribution from heat pump

Table 2 Energy prices and schedule of each tariff period

Time Periods Prices (€/kWh)			
Peak	Floods	Normal Empty	Super Empty
	08:00 – 10:30	00:00 – 02:00	
10: 30 – 13:00	13:00 – 19:30	06:00 – 08:00	02:00 – 06:00
19:30 – 21:00	21:00 – 22:00	22:00 – 24:00	
0.2162	0.1329	0.0918	0.0818

Results

Summer

The weather data used in this simulation is referent to a week in the middle of of the week, due to the cooling needs of the space. The heat pump totals a weekly consumption of 40.72 kWh. The generation (Fig. 4) encompasses both the energy-generated values of power by the PV and the wind power systems.

The total energy consumption during the week, 136.3 kWh, represents only 57% of the total energy generated by the MG, 238.8 kWh. Nonetheless, due to the maximum discharging power limit of the battery (4.2 kW), there is the need to extract power from the utility grid. These explain the peaks observed in Fig. 5 that are characterized by a maximum power of 1.87 kW. The total energy consumption from the grid is 1.35 kWh, making up 0.99% of total energy consumption. The energy bill at the end of the week amounts to 5.76€.

Winter

The weather data used in this simulation is referent to the first week of January. As illustrated in Fig. 6, the heat pump behaves quite differently in this season. Instead of switching on and off several times during the day, it does so only four times during the whole week to heat the space. When it does switch on, it reaches a higher maximum power. The weekly consumption of the heat pump is 34.93 kWh. The total energy consumption of the building during the week is 130.5 kWh. Henceforth, mentions of power generation for the winter will always refer to Fig. 7. The energy

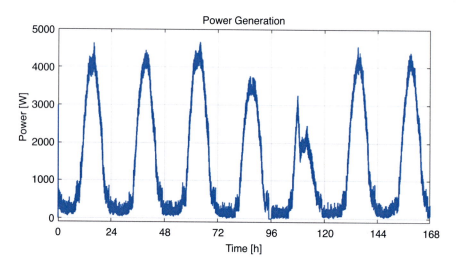

Fig. 4 Power generated by the photovoltaic and wind power systems in a week in July

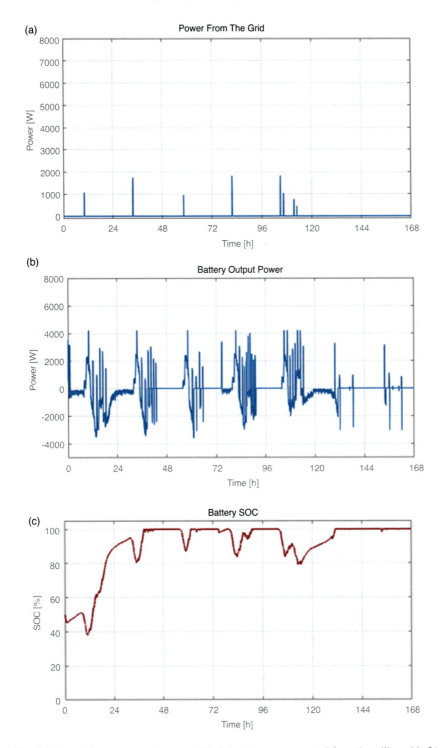

Fig. 5 Microgrid power output for a week in July, (**a**) power extracted from the utility grid, (**b**) battery output power, (**c**) battery SOC

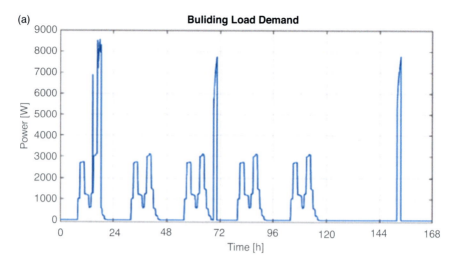

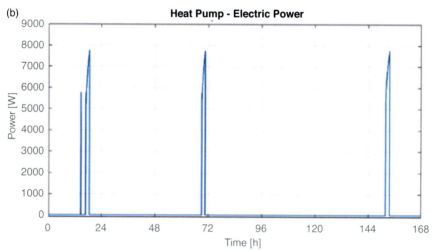

Fig. 6 Building load demand for a week in January: (**a**) total, (**b**) contribution from heat pump

generated by the MG during the entire week is 101.9 kWh, indicating a reduction of 57.3% in relation to the summer values.

From Fig. 8, one notes that the battery SOC is close to its minimum value of 15% for a significant part of the week. The energy extracted from the grid is 45.98 kWh, making up only 35.2% of the total energy consumption. The energy bill increases in relation to the summer to 13,60€.

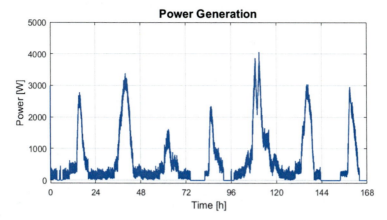

Fig. 7 Power generated by the photovoltaic and wind power systems in a week in January

Modifications to the Case Study

To test some energy management strategies in a case where there is more stress in the battery, i.e., the energy storage is smaller and there are more loads, some modifications to the initial study case were made.

The energy capacity of the battery was reduced from 31,680 Wh (660 Ah, 48 V) to 18,000 Wh (375 Ah, 48 V).

The pilot plant building is part of a main building. It was added to 30% of this building consumption to the pilot plant office equipment power consumption.

Summer

The power consumption curve for the modified study case during the summer week is presented in Fig. 9.

The power consumption from the grid, battery output power, and battery SOC curves are presented in Fig. 10.

During this week, the total energy consumption of the building is 312 kWh, which is 30.65% greater than the energy generation (238.8 kWh). The energy extracted from the utility grid is 97.91 kWh, which is 31.38% of the building load demand. The battery SOC is below 40% for the whole work week. The energy bill at the end of the week is 20.55€.

Winter

Fig. 11 presents the building power consumption curve.

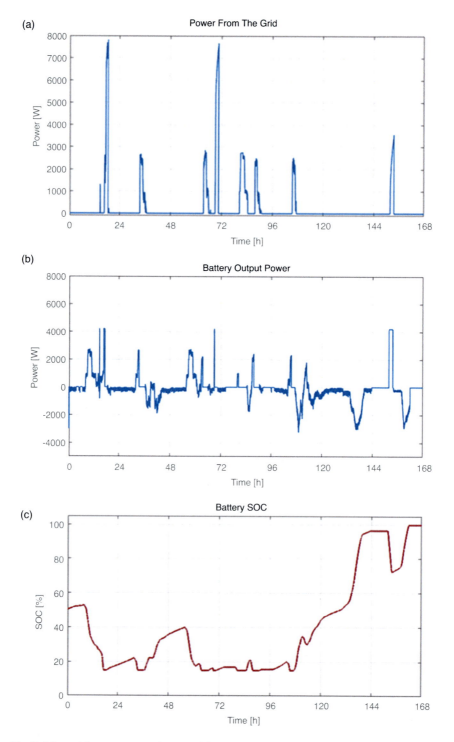

Fig. 8 Microgrid power outputs for a week in January: (**a**) power extracted from the utility grid, (**b**) battery output power, (**c**) battery SOC

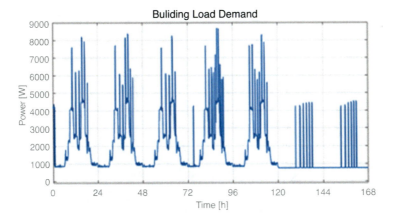

Fig. 9 Microgrid building load demand for the modified case in summer

The power consumption from the grid, battery output power, and battery SOC during the first week of January are presented in Fig. 12.

The total energy consumption during this week is 306.2 kWh, which is three times greater than the energy generated by the MG, and the energy extracted from the grid is 199.8 kWh, which is 65.28% of the energy consumption.

From the battery output power and battery SOC graphs, it can be concluded that the battery is, for most of the time, at its minimum SOC. The energy bill for this case is 38.19€.

Energy Management System

Two different control strategies following a rule-based control approach were developed. The base case, from section "Thermal and Power Microgrid Integration", from now on will be referred to as case without EMS, because the EMS was not integrated.

Control Strategy 1

The first management algorithm developed was based on the real-time electricity pricing signal and the SOC of the battery. It was focused on the economic dispatching of the storage system, charging the battery with power from grid when the energy prices are lower, storing this energy so it can be used when the price value increases.

Thus, when the SOC is between 30% and 90%, if the price of energy is equal or smaller than the quartile of energy prices during that day, the battery charges with

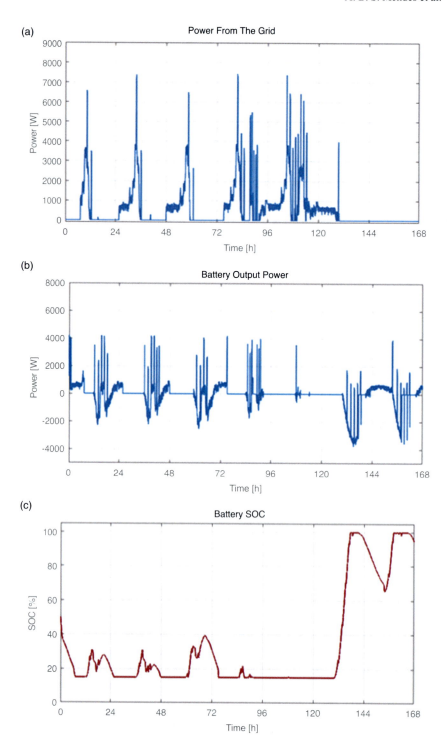

Fig. 10 Microgrid power outputs for the modified case in a week in July: (**a**) power extracted from the utility grid, (**b**) battery output power, (**c**) battery SOC

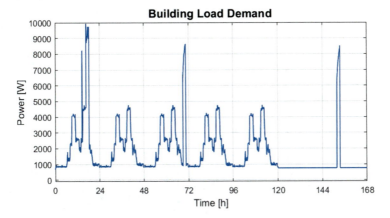

Fig. 11 Microgrid building load demand for the modified case in winter

power from the utility grid. When the SOC is smaller than 30%, the energy storage systems charge with power from the grid if the price is smaller than the median of the energy prices during the day. The battery charges from the grid with a constant power of 1500 W.

To avoid oscillations in the battery during this charging process, a minimum time of 5 h was set for the battery to stay without charging from the grid. It should be noted that these conditions only apply before the end of the work week.

Case Study – Summer

As can be seen from Fig. 13, in this case, only in the first hours of the day the conditions for the battery to charge from the grid are met. Afterward, the battery SOC rapidly increases to values above 90%, and in the few occasions where it goes below this value, the energy prices are greater than the quartile 25 of the prices during that day.

The total energy consumption from the grid had a significant increase relative to the case without EMS, from 1.35 kWh to 13.35 kWh. This increase is due to the interval, in the beginning of the week, when the battery is charging with energy from the grid. The energy bill also increases from 5.76€ to 6.81€.

Case Study – Winter

As can be seen from Fig. 14, a considerable reduction in the maximum of power extracted from the grid is observed. With control strategy 1, this value decreases to 3.62 kW, a difference of 4.2 kW. Because the battery SOC increases substantially due to the process of charging from the utility grid during the peaks in the demand

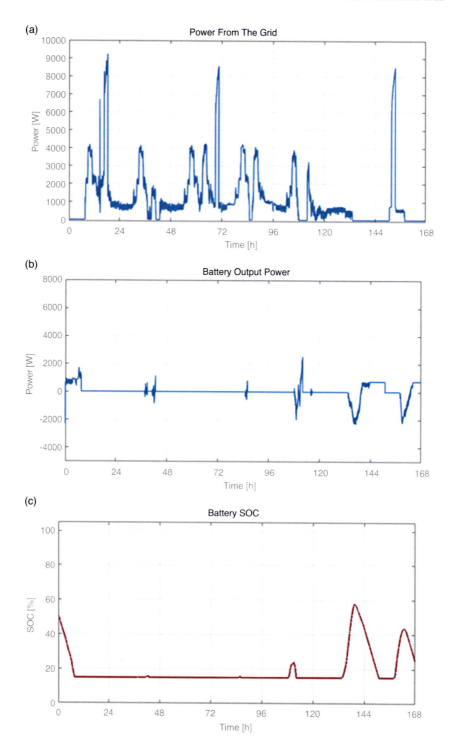

Fig. 12 Microgrid power outputs for the modified case in a week in January: (**a**) power extracted from the utility grid, (**b**) battery output power, (**c**) battery SOC

Toward NZEB in Public Buildings: Integrated Energy Management Systems of Thermal... 269

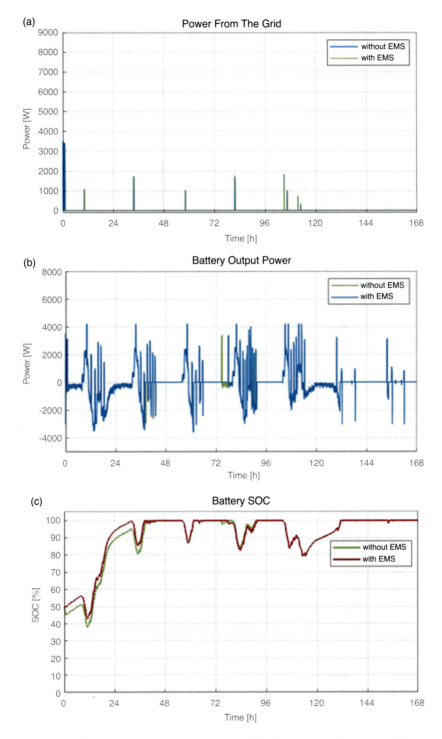

Fig. 13 Microgrid power outputs for the case in a week in January, control strategy 1: (**a**) power extracted from the utility grid, (**b**) battery output power, (**c**) battery SOC

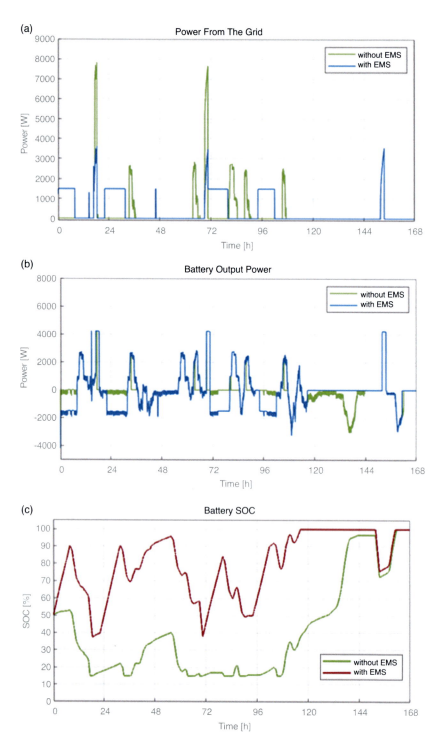

Fig. 14 Microgrid power outputs for the case in a week in January, control strategy 1: (**a**) power extracted from the utility grid, (**b**) battery output power, (**c**) battery SOC

(when the heat pump switches on), the battery has enough energy to fulfill the load until its maximum power output limit.

With this control strategy, the total energy consumption from the grid increases from 45.98 kWh to 65.66 kWh. Nonetheless, the energy bill sees a reduction from 13.60€ to 12.14€.

Modified Case Study – Summer

As can be seen from Fig. 15, the maximum power extracted from the grid suffers a slight reduction of 1 kW, and the battery, as expected, also attains higher levels than the case without EMS, due to the intervals when it was charging from the grid.

The total energy consumption from the grid only increases to 0.01 kWh in relation to the case without EMS. The energy bill reduced considerably, from 20.55€ to 17.01€. This decrease is mainly due to the allocation of power consumption from the utility grid when the energy prices were lower.

Modified Case Study – Winter

Figure 16 shows a very different behavior of battery SOC relative to the case without EMS. This control strategy presents six peaks during the working days, which occur when the battery is charging with power extracted from the utility grid. The total consumption of energy from the grid and the maximum of the power extracted from the grid had the same value as the case without EMS. The energy bill decreased from 38.19€ to 35.80€.

Control Strategy 2

This control strategy builds upon strategy 2. The on-off control described before was kept, and a new control block was added. Thus, when the result of the first part of the heat pump control is to switch on, the second block checks if the average temperature of the five condition spaces is within the comfort range (between 20 °C and 25 °C). If it is, the heat pump is turned off. If not, the SOC of the battery is evaluated. In case the SOC is lower than 50% and the energy price is greater than the daily median, the heat pump switches/stays off; otherwise it switches on. Because the effect of this control strategy in both cases for each season is similar, only the modified case study will be discussed in detail here (Fig. 17).

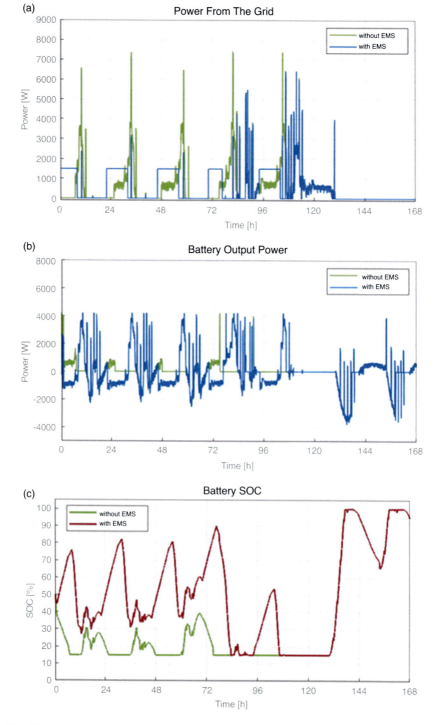

Fig. 15 Microgrid power outputs for the modified case in a week in July, control strategy 1: (**a**) power extracted from the utility grid, (**b**) battery output power, (**c**) battery SOC

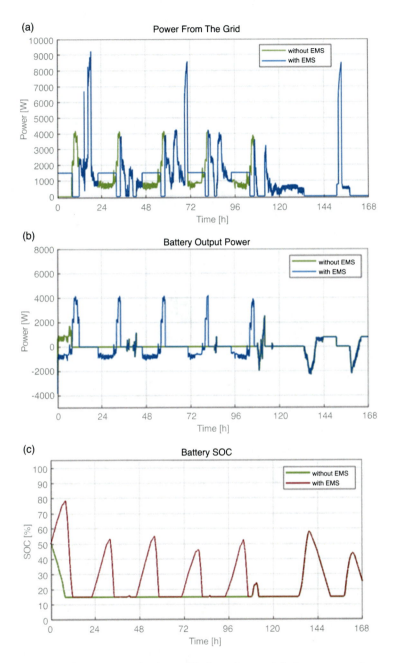

Fig. 16 Microgrid power outputs for the modified case in a week in January, control strategy 1: (**a**) power extracted from the utility grid, (**b**) battery output power, (**c**) battery SOC

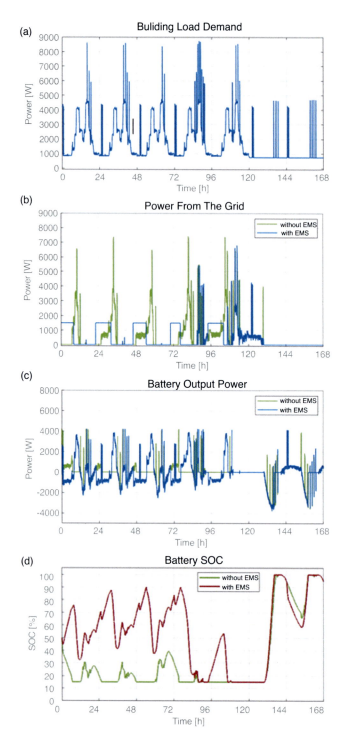

Fig. 17 Microgrid power output for the modified case in a week in July, control strategy 2: (**a**) building load demand, (**b**) power extracted from the utility grid, (**c**) battery output power, (**d**) battery SOC

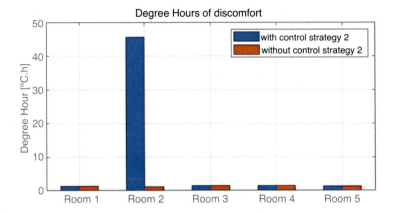

Fig. 18 Degree hours of discomfort for the modified case study for a week in July with control strategy 2

Modified Case Study – Summer

The heat pump switches on fewer times but with a slightly higher power than the case without EMS. The total energy consumed by the heat pump decreased by 28.5%, from 40.72 kWh to 29.12 kWh.

Both the maximum of the power and total energy consumption from the grid decreased to 6.8 kW and 89.73 kWh, respectively. The latter represents a reduction of 8.4% relative to the case without EMS.

The energy bill reduced by 24.4% in relation to the case without EMS, from 20.55€ to 15.53€.

Figure 18 demonstrates that except for room 2, the values of the degree.hours of discomfort are similar for both cases.

In room 2, this value increased from 1.216 °C.h to 45.61 °C.h, representing an average difference of 0.83 °C from the comfort temperature range.

This discomfort can be explained by the fact that as seen in [22], this space is the one with the worst cooling performance due to an excess of thermal loads in the room.

The new control block considers only the average temperature of all rooms. If the latter remains within the comfortable range, even if room 2 by itself is not within that range, the heat pump will stay off. This causes room 2 to reach even higher discomfort levels.

Modified Case Study – Winter

The heat pump switches on several times with control strategy 2. Nonetheless, the total energy consumption of the heat pump decreased from 34.92 kWh to 25.98 kWh (less than 25.6%).

Figure 19 presents the new load demand curve for the modified case study during a week in January and the power consumption from the grid, battery power output power, and battery.

SOC values with and without control strategy 2.

One can see that the maximum of the power consumption from the grid slightly decreased with this algorithm. The total consumption of energy from the utility grid is 190.9 kWh, and it represents a decrease of 5% relative to the case without EMS (199.8 kWh). The energy bill at the end of this week is 34.33€, a decrease of 10.1%.

The levels of thermal discomfort with and without this control strategy are similar and approximately zero.

Discussion

For the *case study* in summer, the lowest energy bill is attained for the case without EMS. This solution is also the one in which the building's energy consumption is lower, making it the preferable solution.

For the case study in winter, control strategy 1 results in a reduction of 11% in the energy bill in relation to the base case, but the energy consumption from the grid increases to 43%. With control strategy 2, the energy consumption from the grid is 26% higher than that for the base case. Still, it is lower than that for control strategy 1 and reduces the energy bill by 7%. Considering that the discomfort levels are zero for strategy 2, it seems to be the best option.

For the *modified case study*, in summer and winter, with control strategy 1, the energy consumption from the grid barely changes from the case without EMS, and the price reduces to 17% in summer and 6% in winter relative to the base case (Table 3).

For the winter case, the preferable control strategy is, again, the second one. It has the lowest energy consumption from the grid (decreases by 5%) and energy bill (decreases by 10%) and the same thermal comfort levels.

In the summer case, the second control strategy would also be the best one if only the decreases of 24% in the energy bill and 8% in the energy consumption from the grid were considered. However, the thermal discomfort levels in room 2 cannot be ignored, even though some things should be noted. First, this room serves as a meeting room that is only used during short periods of time and not in a daily basis. Second, the temperature is never more than one degree above the thermal comfort limit.

It is important to note that the relatively good thermal and lighting performance is largely due to the bioclimatic conception of the building – which is an essential first step to achieve the desired nZEB goals.

Finally, it is observed that the improvement obtained from applying these strategies is more significant for the modified case, where the loads are higher and the battery capacity is lower.

Toward NZEB in Public Buildings: Integrated Energy Management Systems of Thermal... 277

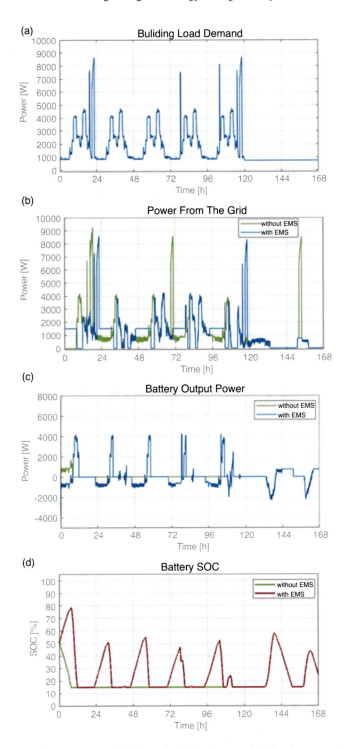

Fig. 19 Microgrid power outputs for the modified case in a week in January, control strategy 2: (**a**) building load demand, (**b**) power extracted from the utility grid, (**c**) battery output power, (**d**) battery SOC

Table 3 Comparison of the results with and without different control strategies for the case study and modified case study in both seasons

Case study	Without EMS		ControlStrategy1		ControlStrategy2	
	Summer	Winter	Summer	Winter	Summer	Winter
Price (€)	5.76	13.60	6.81	12.14	6.57	12.69
Energy from grid (kWh)	1.35	45.98	13.35	65.66	12.18	58.06
Pmax from grid (kW)	1.83	7.82	1.84	3.62	1.5	3.49
Heat pump energy (kWh)	40.72	34.92	40.72	34.92	29.15	26.75
Energy consumption (kWh)	136.3	130.5	136.3	130.5	124.7	122.3
Energy generation (kWh)	238.8	101.9	238.8	101.9	238.8	101.9

Modified case study	Without EMS		ControlStrategy1		ControlStrategy2	
	Summer	Winter	Summer	Winter	Summer	Winter
Price (€)	20.55	38.19	17.01	35.8	15.53	34.33
Energy from grid (kWh)	97.91	199.8	97.92	199.8	89.73	190.9
Pmax from grid (kW)	7.43	9.24	6.42	9.24	6.80	8.60
Heat pump energy (kWh)	40.72	34.92	40.72	34.92	29.12	25.98
Energy consumption (kWh)	312	306.2	312	306.2	300.4	297.3
Energy generation (kWh)	238.8	101.9	238.8	101.9	238.8	101.9

In future, it would be worthwhile to test these strategies on the real building and compare the resulting experimental data with that of the simulations. An enhancement of the proposed strategies should address the issue found in room 2, where the current approach fails to keep the temperature within comfortable levels. The analysis done in this study suggests that this could be achieved by controlling the heat pump with room temperatures, rather than their average, as input.

Appendix

Fig. 20
Fig. 21
Fig. 22
Fig. 23

Fig. 20 The LNEC. South façade (above), East façade (below)

Fig. 21 The atrium: bioclimatic design for daylight and natural ventilation

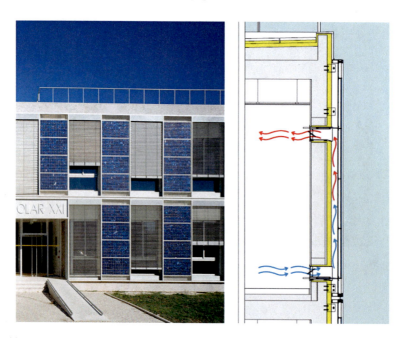

Fig. 22 Integration of PV in the South Façade, doubling as Trombe walls for winter heating

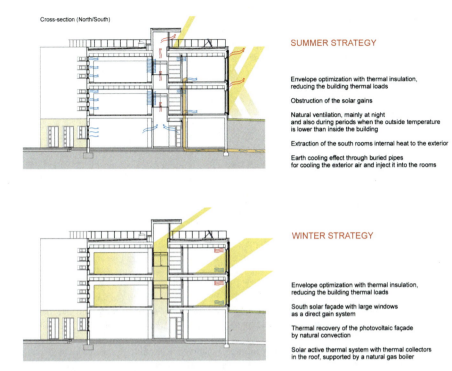

Fig. 23 LNEC: winter and summer bioclimatic design strategies

References

1. National energy and climate plans (NECPs) | Energy. https://ec.europa.eu/energy/topics/energy-strategy/national-energy-climate-plans_en. Accessed 27 Dec 2021.
2. Energy performance of buildings directive | Energy. https://ec.europa.eu/energy/topics/energy-efficiency/energy-efficient-buildings/energy-performance-buildings-directive_en. Accessed 27 Dec 2021.
3. National energy and climate plan 2021–2030 (NECP 2030), 2019, [online]. https://ec.europa.eu/energy/sites/default/files/documents/pt_final_necp_main_en.pdf. Accessed 27 Dec 2021.
4. Desempenho energético de edifícios. https://www.dgeg.gov.pt/pt/areas-setoriais/energia/energias-renovaveis-e-sustentabilidade/desempenho-energetico-de-edificios/. Accessed 7 May 2022.
5. Improvement | lneg Laboratório Nacional de Energia e Geologia. https://www.lneg.pt/en/project/improvement-2/. Accessed 3 Jan 2022.
6. Ton, D. T., & Smith, M. A. (2012). The U.S. department of energy's microgrid initiative. *The Electricity Journal, 25*(8), 84–94. https://doi.org/10.1016/j.tej.2012.09.013
7. Abdelgawad, H., & Sood, V. K. (2019). A comprehensive review on microgrid architectures for distributed generation. In *2019 IEEE electrical power and energy conference (EPEC)* (pp. 1–8). IEEE. https://doi.org/10.1109/EPEC47565.2019.9074800
8. Dorf, R. C. (2017). Energy management. In *Systems, controls, embedded systems, energy, and machines* (pp. 261–270). CRC Press.
9. Su, W., & Wang, J. (2012). Energy management systems in microgrid operations. *The Electricity Journal, 25*(8), 45–60. https://doi.org/10.1016/j.tej.2012.09.010

10. Khan, M. W., Wang, J., Ma, M., Xiong, L., Li, P., & Wu, F. (2019). Optimal energy management and control aspects of distributed microgrid using multi-agent systems. *Sustainable Cities and Society, 44*, 855–870. https://doi.org/10.1016/j.scs.2018.11.009
11. Fontenot, H., & Dong, B. (2019). Modeling and control of building-integrated microgrids for optimal energy management – A review. *Applied Energy, 254*, 113689. https://doi.org/10.1016/j.apenergy.2019.113689
12. Yamashita, D. Y., Vechiu, I., & Gaubert, J.-P. (2020). A review of hierarchical control for building microgrids. *Renewable and Sustainable Energy Reviews, 118*, 109523. https://doi.org/10.1016/j.rser.2019.109523
13. Grosan, C., & Abraham, A. (2011). Rule-based expert systems. *Intelligent Systems Reference Library, 17*, 149–185. https://doi.org/10.1007/978-3-642-21004-4_7
14. Elmouatamid, A., Ouladsine, R., Bakhouya, M., El Kamoun, N., Khaidar, M., & Zine-Dine, K. (2021). Review of control and energy management approaches in micro-grid systems. *Energies, 14*(1). https://doi.org/10.3390/en14010168
15. Naidu, D. S. (2003). Optimal control. In *Optimal control systems* (1st ed.). CRC Press.
16. Dou, C., Lv, M., Zhao, T., Ji, Y., & Li, H. (2015). Decentralised coordinated control of microgrid based on multi-agent system. *IET Generation, Transmission & Distribution, 9*(16), 2474–2484. https://doi.org/10.1049/iet-gtd.2015.0397
17. Garcia-Torres, F., Zafra-Cabeza, A., Silva, C., Grieu, S., Darure, T., & Estanqueiro, A. (2021). Model predictive control for microgrid functionalities: Review and future challenges. *Energies, 14*(5), 1296. https://doi.org/10.3390/en14051296
18. Parejo, A., Sanchez-Squella, A., Barraza, R., Yanine, F., Barrueto-Guzman, A., & Leon, C. (2019). Design and simulation of an energy homeostaticity system for electric and thermal power management in a building with smart microgrid. *Energies, 12*(9), 1806. https://doi.org/10.3390/en12091806
19. Chapaloglou, S., et al. (2021). Microgrid energy management strategies assessment through coupled thermal-electric considerations. *Energy Conversion and Management, 228*, 113711. https://doi.org/10.1016/j.enconman.2020.113711
20. Biyik, E., & Kahraman, A. (2019). A predictive control strategy for optimal management of peak load, thermal comfort, energy storage and renewables in multi-zone buildings. *Journal of Building Engineering, 25*, 100826. https://doi.org/10.1016/j.jobe.2019.100826
21. Pombeiro, H., Machado, M. J., & Silva, C. (2017). Dynamic programming and genetic algorithms to control an HVAC system: Maximizing thermal comfort and minimizing cost with PV production and storage. *Sustainable Cities and Society, 34*, 228–238. https://doi.org/10.1016/j.scs.2017.05.021
22. Guedes, M. C., & Cantuaria, G. (Eds.). (2019). *Bioclimatic architecture in warm climates: A guide for best practices in Africa*. Springer. ISBN 978-3-030-12035-1.
23. da Silva, M. A. G. (2021). *Renewable based thermal systems for microgrids*. Instituto Superior Técnico, Universidade de Lisboa.
24. Coelho, M. Q. S. (2021). *Renewable power systems for microgrids in public buildings*. Instituto Superior Técnico, Universidade de Lisboa.
25. Tabela de Preços 2022 – Mercado Regulado. https://www.celoureiro.com/pdf/precos/2022/Tabela de Precos 2022 – Tarifas Transitoria – Mercado Regulado.pdf. Accessed 12 May 2022.
26. ERSE – Tarifas e preços – eletricidade. https://www.erse.pt/atividade/regulacao/tarifas-e-precos-eletricidade/#periodos-horarios. Accessed 12 May 2022.

The Missing Link in Architectural Pedagogy: Net Zero Energy Building (NZEB)

Maryam Singery

> During COVID we learnt that no one is safe until everyone is safe, so we need to act urgently to protect the most vulnerable in society – in schools, hospitals, and homes for the elderly – in a movement that elevates buildings beyond being simply tradable assets, to seeing them as part of a built environment that is at the vanguard of our fight for a safer future for all. –Sue Roaf, Emeritus Professor of Architectural Engineering, Heriot-Watt University

Introduction

By 2050, Texas will have 115 heat days a year, 55 days more than on average in 2022 [1]. According to the US Census Bureau report 2022, San Antonio, Texas, topped the list of the most significant numeric gainers in the USA, increasing by 13,626 people between 2020 and 2021 [2]. This population growth increases the demand for space and energy. Ed Mazria, a founder of Architecture 2030, believes that the built environment generates nearly 50% of annual global CO_2 emissions. Building materials and construction account for 20% of this amount and building operation to 27% [3]. Both the NZEB design and construction play crucial roles in this region, therefore spotlighting architects.

Academics typically emphasize theorizing the design process in reverse order by setting goals, identifying evidence of learning outcomes, and designing the

---

M. Singery (✉)
School of Architecture + Planning, Klesse College of Engineering and Integrated Design, University of Texas at San Antonio, San Antonio, Texas, United States
e-mail: maryam.singery@utsa.edu

instructional plan as goals set forth according to the Paris Accord.[1] These goals are as follows:

- State current environmental circumstances;
- Select international sites to familiarize students with overseas climatic changes;
- Recognize future climatic changes;
- Extrapolate future resources, demands, and technologies;
- Implement advanced performance modeling to reduce energy consumption;
- Cooperate with construction industries, community, and other stakeholders;
- Design a net zero energy building with emphasis on maximizing the passive system through different architectural design concepts;

According to the New Building Institution (NBI), out of 56 verified NZE offices in the USA, only one branch exists in Texas[2] [5] (Fig. 1). The 17 third- and fourth-year undergraduate architecture students enrolled in the ARC 4156 Building Design studio course, entitled: Net Zero Energy US Embassy, US Consulate architectural design in different countries like Colombia, Dominican Republic, Greece, Japan, Norway, and Mexico.

Most of the students taking this studio course also took the ARC 4183 lecture course on environmental systems, which primarily focused on designing environmentally responsive buildings and the natural and artificial systems supporting them, i.e., embedded energy, active and passive heating cooling systems, and so on.

Fig. 1 HARC's energy manager explains how beekeeping helps the local ecosystem (left). Students promised to design NZEB at the end of the field trip to Houston and Woodlands at the beginning of the semester, HARC, Woodlands, TX (right) (Image represents the students' engagements and beginning potential for designing NZEB, and it stayed by the end of the semester). (Source: Author (2022))

---

[1] The Paris Agreement is a legally binding international treaty on climate change. It was adopted by 196 Parties at COP 21 in Paris on 12 December 2015 and entered into force on 4 November 2016. Its goal is to limit global warming to well below 2, preferably to 1.5 °C, compared with pre-industrial levels [4].

[2] The first NZE office in Texas is named HARC and is located in Woodlands. The studio had a site view at HARC.

Therefore, the students taking the theory course could practically apply those theoretical discussions in their studio projects.

Passive Design Strategies

Vernacular architecture presents perfect examples of inspiring passive design strategies to solve climate conditions without using fossil fuel energies and, in some cases, no modern construction machines or technologies (Fig. 2).

Every two or three students chose one country and conducted research on their climate zones. Using sources like Google Map and Google Earth, they discovered and analyzed their sites. Variable sites and climatic conditions in different countries provided a pedagogical opportunity for students to familiarize themselves with opportune climatic zones and their respective climate changes.

Premised upon passive design, resource conservation, and passive and active cooling and daylighting, conserving embodied energy base can significantly reduce gas and fuel consumption. Depending on their project site selection, each student in this design studio chose specific strategies adopted and used in various parts of the world. One of these solutions had to do with spatial scarcity predicting a significant reduction in spatial needs over the foreseeable future.

Sustainable Landscape

Students paid particular attention to using and enhancing public transportation, worked on the pedestrian paths surrounding bus stations, designing the necessary street furniture, and bicycle racks, and social and welfare amenities, i.e.,

Fig. 2 The student worked at the NZE US consulate at Barranquilla, Columbia, and used wind captures to provide natural ventilation inspired by Middle East countries' vernacular architecture. (Source: Erik Ortega (2022))

Fig. 3 All students focused on their site interventions and their impacts on local ecosystems. For example, one student designed the US consulate in a hot and humid climate zone (Santo Domingo Republic). He incorporated using low albedo materials, fauna, and flora in his design, where his passive design strategies provide appropriate shading areas and natural ventilation methods. (Source: Jhardon Small (2022))

supermarkets and food stores near them. Depending on the climate, flora and fauna play essential roles in creating shades and cooling systems apropos of designing sustainable landscapes, where rainwater serves irrigation purposes (Fig. 3).

Reducing the albedo effect is responsible for 23% of the global emissions – most of which are used in the built environment. This prompts incredible opportunities for embodied carbon reduction in high-impact materials [3].

Conceptual Architectural Design

A variety of reactions regarding architectural concepts had been seen in this stage of the design process: biomimicry, biophilic, vernacular, and sustainable architecture, and sometimes a combination to achieve better results (Figs. 4, 5, and 6).

When the time came to conserve historical buildings and save embodied energy, another student gave a new face to a historical US consulate building in Tokyo, Japan, by remodeling the building and using passive design strategies.

The Missing Link in Architectural Pedagogy: Net Zero Energy Building (NZEB)

Fig. 4 Student work: A bird perspective (left), floor plan (right); one student represented a biomimicry design using golden ratio geometry as his concept. He was also inspired by Greek theater forms for designing the US embassy at Athens, Greece. (Source: Covey Johnson (2022))

Fig. 5 Students work at Ciudad Juarez, Mexico, with a hot and dry climate zone; one student, inspired by some vernacular architecture strategies, used arches that provide a passive cooling system. Thanks to the arch form, half of the arch would be shaded during the day except at noon. (Source: Samuel Ruan (2022))

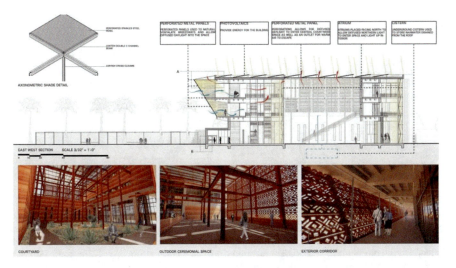

Fig. 6 Students work at Ciudad Juarez, Mexico; one student's passive cooling strategy included a courtyard and covered pedestrian path to provide a thermal landscape. (Source: Noel Parra (2022))

Sustainable Materials

Students were instructed to use recycled, local, low-carbon, and carbon-neutral materials. One student used plant-based concrete because it is eco-friendly and Avi-protect glass. Avi-Protect glass is an eco-friendly solution to saving flora and fauna. Birds lose their lives flying into reflective buildings by hundreds, so to counter that, the glass is acidly etched, allowing the birds to see it while not distracting by the facade. Another student reached out to use autoclaved aerated concrete, known as cellular concrete, and 100% natural materials.

Active Design Strategies

The studio familiarized the students with the world's renewable energy sources, including the photovoltaic cells on rooftops and building facades geared toward generating clean energy. The students used some of these devices as canopies or roof overhands in designing covered parking spaces. One student used a composition comprising PV cells and green surfaces that, in addition to natural ventilation and daylight, generated electricity as part of a green façade and collected the rainwater for watering the landscape and green panels (Fig. 7). Other students used the following design methods in their projects: recycling and reusing wastewater, flexible or movable walls for interior design as needed. Geothermal energy, especially in Oslo, Norway, inspired yet another renewable source that students used.

Fig. 7 Student work: One student practiced designing the NEZ US consulate in Fukuoka, Japan, and minimized the ratio of surface/volume; four facades have different designs based on their orientations. They provide natural ventilation and daylight. They combined PV panels with green panels. The façade panels are kinetics. The corridor in the center of the pyramid acts as a chimney effect. During our field trip, the student inspired the façade pattern from one of Houston's Fine Art Museum artworks. (Source: Udal Kosta (2022))

Results and Discussion

Luckan (2014) discussed embedding "sustainable design principles" into architectural design studio curricula [6]. Oliveira (2017) also shared the benefits of interdisciplinary or cross-disciplinary approaches to fill the existing theory-practice gap in architectural education. These strategies show how to incorporate sustainable development into architectural design [7]. Similarly, Azari and Caine (2017) demonstrated how to weave technical expertise – especially adding different methods of building performance evaluation – into architectural studio pedagogy [8]. In another research, Malini (2020) acknowledged the unpreparedness of architectural students in addressing the negative consequences of climate change and believes that "traditional design studio pedagogies" do not promote the culture of "cooperation and collaboration" among students. The undeniable fact remains that students learn from co-learning and sharing works, which, in some cases, helps them design more efficient buildings closer to net-zero goals, i.e., reducing energy demands to about 70% (ibid.) [9].

Therefore, according to Moosavi and Bush (2021), educating sustainability should be incorporated into architectural education. In their experience, the students simulated real-world scenarios in creating sustainable design strategies in their studio projects [10]. Mohamed (2022) also emphasized similar recommendations in incorporating sustainable design strategies into the architectural design curricula [11].

The word integration in architectural education has witnessed some changes since 2014. Integrating studio courses with sustainable design principles between 2014 and 2020 is a case. The 2050 achieve zero-carbon emission goals have incorporated even more pedagogical techniques in recent years. Combining these skills goes beyond teaching the students how to design NZE buildings and creates broader links for connecting the construction industry and community partnerships. Such new and integrative linkages can help thrive and develop new construction skills.

Conclusion

The learning outcomes of NZEB in the UTSA design studio pilot project include the following:

1. Innovation integration model between the academia, profession, and community (Fig. 8)

    1. Link between A and P

        - Expediting and expanding the NZEB projects.
        - Reducing the time to reach the set goals.

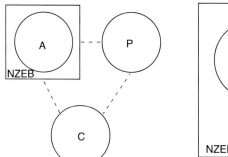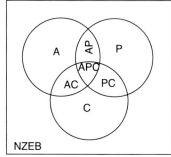

Fig. 8 Compares the unlinked (left) and linked (right) connections or relationships between the academia, the profession, and the local community in the area of NZEB studio. Specifically, the point shows two connections that seem more realistic and suitable to students and their communities (Source: Author (2022)). Note: A = Academia; P = Profession; C = Community; AP = Academia and Profession; AC = Academia and Community; PC = Profession and Community; APC = Academia, Profession, and Community

- The jury feedback shows more enthusiasm toward these projects than the students' purely conceptual projects, which ultimately boosts interdisciplinary professional creativity.

2. Linking A and C relationships leads to optimization of the NZE building

    - The community part of the studio jury who came from the Mexican Consulate invited us to have an exhibition in the Consulate building.
    - Social media can also promote this building construction approach within the community networks.
    - The final project review also further strengthened the community, Mexican Consulate relationship.

The students learned a lot from the professional feedback for their projects and self-reflection. This relationship also helped them find summer internship employment opportunities in San Antonio construction companies.

The final jury demonstrated student satisfaction with this project. With its "international" concentration, this studio encouraged some students to pursue UTSA study abroad in Urbino, Italy, for the next semester (Fig. 9).

Acknowledgments The author would like to sincerely thank international and local (San Antonio, Texas Area) academics, professionals, diplomats, and community listed below for their time, comments, and participation as jury members during the design review and for providing job/career opportunities for my undergraduate ARC 4156: Building Design Studio students: The University of Texas at San Antonio, I. Azad University of Tehran, Institute of Photovoltaics d'Ile-de-France (IPVF), Technical University of Munich (TUM), Houston Advanced Research Center (HARC), PBK Sports Rio Grande Valley, WJE (Wiss, Janney, Elstner Associates), Vaquero Group, IBTX, Huxton Group, and The Mexican Consulate in San Antonio.

Fig. 9 Final review day of the studio at UTSA. The image represents the integration of the academics, professionals, community members, and students who are aiming to "build an environment that is at the vanguard of our fight for a safer future for all." (Source: Author (2022)). Note: The quote is from Sue Roaf

References

1. States at Risk. (2022). Texas. https://statesatrisk.org/texas/all. Accessed 28 May 2022.
2. The U.S. Census Bureau. (2022). https://www.census.gov/construction/nrc/pdf/newresconst.pdf. Accessed 18 May 2022.
3. Architecture 2030. https://architecture2030.org/enews/. Accesses 29 May 2022.
4. United Nations, Climate Change. (2022). https://unfccc.int/process-and-meetings/the-paris-agreement/the-paris-agreement. Accessed 30 May 2022
5. New Building Institution (NBI). https://newbuildings.org/resource/getting-to-zero-database/. Accessed 28 May 2022.
6. Luckan, Y. (2014). *Analysis and perception: Architectural pedagogy for environmental sustainability*. W I T Press.
7. Oliveira, S., & Marco, E. (2017). Preventing or inventing? Understanding the effects of non-prescriptive design briefs. *International Journal of Technology and Design Education, 27*, 549–561.
8. Azari, R., & Caine, I. (2017). Applying performative tools in academic design studio: A systemic pedagogical approach, ARCC 2017 conference: Architecture of complexity, Salt Lake City, University of Utah.
9. Srivastava, M. (2020). Cooperative learning in design studios: A pedagogy for net-positive performance. *Buildings & Cities, 1*(1). https://doi.org/10.5334/bc.45
10. Moosavi, S., & Bush, J. (2021). Embedding sustainability in interdisciplinary pedagogy for planning and design studios. *Journal of Planning Education and Research*. https://doi.org/10.1177/0739456X211003639
11. Mohamed, K. E. (2022). An instructive model of integrating sustainability into the undergraduate design studio. *Journal of Cleaner Production, 338*, 13059.

Environmental Dimensions of Climate Change: Endurance and Change in Material Culture

Mona Azarbayjani and David Jacob Thaddeus

Background

M. Azarbayjani (✉)
University of North Carolina at Charlotte, Charlotte, NC, USA
e-mail: mazarbay@uncc.edu

D. J. Thaddeus
School of Architecture, College of Arts + Architecture, UNC Charlotte, Charlotte, NC, USA
e-mail: thaddeus@uncc.edu

In her seminal 1962 book the *Silent Spring* [1], Rachel Carson declared, "at times, technological progress is so fundamentally at odds with natural processes that it must be curtailed." She wrote: "Man's attitude toward nature is today critically important simply because we have now acquired a fateful power to alter and destroy nature. But man is a part of nature, and his war against nature is inevitably a war against himself?" Carson's activism inspired the environmental movement that led to the creation of the US Environmental Protection Agency in 1970 (EPA).

For many, "man's war against nature" began with the industrial revolution that has since brought much progress and prosperity to industrialized nations but comes as a threat to man's existence in the form of critical worldwide climate change. The 2015 Paris Agreement [2] is the international treaty on climate change that has set a goal of keeping global warming to less than 2 °C (preferably 1.5 °C) compared to preindustrial levels. This is the goal that 193 parties agreed to accomplish by 2050 by each proposing a Nationally Determined Contribution (NDC) and a path to reach it.

In its Executive Summary, the McKinsey Global Institute [3] stated that "As of December 2021, more than 70 countries accounting for more than 80% of global CO_2 emissions and about 90% of global GDP have put net-zero commitments in place, as have more than 5000 companies, as part of the United Nations' Race to Zero campaign." This underscores the importance of collective and immediate global action to confront climate change.

A framework to measure environmental impact is known as the I = f (P, AT), created by Paul Ehrlich and John Holdren in 1971 [4]. It concludes that "there are three main factors affecting environmental impact: (1) Population (P), (2) Affluence Level (A), and (3) Technology," and as all these continue to rise and develop, so will their collective impact (I). These three factors contribute to urbanization rates, which are also on the rise. The United Nations projection is that by 2050, two-thirds of the world's population will live in urban areas. The surge in construction projects to meet this demand is why net-zero buildings are more relevant [5]. The projected growth will require increased infrastructure and the construction necessary to absorb the additional 2.3 billion urban migrants. This underscores the gravity of trending urban development and the dramatic increase in carbon emissions that it will undoubtedly entail, in addition to impacting land use, air and water pollution, resource shortages, the heat island effect, and the resulting exponential increase in energy consumption.

It is now widely accepted more than ever before that it is imperative to lower carbon emissions to avert a major climate calamity. It is evident that there is no silver bullet or a single solution to the environmental crisis. Global cooperation is essential to solve this global quandary. Everyone has a role to play in the fight against climate change, but there is only so much we can do as average citizens. On the other hand, the building sector can have a much greater impact [6]. According to the US Department of Energy (DOE), buildings are responsible for 38% of all energy-related greenhouse gas (GHG) emissions each year. This is a considerable percentage, and by targeting the building sector specifically, significant progress could be made in the global effort to reduce carbon emissions. The Paris Agreement calls for the building sector to reduce the primary GHG emissions by 50% by 2030 and net-zero by 2050 [7].

This goal may seem lofty, but countries like the UK demonstrate that it is possible. The UK was the first to sign "a legally binding target to achieve net-zero carbon emissions by 2050." The ARUP Group, located in London, England, conducts research and provides detailed reports to be used as a framework to empower building sectors and other countries to strive for their net-zero aspirations. The authors of the ARUP publication titled *Net-zero Carbon Buildings: Three steps to take now* report [8] that they saw surprisingly widespread and fast action in the property sector. "Twenty-three leading property organizations signed a climate change commitment agreeing to set a pathway to net-zero carbon for their organizations." Others in the UK have also joined this movement and have begun implementing changes to achieve this goal, which, until recently, was something that seemed impossible. In another publication, the ARUP Group further promotes using building incentives, conducting complete lifecycle analyses, and defining a clear net-zero energy definition to move the property sector toward reaching the goal of carbon neutrality and net-zero buildings. They believe that achieving net-zero carbon for buildings is possible for both existing and new buildings, but to achieve this goal, new approaches must be taken for buildings of all types and scales [9].

This type of large-scale change will not happen overnight, and it is important to understand how to "encourage organizations to continue on the journey to achieving net-zero." One way to do this is through incentives and short-term recognition for progress. It is important to communicate the importance and value of making these changes with companies because "improving emissions can be a costly project and will often see its biggest returns in the long term. Therefore, short-term recognition can be the added boost that many companies need to push them to engage fully with greener practices."

The Leadership in Energy and Environmental Design (LEED) is an example of one of these recognitions that constitute an "international verification of a building's green features." [5] The Building Research Establishment Environmental Assessment Method (BREEAM) is another leading sustainability assessment for buildings and infrastructure. These incentives and the organizations they represent also participate in research and create the industry standards that the ARUP Group calls for.

These are just two of the main recognition certifications already in place that can distinguish built projects and serve as the motivation a company needs to make greener choices that lower its carbon emissions [5]. Along with these incentives, certain cities lead the way in getting sustainable accredited buildings by making cities broad goals and requirements for new construction. To lay out some of the leading states' approaches and targets, Boston was ranked the number one energy-efficient city in the country as of 2013–2019. A few of the city's targets are to be carbon neutral by 2050 and that every new building must meet the minimum requirements of the US Green Building Council (USGBC) and be LEED-certified. Many other states also make great strides toward creating a greener future, and these show the importance of policy-making in a city toward incentivizing and requiring greener building practices.

Another way to reach similar outcomes has been observed in Australia, where there has been a clear link between operational energy/carbon performance and overall value. There, an energy rating scheme called the National Australian Built Environment Rating System (NABERS) is "credited with halving the average energy intensity of commercial property." This change has happened because the majority of "commercial properties that have a high NABERS rating in Australia benefit from a value premium of approximately 20%." [8] Linking lower carbon emissions to a direct monetary value has resulted in greater industry participation. The UK has just launched a similar rating scheme. While these types of efforts are making significant changes, they mainly address operational emissions. Currently, there is no framework that effectively links "embodied carbon and asset value," which is needed "if we are to empower the market to move toward true net-zero." This example illustrates that "transformation is driven by incentives," and having accreditations that "define clear markers" and clear benefits will produce results.

While these are promising examples, there is still much more to be done. Two-thirds of countries do not even have building codes let alone incentives to reach net-zero carbon emissions. This means that in 2019, approximately 55 billion square feet of building area were constructed without energy standards. This is really important because, as stated earlier, the world's overall population is growing exponentially, and 90% of urban growth is concentrated in countries in Africa and Asia. Calculating an individual country's carbon impacts can get tricky, but the main point is that there needs to be regulations put in place to ensure that new construction is bringing us closer to net-zero rather than farther from it.

Globally and across all sectors of the economic and political spectra, there needs to be a multifaceted, gradual yet deliberate, and orderly transition to reducing carbon emissions to meet the goals of the Paris Agreement by 2050. In the industrial manufacture of building products and components, Life Cycle Analysis (LCA), Design for Disassembly (DfD), component reuse, and other parameters and benchmarks all need to be mandated to slow down global warming to an acceptable level compared to preindustrial standards. In the construction and operation of buildings, decarbonization through the use of clean energy and sustainable alternatives will collectively reduce emissions by appreciable margins.

Definitions

Net zero refers to the state of overall equilibrium between GHG emissions into and removal from the atmosphere. Although the concept is simple and the objective is righteous, many believe that the net-zero equation has not yet been solved. Not unlike a *steady state* condition in thermodynamics, net zero involves many sources of emissions (carbonization) and unequal sources of negative emissions (decarbonization) that need to be balanced.

The following definitions are from the 2020 McKinsey Article: Data to the rescue: Embodied carbon in buildings and the urgency of now.

> **Embodied carbon** consists of ALL the GHG emissions associated with building construction, including those that arise from extracting, transporting, manufacturing, and installing building materials on-site, as well as the operational and end-of-life emissions associated with those materials. [10]
>
> **Cradle to gate** embodied carbon refers to the emissions associated with only the production of building materials, from raw material extraction to the manufacturing of finished products; it can be thought of as supply-chain carbon, and it accounts for the vast majority of a building's total embodied carbon. [10]

Operational Carbon, on the other hand, is what ensures user comfort in a building and includes carbon emissions from building operation and maintenance such as heating, cooling, ventilation, lighting, power, etc.

Design Decisions Based on Embodied Versus Operational Carbon

> Buildings are currently responsible for 39% of global energy-related carbon emissions: 28% from operational emissions, from the energy needed to heat, cool and power them, and the remaining 11% from materials and construction. [11]

The International Building Code (IBC) has required a 50% decrease in energy use in *new buildings* since it was first adopted in the USA in 2000. This has led to significant reductions in operational carbon emissions but not in embodied carbon emissions. Also, the IBC does not require upgrading energy efficiency for *existing buildings* unless they are scheduled for "substantial" renovations.

In contrast to operational carbon, embodied carbon is finite rather than perpetual, can only be reduced in the initial building design and construction stages, and cannot be removed from existing buildings. A large part of a building's embodied carbon emissions lies in the choice of building materials and components. In general, carbon emissions can be reduced firstly by choosing locally sourced materials that will require nearer transportation and secondly by choosing products with longer life spans that will not require frequent replacement.

To reach net zero, it is critical that building design and construction follow a sequential approach to critical decision-making from the inception and at every stage of development and implementation. The greatest impact on a building's carbon emission trajectory begins in the earliest design phases. Fundamental decisions regarding massing, ceiling heights, facade designs, and passive design strategies all impact a building's full carbon lifecycle. Implementing efficient systems, using on-site renewable energy, minimizing waste by reusing materials, and choosing products manufactured nearby are all to be considered early on in a designer's efforts to mitigate carbon emissions. By taking a whole life-cycle approach, tradeoffs will be more accurately considered at every stage of a project [8].

While this is all crucial to reaching net-zero, it is only really "delivered in operation." One of the major hindrances to buildings reaching their carbon goals is the enduring conflict between "building occupants and their operational expectations." To have a building reach net zero, there will inevitably be greater restrictions on energy use, which will impact the occupants who may not be used to limiting their energy use. This disconnect can be represented through the distinction between a net-zero-*enabled* building and a net-zero-*achieved* building. This means that a building may be enabled to become a net-zero building, but it is up to the occupants and its operational usage to achieve this goal. Reaching full net zero will require a method to measure energy usage, and one such format is known as the energy use intensity (EUI), which gives a building a floor-by-floor calculation of its energy usage and helps give a definite tracking system for energy usage. In addition to clearly understood markers, early involvement and collaboration from every stakeholder throughout the building's lifecycle are needed, along with cooperation and clarity of expectations with all involved to accomplish a net-zero full lifecycle building [8].

Whether it is a net-zero retrofit, or a new build, designing buildings to achieve net zero requires a fundamental change in approach. Design success has been measured by compliance outcomes but needs to shift to a focus on designing for operational performance and carbon reduction in every design decision. Demonstrating a strong link between value and performance will advance this data-driven approach vital to accomplish these changes. Some strategies to design with this approach include using advanced energy modeling systems, strategically locating apertures in the envelope to exploit natural ventilation, deploying the appropriate energy systems, and smart building controls that adapt to the occupancy of the building while minimizing energy waste. The benefits of each of these strategies need to be communicated so that these practices will become accessible and lead to widespread change in the industry.

In general, designing a building by following the "long life/loose fit" strategy will help ensure that new construction will stand the test of time. Having digital data on building materials thoroughly documented and readily available will help future designers learn how to best reuse and retrofit the buildings that are currently in use. After all of the design, decisions have been made, and the operational consumption has been addressed "to minimize a building's whole life carbon emissions, the final step to net-zero is to offset what's left." [8] The first steps will make the most impact, but investing in renewable energy is a positive step toward reaching net zero for the building industry and the rest of the world.

Many measures to reduce operational carbon emissions have been implemented across the USA and the world. These include replacing incandescent light bulbs with compact fluorescent bulbs since they last longer, produce less heat, and are generally more efficient. Zero emissions of clean energy from sources such as nuclear, wind, solar, hydro, geothermal, etc., are replacing energy generated from fossil fuels such as coal, oil, and gas. Although electric heat pumps are more expensive than traditional gas hot water boilers, they are more efficient and produce less CO_2 emissions as they run on clean energy instead of fossil fuels. Local jurisdictions in large cities are trying to reduce emissions in buildings in several ways. In New York City (NYC), there will be a ban on gas-powered stoves and water boilers for a new building under seven stories starting in 2023 and for taller new buildings beginning in 2027. This targets 6% CO_2 emissions that are traced back to residential gas heating and cooking [12]. Another example of NYC leading in sustainable development nationally and internationally by restricting operational carbon is the Local Law 97 (LL97) [13], which places caps on carbon emissions on larger residential and commercial properties in the city.

On average, the embodied carbon in a typical building is 50% in its structure, 30% in its envelope, and 20% in the interior [11]. The efforts to limit embodied carbon in buildings appear to be less stringent than those of operational carbon. For example, Architecture 2030 [14] has set voluntary targets for embodied carbon reductions; these are an immediate reduction of 40%, then 50–65% by 2030, and zero emissions from materials by 2040. Similarly, the SE2050 [15] is the commitment by the Structural Engineering Institute of the American Society of Civil Engineers to meet the transitional embodied carbon reductions that reach zero by 2050. There are many similar efforts in the UK, Europe, and Australia, and cities and countries across the world are drafting commitments to reduce carbon emissions in building materials.

Concrete is the second most used material on the planet after water [16]. It is, however, the greatest contributor to embodied carbon emissions because of the energy-intensive process of producing cement and other sources of emissions in the concrete industry as a whole. In addition to the energy-intensive process of burning limestone to make cement, there is also the extraction and transportation of fine and coarse aggregates. There are several strategies being deployed to improve concrete's environmental performance. These include carbon capture, in which CO_2 is injected into a concrete mix to sequester the carbon while providing additional strength and durability to a concrete mixture. Other strategies involve using Supplementary Cementitious Materials (SCMs) such as fly ash, slag cement, and silica fumes as a partial replacement for cement. There is self-consolidating concrete (SCC) which eliminates the need for vibrating the concrete mix in the formwork while reducing emissions. Photocatalytic concrete uses titanium oxide in the mix to keep the concrete clean while also healing any potential cracks.

This is in addition to many promising technologies that are in the research and development stage to scale up production to industrial levels. Graphene concrete, for example, has much greater compressive and flexural strength while also significantly improving the impermeability of the cured concrete, which implies that less volume of concrete and less reinforcing would be needed. Also, the stronger the concrete, the smaller and lighter the member, which would result in smaller support members and a resulting smaller foundation.

The carbon emissions from the production of steel are primarily attributed to the use of blast oxygen furnaces that burn fossil fuels. The making of steel from iron ore requires the use of this type of energy-intensive furnace. Significant reductions of emissions in the steel-making process are being realized through the use of electric arc furnaces, which are used to melt down steel and recycle it into other applications of the material. Steel does not lose any of its properties when it is recycled from a soup can or a car into a wide flange beam.

The Profound Impact of the Envelope Is Anything But Skin-Deep

One of the most impactful ways to reduce both the embodied and operational carbon levels in a building is by considering one of the most dominant systems of a building, namely, the building envelope. The facade design, building orientation, and envelope's mechanics are all integral parts of a building's carbon emissions and should be addressed in the following crucial strategies:

The first step to ensure the maximum utilization of natural resources is to purposefully study the placement and orientation of the building on the site. For example, harnessing the sun and providing access to natural light will reduce energy consumption while also reducing the demand for electrical lighting. Shading devices on the facade are also an important factor because they keep unnecessary heat out of the building so that less energy is used to keep it cool. Also, when considering the overall building massing and orientation, the phrase "long life loose fit" is often suggested [8]. This implies that design for the future with long-lasting and durable materials along with a loose fit may be adapted in the future to reduce the likelihood of premature demolition and the associated carbon release. Bolting steel members to each other, as an example of "loose fit," instead of welding them would produce less carbon release during the disassembly process.

The envelope should be designed or refurbished to ensure airtightness with good U-values. As mentioned earlier, currently, one of the highest potentials for energy savings is by achieving greater efficiency in heating and cooling loads. The facade has a direct link to the effectiveness of these systems based on their performance qualities. The Locker Group promotes its product as being able to help reduce a building's utility bills and the impact on the environment through its creative façade solutions [17]. With proper insulation, air-conditioned and heated air will stay in the building, and the lower energy consumption is attributed to leakage loss. Thermal readings can help identify areas where excessive thermal bridging exists in the facade that should be considered in new buildings and when updating existing buildings.

The common approach to designing a building to have a completely sealed facade with access only to artificial heating and cooling often results in dissatisfied occupants and higher energy consumption. One way the façade condition can be altered is to utilize hybrid passive strategies, including natural and mixed-mode ventilation. Movable components that allow occupants to access natural ventilation are a beneficial consideration in terms of both energy use and occupant satisfaction.

In addition to these overarching strategies, there are many specific technologies that have been designed to help support a net-zero building through the design of its facade. The climate emergency has motivated net-zero energy practices that have produced a wide range of new technologies. Among these are advanced glass window technologies that revolutionize energy efficiency in building facades. These include technologies that not only manage the energy transfer between the interior and exterior spaces but can also serve as a generator of power. The skin of a building is the foremost location to harvest solar energy. Photovoltaic panels are also among the new technologies available for use directly in the façade of a building. Harnessing solar power in the façade may facilitate reaching net-zero by providing a building with supplemental energy to offset operational demands.

There are a few avenues for upgrading the performance buildings that do not require complete demolition. Updating the facade is one of these. Companies like Pic Perf are suggesting a new facade to reach lower carbon footprints and net-zero emission rates because of the large impact the facade has on rendering a building net zero. A product like Pic Perf can make a building envelope more efficient by blocking the sun, improving wind resistance, and more [6]. Another way to upgrade the facade is through an insulation-retrofit, which will save energy and cut carbon in the long run, although it will increase the embodied carbon in the short run. This increase will be insignificant compared to the carbon impact of demolishing and rebuilding the structure. Retrofit measures extend the life of any building and will thus contribute significantly to reaching net-zero objectives.

An example of this type of retrofit is Triton Square in London [17]. Originally built in the 1990s, the redesign shows "what is possible through imaginative reuse, demounting, refurbishing and re-erecting the existing facade." The Arup Group took a marginal gains approach to this redesign and called it a revolution. "Team Triton chipped away at every aspect to save carbon, cut waste, and deliver the best working environment possible. Through a marginal gains approach, the team has refined and optimized dozens of systems, components, and strategies to deliver a highly sustainable building." This building was a huge success with 43% cost saving compared to typical commercial buildings, 40,000 tons of carbon saved, and 30% faster to completion versus a typical new build. Arup is leading the way in how we can approach new building projects to achieve net zero and particularly for retrofitting buildings. Updating existing facades with this type of strategy will be revolutionary in the building industry and the global efforts to reach net zero.

In this chapter, a comparison of different case studies is conducted with a focus on the facades and building envelope. In each case study, the façade was used anywhere between 6% and 21% of the whole life cycle carbon assessment. Comparative studies such as these are imperative in moving toward a net-zero future. Companies across the globe are calling for more data-driven, transparent approaches to creating clear targets that will help provide the framework needed to realize the net-zero carbon emissions goal by 2050.

Case Study: 888 Boylston Street

Architect: FXCollaborative

Introduction

Completed in 2016, FXCollaborative's 888 Boylston Street serves as a unique example of a cold climate high rise with a LEED Platinum certification. 888 Boylston Street is located in the city of Boston, MA, and was designed for the well-known sustainably interested Boston Properties, Inc. (BXP). While much of the attention given to this building is often centered around the energy generation of the design, this writing aims to examine a closer look at the envelope materials utilized and their contribution to the sustainable characteristics of the whole. While 888 Boylston Street does not set any records for its size or height, the building sets a precedent for the possibilities of sustainable buildings in the USA and serves as a model where educational tours on sustainable design are held. This writing primarily investigates the building's use of glazing as an envelope material and its relationship to the cold climate of the northeastern USA (Fig. 1).

Identified as a cold continental climate, Boston's weather includes warm summers and very cold and snowy winters. Boston is known to have a relatively unstable climate with alternating days of stormy and clear weather due to different air masses colliding from different directions [18]. Similar to Chicago, Boston's location along the water can result in relatively cool temperatures in summer due to cold current flows above the sea. Boston averages 47 inches of rain per year, 48 inches of snow per year, and slightly below the national average of sunny days per year [19]. The average temperature in Boston ranges from a high of 82 °F in summer to a low of 19 °F in winter [19]. As heating consumes more energy than any other building system in Boston's climate, passive techniques can and have been utilized to better provide thermal comfort indoors [20]. Air barriers and continuous thermal insulation are two of the techniques that can help mitigate thermal bridging. In some areas in Boston, water levels are also a concern. Many historical buildings contain elevated entryways to help combat possible flooding [21]. While many of the climatic characteristics of Boston can be combated with mechanical systems, these methods are not sustainable when compared to smart passive systems and other efficient sustainable strategies (Fig. 2).

Sustainable Features

888 Boylston Street is located among several other mid- and high-rise buildings near Boston's downtown and the Charles River Basin. The building stands as a mixed-use office and retail building at 17 stories high and includes an area of over

Fig. 1 888 Boylston Street. FXCollaborative Architects LLP

425,000 square feet [22]. In addition to the two floors of below-ground parking, the design includes 14 floors of office space and 3 floors of retail with additional varying occupancies scattered at alternate levels. Sustainable features of the building from the architect as illustrated in Fig. 3 are most notably recognized through the green roof, green terrace, energy-efficient lighting, chilled beam system, energy-generating wind turbines, sky gardens, rainwater-harvesting system, high-performance envelope, PV panels, bike storage, and elevated equipment [22]. Compared to traditional offices of similar size, the design consumes 47% less energy and 37% less water [23]. This results in an annual saving of $650,000 [24]. 888 Boylston Street serves as one of the highest-performing buildings in the

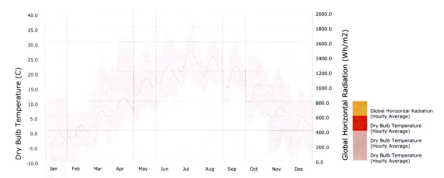

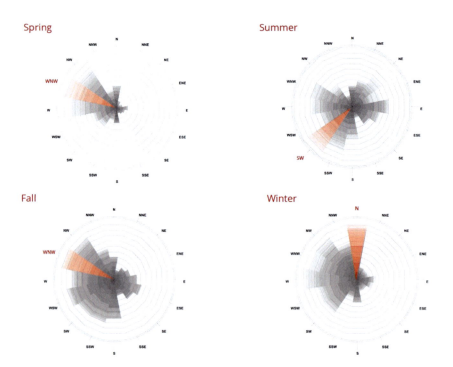

Fig. 2 Climatic data, dry-bulb temperature, wind roses, and psychrometric chart

Environmental Dimensions of Climate Change: Endurance and Change in Material...

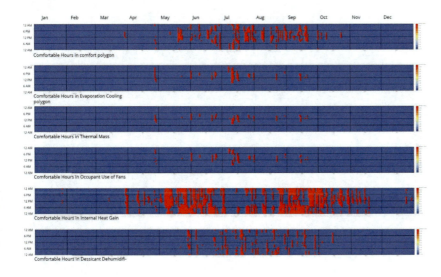

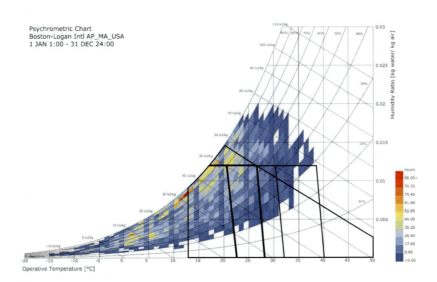

Fig. 2 (continued)

northeast and has the lowest EUI of any building in Boston at 40 kBTU per square foot [23].

The sustainable features in place at 888 Boylston Street reduce the energy use of the building by nearly 34.6% when compared to a traditional building of similar location and size [25]. One of the strategies in use that contributes to this amount is the rainwater-harvesting system. The rainwater-harvesting system, located on the

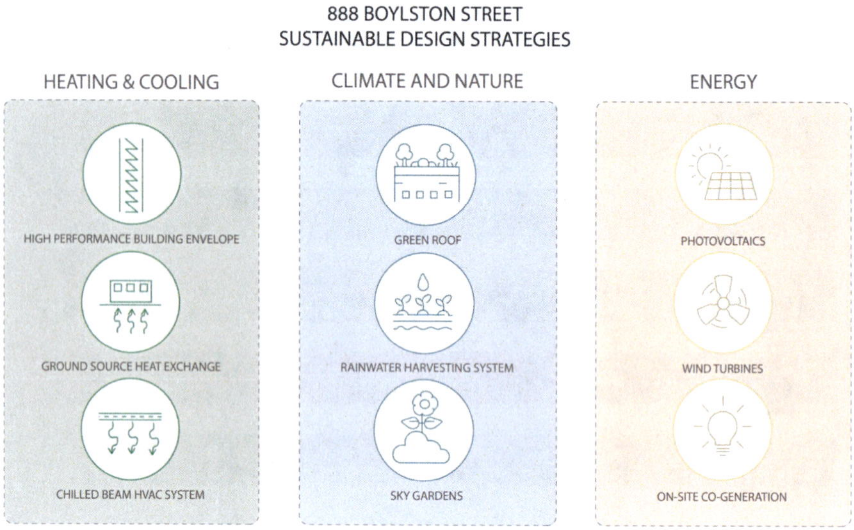

**888 BOYLSTON STREET
SUSTAINABLE DESIGN STRATEGIES**

HEATING & COOLING

HIGH PERFORMANCE BUILDING ENVELOPE

GROUND SOURCE HEAT EXCHANGE

CHILLED BEAM HVAC SYSTEM

CLIMATE AND NATURE

GREEN ROOF

RAINWATER HARVESTING SYSTEM

SKY GARDENS

ENERGY

PHOTOVOLTAICS

WIND TURBINES

ON-SITE CO-GENERATION

Fig. 3 Sustainable strategies. (Credit: https://www.usgbc.org/articles/boston-properties-pushes-boundaries-sustainable-design)

roof, utilizes collected water for cooling and irrigation. In the system in place, water is first stored in an underground tank where pollutants are removed. Nearly 20% of the total water used in the building is a result of the rainwater collection system [26]. In addition to water used in a chilled beam system for thermal comfort, the building efficiently utilizes a dedicated outdoor air system (DOAS). The system for air circulation uses only fresh outdoor air. Comparatively, a traditional HVAC system uses up to 75% stale recycled air. The DOAS system provides around 30% more fresh air and 50% more air changes per hour than a traditional HVAC system [24].

In addition to water and air, other natural elements such as light and plants are included in the sustainable design strategies in place. Biophilic elements, such as the rooftop garden and living walls, bring natural plant life to the building. The two common area walls display 13-feet-high green walls that provide connections with nature while in central Boston [25]. On the northern facade, 13′–6″ floor-to-ceiling heights provide vertical views to the outdoors. This viewing height, often called the "visual zone," is estimated to be 145% larger than most office buildings [26]. The structural design was even included in the daylighting to reduce the amount of columns that might intervene with natural light exposure.

Envelope

It is estimated that nearly 70% of the facade of 888 Boylston Street is comprised of glass. (Fig. 4) [25]. While this might not initially appear to be the most sustainable choice for daylighting and interior comfort, often requiring additional tools to control thermal comfort, the envelope of the design was analyzed extensively to

Fig. 4 Green roof. (Credit: FXCollaborative Architects LLP)

determine the most efficient system for the building's use. The chosen double-pane insulated glass reduces artificial lighting runtime in the building by nearly 60% to the baseline [27]. The glass itself is not entirely vertical. In this manner, the envelope system on the northern orientation curves upward to allow a larger quantity of light to enter the building's interior spaces (Fig. 5).

Operational Versus Embodied Energy

888 Boylston Street reaches a LEED Platinum status through both operational strategies in place as well as material choices that lower the total embodied energy of the structure. Operational energy is most notably seen through the visually present wind turbines and solar panels on the roof. However, additional mechanical and electrical systems in place help lower the operational energy requirements. Both the chilled beam system and DOAS help condition interior spaces without wasteful traditional HVAC techniques. Similarly, the building includes high-efficiency chillers that get rid of ozone-depleting chemicals and refrigerants [25].

The roof itself and subsequent wind turbines are visible from the exterior and not only contribute to the energy efficiency of the building but also provide a dynamic

Fig. 5 Envelope. (Credit: AntyDiluvian)

crest of the building. The design includes 14 vertical axis wind turbines and a 134-kW photovoltaic (PV) system. As shown in Fig. 6, together, the roof system generates enough energy to run an estimated 15 homes in the state [28]. The roof system additionally includes garden areas and beehives. The beehives provide a safe home for the bees which in turn help pollinate the native plants on the rooftop. The plants themselves help to reduce the heat-island effect, absorb carbon dioxide, and reroute water to the rain collection tanks (Figs. 7 and 8).

The diagrams below display data obtained by using the ATHENA® Impact Estimator for Buildings. Located in Ontario, Canada, the ATHENA® Impact Estimator for Buildings was developed by the Sustainable Materials Institute. As part of the institute's mission, it leverages the life-cycle assessment in North America to promote sustainability in the built environment [23]. According to the developers, "robust life cycle inventory databases provide exact scientific cradle-to-grave information about building materials and products, transport, and construction and demolition activities" [28]. The Athena Institute connects designers to the power of life-cycle analysis without requiring them to become LCA experts themselves [28].

Any part of a building has the potential to be modeled using the Impact Estimator when the bill of materials has been provided. Using simple inputs, the Impact Estimator can create a bill of materials for users who do not have one. Examples include [24] foundations, footings, slabs, all below- and above-grade structure and envelope, windows and doors, and building interiors. Based on a 60-year life cycle, the study examines the overall building's life cycle. According to ISO 14040, we

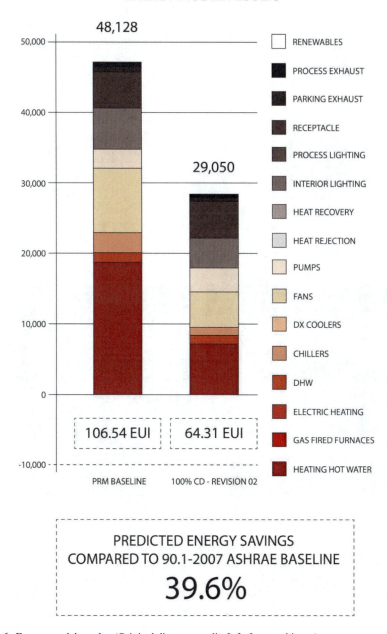

Fig. 6 Energy model results. (Original diagram credit: Info from architects)

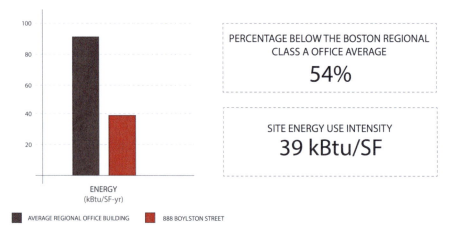

Fig. 7 Performance comparison to average regional office building.

Fig. 8 Facade, photovoltaics, and wind turbine integration. Credit: FXCollaborative Architects LLP

can compare up to five design scenarios according to the US Environmental Protection Agency's environmental impact categories [23]. In this study, the following environmental metrics were used: Global Warming Potential, Smog Potential, Acidification Potential, Non-renewable Energy, Eutrophication Potential, and Ozone Depletion Potential. As inputs to the Impact Estimator, a series of factors related to the building are considered in order to calculate the life-cycle impact of each factor on the above categories. There are five assemblies that consist of information on the project: foundations, floors, columns and beams, roofs, and walls (Figs. 9 and 10).

Environmental Dimensions of Climate Change: Endurance and Change in Material... 311

Fig. 9 Envelope materials and integrated sustainable strategies. Credit: FXCollaborative Architects LLP

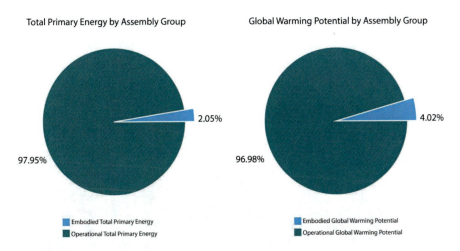

Fig. 10 Left: Operational versus embodied energy; right: global warming potential

A comparison is made between operating and embodied in both primary energy and global warming potentials. Operating accounts for a greater share in both charts.

Figure 11 displays the comparison between different constructional categories that are used in this building. The report compares the amount of CO_2 emissions that each category can have on the environment.

Figure 12 displays the comparison between different constructional categories that are used in the case study. The report compares the amount of O_3 emissions that each category contributes.

Conclusion

888 Boylston Street sets a precedent for the possibilities of sustainable buildings in US cold climate regions and serves as a model where educational tours on sustainable design are held. The design includes natural elements such as light and air in addition to biophilic elements such as the rooftop garden and green wall. As a mid-high-rise building that is comprised of 70% glass, 888 Boylston Street showcases just how environmentally friendly and sustainable large-glazed buildings can be. While the roof wind turbines might be the element of the building that catches the attention of most passersby, 888 Boylston Street has much more to offer in regard to its envelope, sustainable strategies, and operational and embodied energy.

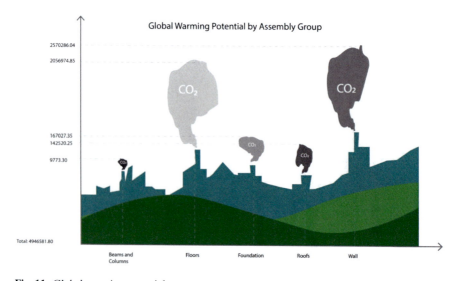

Fig. 11 Global warming potential

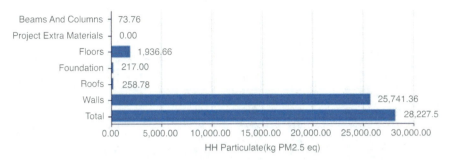
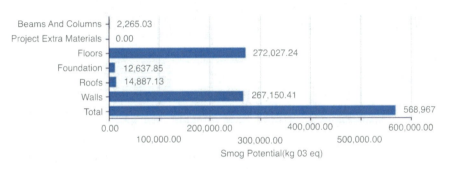

Fig. 12 Human health particulate and smog potential

Case Study: CLT Passivhaus

Architect: Generate Architects

Introduction

Generate, in collaboration with Placetailor, delivered the world's first fully integrated Cross-Laminated-Timber (CLT) Passivhaus demonstration project. The project is set to be located in the Roxbury neighborhood of Boston, MA. This project is a response to both global warming and urban density issues. The collaboration between Placetailor and Generate will make it possible to propose sustainable construction for mid-size and, in the future, high-rise buildings. Generally, this is a typology of housing delivery method that focuses on climate and community [29]. In terms of carbon footprint, the building aims to reach Passivhaus standards and is expected to be net-zero carbon in operation. Boston has a cold climate, which makes well-insulated exterior walls quite practical in this setting (Fig. 13).

Fig. 13 CLT Passivhaus. (Credit: Generate | Placetailor)

Roxbury is one of the 23 official neighborhoods in Boston, MA, located south of the central business district. Identified as a cold continental climate, Boston's weather includes warm summers and very cold and snowy winters. Boston is known to have a relatively unstable climate with alternating days of stormy and clear weather due to different air masses colliding from different directions [18]. Similar to Chicago, Boston's location along the water can result in relatively cool temperatures in summer due to cold current flows above the sea. Boston averages 47 inches of rain per year, 48 inches of snow per year, and slightly below the national average of sunny days per year [19]. The average temperature in Boston ranges from a high of 82 °F in the summer to a low of 19 °F in winter [19]. As heating consumes more energy than any other building systems in the Boston climate, passive techniques can and have been utilized to better provide thermal comfort indoors [20]. Air barriers and continuous thermal insulation are two techniques that can help with thermal bridging. In some areas in Boston, the water levels are also a concern. Many historical buildings contain elevated entryways to help combat possible flooding [21]. While many of the climatic characteristics of Boston can be combated with mechanical systems, these methods are not sustainable when compared to smart passive systems and other efficient sustainable strategies (Fig. 14).

Sustainable Features

The CLT Passivhaus includes a variety of spaces and techniques related to sustainable features and methods of construction. In addition to 14 residential units, the building includes a co-working space accessible to the local community on the ground floor (Fig. 13). A mix of housing types will be available in the 14-unit Model C building, ranging from studio apartments to 3-bedroom apartments. On the ground floor, there will be affordable commercial space for local businesses. Because the building is located in the city of Boston and close to public transportation, Placetailor was not required to provide parking [30]. The building will highlight the unique benefits of a prefabricated kit-of-parts for developing workforce housing that is both healthy and carbon positive (Figs. 15 and 16).

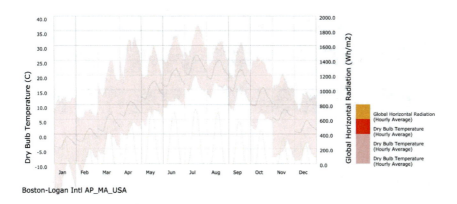

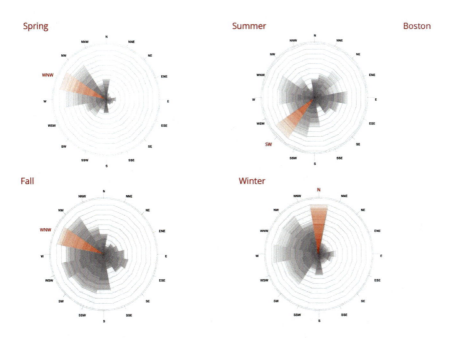

Fig. 14 Climatic data, dry-bulb temperature, wind roses, and psychrometric chart

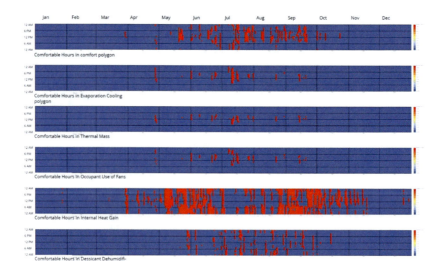

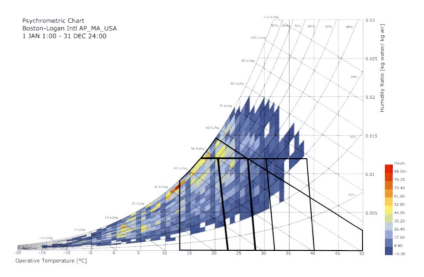

Fig. 14 (continued)

Despite the concrete foundation, all other elements of the building are made of CLT panels – an engineered wood made of laminated timber sections [31]. As illustrated in Fig. 17, CLT panels of varying thicknesses make up the floor, interior partitions, exterior walls, and roof assemblies of the Model C [30]. A high-density cellulose thermal panel is to be installed on the interior of the exterior assembly, and

Environmental Dimensions of Climate Change: Endurance and Change in Material... 317

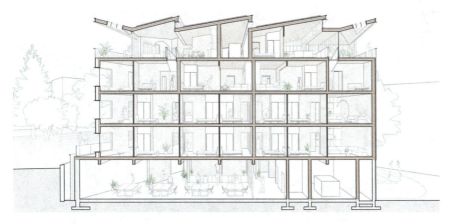

Fig. 15 The building section displays the array of interior spaces including the 14 residential units and co-working space. Credit: Generate | Placetailor

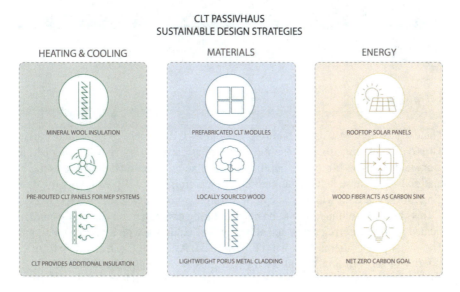

Fig. 16 Sustainable strategies

a wood-fiber board must be used on the exterior. Clean and modern, but warm and spacious, CLT walls give interior spaces a sense of comfortable living [32]. The building is also designed to reduce the amount of radiation it receives and to insulate against heat and cold. Mineral wool and CLT will be used as insulation on the walls [31].

Fig. 17 CLT elements such as beams, columns, and ceilings can be seen from this render view. (Credit: Generate | Placetailor)

Envelope

The exterior elements of the building are mainly CLT panels, glazing, and frames. The CLT's cellular structure and envelope act as carbon sinks by capturing carbon dioxide during the life of the building and replacing traditional concrete and steel materials that cause significant carbon dioxide emissions [29]. Additionally, cladding for the building will be made of lightweight porous metal. As shown in Fig. 18, in this case, scaffolding will not be needed during construction because it is prefabricated [31]. The corrugated metal siding will be equipped with a rainscreen air gap. The metal panels themselves will be perforated. As a result, select views from the exterior of the building will allow for glimpses of the exposed wood inside of the building. Additionally constructed of CLT, the sawtooth roofline is oriented toward south for maximum solar PV exposure [30]. Rooftop solar panels are mounted easily on this system due to the CLT roof canopy.

Operational Versus Embodied Energy

A Model C demonstration project was designed to generate net-zero carbon emissions by measuring both the embodied energy of the building and its operating energy. Any excess energy was compensated for by carbon offsets [32]. When

Fig. 18 Lightweight porous metal cladding will be used as the envelope of the building. (Credit: Generate | Placetailor)

compared to traditional buildings constructed with conventional steel or concrete, the Model C project reduced total embodied carbon emissions by less than 50% [30]. Due to the CLT construction, fireproofing materials can be reduced, along with the use of dyed plaster or drywall. Additionally, exposed CLT walls and ceilings can help reduce the use of harmful materials [31]. The CLT panels are in fact a carbon sink due to the high amounts of wood fiber, which allows them to address both operational and embodied energy with one solution [31]. As shown in Fig. 19, the CLT panels which will be used in construction are set to be locally sourced from Montreal, Canada, and trimmed locally by panel manufacturer Bensonwood in Keene, New Hampshire [30].

The operational energy requirements of the building rely on a heat pump for cooling and cost-effective electric-baseboard system for heating. The mechanical ventilation for air circulation is set to be supplied by a semi-centralized system where one system supplies four housing units. The building's source of hot water and part of the electricity will be supplied from a gas-fueled combined heat and power (CHP) plant. Current calculations show that the use of the CHP will lower greenhouse gas emissions when compared to using the heat pump for traditionally heated water [30]. The building also includes off-site prefabricated modular

Fig. 19 Locally sourced and trimmed CLT will be exposed in indoor spaces. (Credit: Generate | Placetailor)

bathrooms that can be hoisted and installed easily, enhancing the project's timeline and reducing construction waste [29]. This material is more energy-efficient. A Passivhaus already has a low energy demand, so in order to reduce the MEP requirements, the CLT panels can be pre-routed to incorporate the system (Fig. 20).

The diagrams below (Figs. 21, 22, and 23) display data obtained by using the ATHENA® Impact Estimator for Buildings. Located in Ontario, Canada, the ATHENA® Impact Estimator for Buildings was developed by the Sustainable Materials Institute. As part of the institute's mission, it leverages the life-cycle assessment in North America to promote sustainability in the built environment [18]. According to the developers, "robust life cycle inventory databases provide exact scientific cradle-to-grave information about building materials and products, transport, and construction and demolition activities" [33]. The Athena Institute connects designers to the power of life-cycle analysis without requiring them to become LCA experts themselves [33].

Any part of a building has the potential to be modeled using the Impact Estimator when the bill of materials has been provided. Using simple inputs, the Impact Estimator can create a bill of materials for users who do not have one. Examples include [19] foundations, footings, slabs, all below- and above-grade structure and envelope, windows and doors, and building interiors. Based on a 60-year life cycle, the study examines the overall building's life cycle. According to ISO 14040, we can compare up to five design scenarios according to the US Environmental

Environmental Dimensions of Climate Change: Endurance and Change in Material... 321

Fig. 20 Envelope materials. (Credit: Generate | Placetailor)

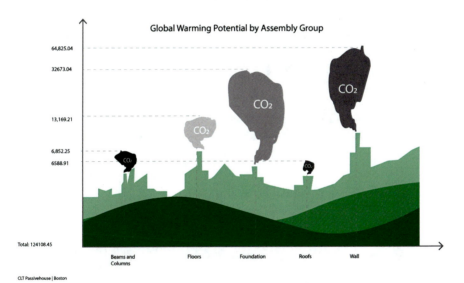

Fig. 21 Global warming potential

Protection Agency's environmental impact categories [18]. In this study, the following environmental metrics were used: Global Warming Potential, Smog Potential, Acidification Potential, Non-renewable Energy, Eutrophication Potential, and Ozone Depletion Potential. As inputs to the Impact Estimator, a series of factors related to the building are considered in order to calculate the life-cycle impact of

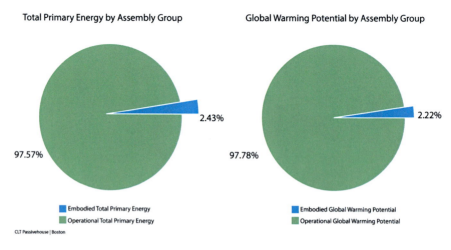

Fig. 22 Left: operational versus embodied energy; right: global warming potential

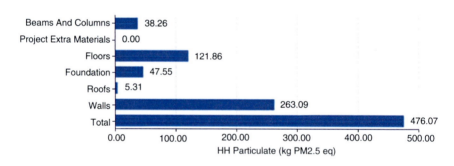

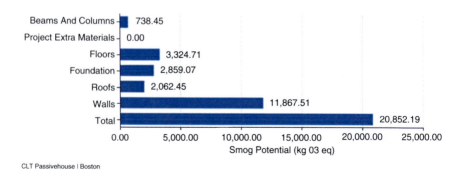

Fig. 23 Human health particulate and smog potential

each factor on the above categories. There are five assemblies that consist of information on the project: foundations, floors, columns and beams, roofs, and walls.

Figure 21 displays the comparison between different constructional categories that are used in the case study. The report compares the amount of CO_2 emissions that each section can have on the environment.

A comparison is made between operating and embodied in both primary energy and global warming potentials. Operating accounts for a greater share in both charts.

Figure 23 displays the comparison between different constructional categories that are used in the case study. The report compares the amount of O_3 emissions that each section can have on the environment.

Conclusion

Not only does the CLT Passivhaus serve as a precedent for net-zero projects in the USA, but it also serves as an example of fully integrated CLT construction. The project is a response to both global warming and urban density issues. Reacting to Boston's cold climate, both passive and active techniques in the building aid in reaching this net-zero goal. When compared to traditional buildings constructed with conventional steel or concrete, the Model C project reduced total embodied carbon emissions by less than 50%. With the addition of locally sourced CLT, it is clear that the CLT Passivhaus prioritizes sustainability and is deserving the title the world's first fully integrated CLT Passivhaus demonstration project.

Case Study: Golisano Institute for Sustainability

Architect: FXFOWLE, New York City, NY.
  Architect of Record: SWBR, Rochester, NY

Introduction

The Golisano Institute for Sustainability (GIS) at Rochester Institute of Technology is a LEED Platinum building that not only serves as a laboratory for scientific research on sustainable technology but itself exemplifies energy efficiency and high-performance systems. Designed by FXFowle (NYC) and SWBR in Rochester, NY, the Golisano Institute for Sustainability is located in Rochester, NY. Designed for a very cold climate region, the building includes a high-performance facade system that helps reduce the overall carbon footprint. Since its completion in 2013, the building has won multiple awards, including the 2014 National Award of Excellence from the Design Build Institute of America (DBIA), and was titled the best project in the Green Project category by Engineering News Record New York [34]. It has earned LEED Platinum certification, which is the highest standard in the certification system. This writing aims to investigate the material involvement with specific attention to the envelope and its relationship to the overall sustainable status of the structure (Fig. 24).

Identified as a cold continental climate, Rochester is characterized by warm summers and snowy freezing winters. Similar to many of the cities in the included case studies, Rochester is located near a body of water. The southern shoreline of Lake Ontario reaches along the northern portion of Rochester. As a result, much of the snow is a direct result of the "lake effect," in which cold air crosses warmer water, resulting in clouds, precipitation, and snow [35]. Rochester averages 33 inches of rain per year, 77 inches of snow per year, and below the national average of sunny days per year [36]. The average temperature in Rochester ranges from a high of 82 °F in summer to a low of 17 °F in winter [36]. With the help from the US Department of Energy, the Passive Solar Industries council has set guidelines for climate-reactive passive strategies employed in the Rochester region [37]. While this information is specifically targeted to residential homes, much of the climate combatant information can also relate to larger structures such as the Golisano Institute for Sustainability. Some of these guidelines include increasing thermal resistance and increasing south-facing glazing up to 7% of the building's total floor

Fig. 24 Golisano Institute for Sustainability. Credit: David Lamb

area. Natural cooling and fully insulated basement walls are additionally suggested. As part of New York State's Genesee-Finger Lakes Region, Rochester has additionally been included in Stockholm Environment Institute's plan for combating climate change, which includes support for structures that are energy efficient and reduce waste [38] (Fig. 25).

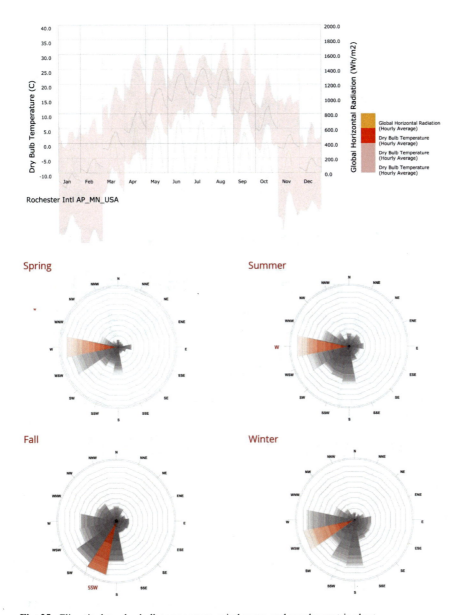

Fig. 25 Climatic data, dry-bulb temperature, wind roses, and psychrometric chart

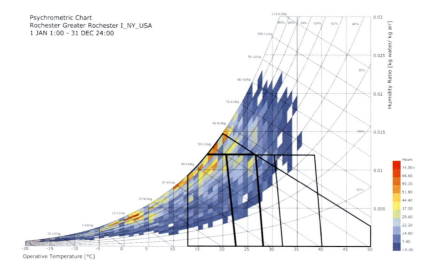

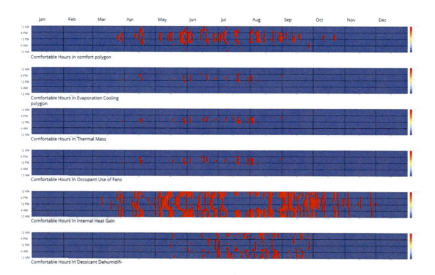

Fig. 25 (continued)

Sustainable Features

The building systems and envelope are both involved in the sustainable features utilized in the design of the Golisano Institute for Sustainability. In a similar manner, both active and passive strategies are central sustainable design features that contribute to the building's LEED Platinum status. Together all sustainable features reduce

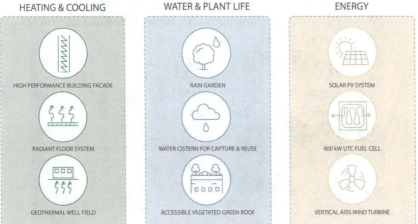

Fig. 26 Sustainable strategies, original diagram. (Credit: http://www.swbr.com/wp-content/uploads/2014/03/rit-gis-case-study_lo.pdf)

the annual carbon footprint of the building by 61%, meeting the AIA 2030 goal of 60% [39]. Other sustainable statistics show that the building has a total annual water saving of 75%, utilizes 88% forest stewardship-certified wood, and recycles nearly 80% of construction waste [39]. Some of the notable features included in the design that contribute to these statistics are vertical-axis wind turbines on-site, radiant flooring, geothermal well, solar shading controls, high-performance facade, green roof and walls, and roof photovoltaic (PV) system (Fig. 26).

Envelope

The envelope of the Golisano Institute for Sustainability alone contributes up to 15% of the total energy savings of the design [39]. The primary material utilized in the envelope of the building is glazing. As shown in Fig. 27, the glazing system consists of factory-assembled large units of glass situated in a thermally broken curtain wall system. A 40% vision glass or less was utilized to provide a triple-pane glazing performance while being only 1-inch thick [40]. In contrast, the spandrel glazing itself includes thermal insulation that is 4.5-inches thick. In areas of the envelope system where occupants are meant to sit near the glass, an innovative system of glazing termed the "perfect window" is utilized to improve thermal comfort. This system involves a double-pane window with a metal coating that is heated by electrical currents. In this system, when temperatures drop below 42 °F, the system

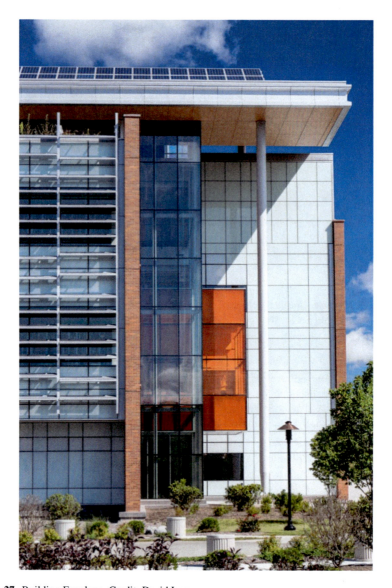

Fig. 27 Building Envelope. Credit: David Lamz

is heated providing optimal thermal comfort while also minimizing heat loss [40]. This electric-powered heated glass is essential for mitigating heat loss in the cold climate of Rochester, NY.

The envelope facade system of the Golisano Institute for Sustainability is different for different orientations. While both south and north facades are primarily glass, the east and west facades are constructed of masonry materials. In addition to

Fig. 28 Sun shading, PV systems, and green roof. Credit: Craig Shaw, Stratus Imaging

glazing, the south facade features solar shades that reduce solar heat gain by 70% [39]. The shading elements help reduce the building's cooling demand and allow a more efficient chilled beam technology. The building roof is equipped with 144 photovoltaic panels that generate an average of 45,000 kWh per year [41]. In addition to these energy-generating panels, the roof also houses 3300 square feet of vegetation and a butterfly habitat, as illustrated in Fig. 28 [39].

Operational Versus Embodied Energy

The Golisano Institute for Sustainability earned its LEED Platinum status by deploying both multiple operational and embodied carbon reduction strategies. The most visible techniques used to contribute to the operational energy of the building are the roof-mounted solar PV panels and the vertical-axis wind turbines located on-site. Currently, 170 solar panels are mounted on the roof of the building. This number of panels generates enough energy to power seven homes in New York. In addition to the wind and solar energy-generation strategies employed, additional strategies utilized include geothermal and a 400-kW fuel cell to supplement energy demands [42]. Additionally, an operational microgrid room incorporates data sensors, feedback loops, and control systems to monitor building performance and efficiency of operation [39]. With this system, a stand-alone power supply source is provided that consists of a lithium ion battery storage bank with a power of 50 kW\50 kWh.

With regard to the building's systems, many different interlocking and separate components and strategies work together to provide maximum comfort for the occupants. Air-handling units and terminal units are situated on the floors above grade while heating, cooling, and water conservation take place below grade. Included at the roof of the building is an exhaust terminal for the expulsion of air. While lighting and plug load controls exist throughout the entire building, the microgrid base exists on the first floor. This microgrid consists of the fuel cell, PV generation, batteries, lighting, and other miscellaneous loads. All of these components together contribute to an annual energy use of 112 kBtu/sf and a predicted energy saving beyond ASHRAE 90.1 of 57% [43].

The diagrams below display data generated using the ATHENA® Impact Estimator for Buildings. Located in Ontario, Canada, the ATHENA® Impact Estimator for Buildings was developed by the Sustainable Materials Institute. As part of the institute's mission, it leverages the life-cycle assessment in North America to promote sustainability in the built environment [39]. According to the developers, "robust life cycle inventory databases provide exact scientific cradle-to-grave information about building materials and products, transport, and construction and demolition activities" [36]. The Athena Institute connects designers to the power of life-cycle analysis without requiring them to become LCA experts themselves [36].

Any part of a building has the potential to be modeled using the Impact Estimator when the bill of materials has been provided. Using simple inputs, the Impact Estimator can create a bill of materials for users who do not have one. Examples include [35] foundations, footings, slabs, all below- and above-grade structure and envelope, windows and doors, and building interiors. Based on a 60-year life cycle, the study examines the overall building's life cycle. According to ISO 14040, we can compare up to five design scenarios according to the US Environmental Protection Agency's environmental impact categories [39]. In this study, the following environmental metrics were used: Global Warming Potential, Smog Potential, Acidification Potential, Non-renewable Energy, Eutrophication Potential, and Ozone Depletion Potential. As inputs to the Impact Estimator, a series of factors related to the building are considered in order to calculate the life-cycle impact of each factor on the above categories. There are five assemblies that consist of information on the project: foundations, floors, columns and beams, roofs, and walls (Fig. 29).

Figure 30 displays the comparison between different constructional categories that are used in the case study. The report compares the amount of CO_2 emissions that each section can have on the environment (Fig. 31).

A comparison is made between operating and embodied in both primary energy and global warming potentials. Operating accounts for a greater share in both charts.

Figure 32 displays the comparison between different constructional categories that are used in the case study. The report compares the amount of O_3 emissions that each section can have on the environment.

Environmental Dimensions of Climate Change: Endurance and Change in Material... 331

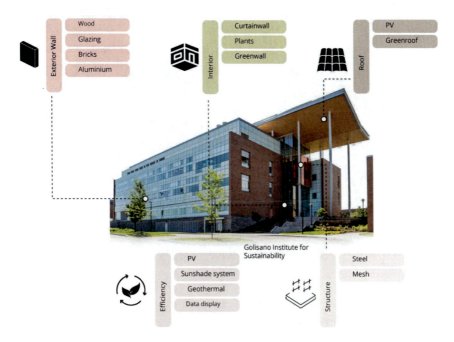

Fig. 29 Envelope materials and integrated sustainable strategies

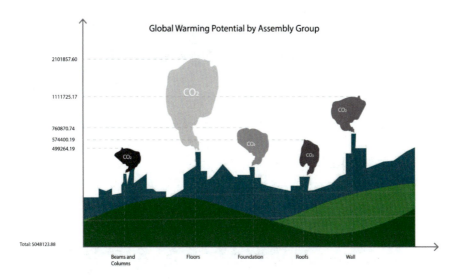

Fig. 30 Global warming potential

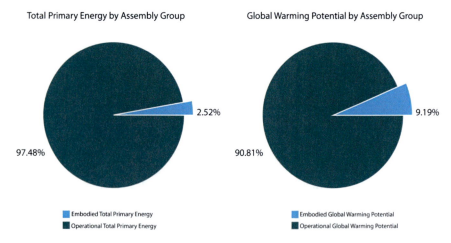

Fig. 31 Left: operational versus embodied energy; right: global warming potential

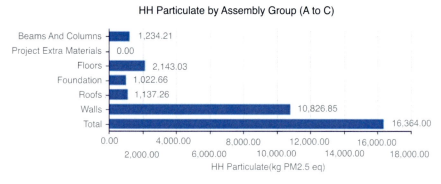

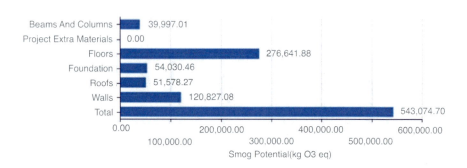

Fig. 32 Human health particulate and smog potential

Conclusion

Since its completion in 2013, the Golisano Institute for Sustainability (GIS) at the Rochester Institute of Technology (RIT) has won multiple awards, including the 2014 National Award of Excellence from the Design Build Institute of America (DBIA), and was titled the best project in the Green Project category by Engineering News Record New York. Along with a LEED Platinum certification, the design of this building exemplifies energy efficiency and regard for the environment. This feat is obtained through both passive and active strategies that relate to the envelope, operational, and embodied energy use in the building. With all strategies in place, the Golisano Institute for Sustainability serves as an ideal precedent of the region for the sustainable possibilities of lab and educational buildings.

Case Study: Orlando McDonald's Flagship

Architect: Ross Barney Architects

Introduction

The McDonald's Disney Flagship, located in Orlando, Florida, not only serves as a precedent for the sustainable possibilities of quick-serve restaurants but responds to the humid subtropical climate of its location. Designed by Ross Barney Architects in 2021, the building pushes the boundaries of ordinary McDonald structures through the use of photovoltaic glass panels, natural ventilation techniques, and on-site energy generation. Aside from energy generation of the building amounting to 100% of the building's needs, the materials used in construction and design strategies in place are paramount to the building's net-zero status. Specific attention is given to the V-shaped roof that responds to Florida's climate as well as the wood louvers and plant-covered walls. This writing aims to investigate this material involvement with specific attention to the envelope and its relationship to the overall net-zero status of the structure (Fig. 33).

Identified as a humid subtropical climate, Orlando winters are mild and short, while summers are hot and sunny. Orlando averages 52 inches of rain per year, 0 inches of snow per year, and above the national average of sunny days per year [44]. The average temperature in Orlando ranges from a high of 92 °F in summer to a low of 49 °F in winter [44]. In June to September, there are also frequent thunderstorms and muggy weather. Hurricanes are likely to hit Florida during the summer and early fall [45]. Prior to the invention of air conditioners and the widespread use of this technology, open and breezy dwellings were the ideal choice of homes in this climate. The air conditioning in modern homes makes them more comfortable despite being tightly insulated [46]. Additionally, there are two envelope concepts

Fig. 33 Orlando McDonald's Flagship. (Credit: Kate Joyce Studios, McDonald's Flagship—Orlando, Ross Barney Architects, 2020)

that are known to be efficient for Florida's climate: (1) a compact shape to reduce the exterior wall area and (2) continuous insulation to ensure that the building envelope has no interruptions. It is imperative that buildings in this climate zone are sealed tightly to avoid cool air indoors releasing outdoors and vice versa (Fig. 34).

Sustainable Features

The McDonald's Disney Flagship is part of a corporate McDonald strategy that aims to spread and implement net-zero energy-certified restaurants internationally. McDonald's global sustainability efforts will be informed by data, including progress toward the company's science-based target to reduce greenhouse gas emissions by 36% by 2030, compared with 2015 [47]. Disney's newly remodeled building is situated on the west side of the Disney property. In response to Florida's climate and its site location, the restaurant is covered in solar panels, creating a sustainable and healthy environment. The design features natural ventilation for about 65% of the time [48]. By doing so, the restaurant takes advantage of the humid subtropical climate.

As presented in Fig. 35, one of the key sustainable elements in the project's design is the use of green walls (Fig. 33). With the help of Florida-based architecture and engineering firm CPH, Ross Barney Architects designed the plant selection utilized in the walls [49]. The plants were chosen according to the subtropical climate of Orlando. They are equipped with a plant care system that works to keep their facades green throughout the year [49]. The plant care system sends automatic updates to an app that keeps those involved in the wall's design and maintenance

Environmental Dimensions of Climate Change: Endurance and Change in Material... 335

updated on the green wall's condition [50]. The web-based system delivers the right amount of nutrients and water to the plants. Because the quantities of nutrients and water are monitored remotely, they can be adjusted easily [49]. In addition to these green walls, the company's logo is also incorporated into a lush garden wall that absorbs additional CO_2 [51]. In total, 1766 square feet of the living green walls are included in the project and help increase local biodiversity (Fig. 36).

Other sustainable strategies in place include paving materials that help reduce the heat island effect as well as pervious surfaces to redirect rainwater. Additionally

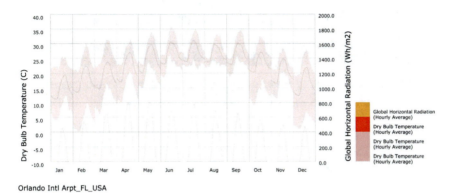

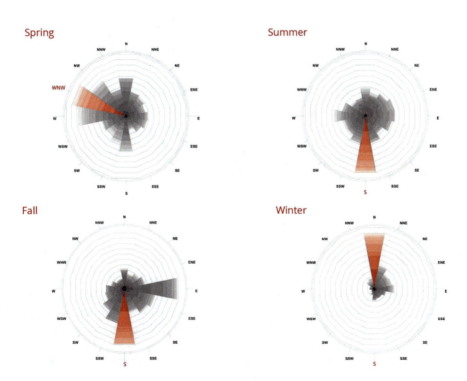

Fig. 34 Climatic data, dry-bulb temperature, wind roses, and psychrometric chart

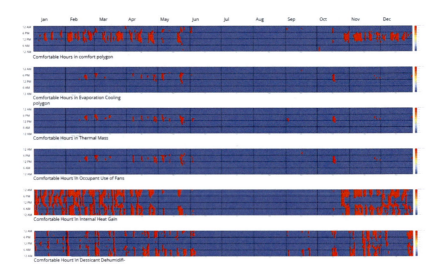

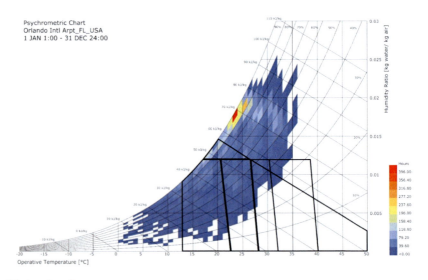

Fig. 34 (continued)

located on the exterior of the building and resembling an outdoor play structure, the building site features stationary bikes that allow users to pedal in place, ultimately harnessing that kinetic energy, and charge their devices [52]. This kinetic energy can also be used to light up a display of the McDonald's logo on one of the green walls. Sustainable and renewable energy generation of this sort represents the design commitment to a net-zero status as well as ingenuity and liveliness for all those who will visit the fast-service restaurant.

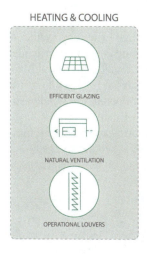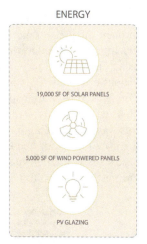

Fig. 35 Sustainable strategies

Fig. 36 A green wall equipped with a plant care system can be seen on the exterior of the building from the parking lot. (Credit: Kate Joyce Studios, McDonald's Flagship—Orlando, Ross Barney Architects, 2020)

Envelope

Although the building takes advantage of Orlando's humid subtropical climate through the use of natural ventilation, it also contains a high-efficiency system of glazing. With wood louvered walls and fans, the indoor dining area can be extended as an outdoor entrance. A specific and efficient type of automatic window, coined "jalousie windows," is utilized as the primary glazing material [48]. As shown in Fig. 37, temperature and humidity sensors in the exterior operate the windows to close automatically, and air conditioning is used instead of natural ventilation. In this way, the building functions similarly to vernacular Florida architecture prior to the invention of air conditioning where natural breezes are taken advantage of and used to cool indoor spaces. However, the dynamic aspect of the sensor system ensures indoor comfort even on the most humid of days.

Fig. 37 Operational "jalousie windows" and wood louvers are combined to provide shade and allow for natural ventilation when desired. (Credit: Kate Joyce Studios, McDonald's Flagship—Orlando, Ross Barney Architects, 2020)

Operational Versus Embodied Energy

The glazing system is not only operable but also integrated with photovoltaic (PV) panels. PV Glass panels with a size of 4809 square feet were manufactured by Onyx Solar and installed on the porch [48]. "Belnor Engineering's Onyx Solar photovoltaic glass" is the name of the specific glazing used in the building. With regard to its benefits, this type of glass can naturally illuminate various spaces with sunlight [53]. Additionally, it avoids UV and IR radiation while generating renewable energy [53]. Using these solar panels helps combine both active and passive properties. These panels both provide thermal and acoustic insulation [53] and consist of two layers of tempered glass, each of which is 14″ in height [51]. While the outermost layer is transparent, the interior layer is light gray [51]. A total of 192 units of 291 watt/unit, gray-finished crystalline silicon glass are utilized in this system, providing a power of 55.80 kWp. Additionally, 66 monocrystalline silicon solar cells are embedded in the glass of each unit [51]. As a result of the PV glass panel system, an average light transmittance of 36% is achieved. This amount provides increased light in the dining area of the porch [51].

Although the PV glazing system is responsible for providing increased light in the dining area of the porch, it is not the only means of energy generation found in the building design. Other energy-generation strategies include 18,727 square feet of PV panels and 25 smart off-grid parking lot lights. The building additionally utilizes low-flow plumbing fixtures and LED lighting. In total, these strategies produce more energy than the restaurant uses [48]. A large portion of the corporate goal to reach net zero has been achieved by using solar energy (Fig. 35). The design includes 19,000 square feet of traditional solar panels on its roof and canopy and 5000 square feet of wind-powered solar panels on the porch [51]. As a result, the building is capable of producing 679,000 kWh each year [51]. This energy generation is particularly important as it is utilized for the consumption of energy in the building's kitchen systems (Fig. 38).

The diagrams below display data obtained by using the ATHENA® Impact Estimator for Buildings. Located in Ontario, Canada, the ATHENA® Impact Estimator for Buildings was developed by the Sustainable Materials Institute. As part of the institute's mission, it leverages the life-cycle assessment in North America to promote sustainability in the built environment [50]. According to the developers, "robust life cycle inventory databases provide exact scientific cradle-to-grave information about building materials and products, transport, and construction and demolition activities" [53]. The Athena Institute connects designers to the power of life-cycle analysis without requiring them to become LCA experts themselves [53].

When the bill of materials has been provided, any part of a building has the potential to be modeled using the Impact Estimator. Using simple inputs, the Impact Estimator can create a bill of materials for users who do not have one. Examples include [51] foundations, footings, slabs, all below- and above-grade structure and envelope, windows and doors, and building interiors. Based on a 60-year life cycle,

Fig. 38 The underside of solar panels can be seen from the exterior dining area. (Credit: Kate Joyce Studios, McDonald's Flagship—Orlando, Ross Barney Architects, 2020)

the study examines the overall building's life cycle. According to ISO 14040, we can compare up to five design scenarios according to the US Environmental Protection Agency's environmental impact categories [50]. In this study, the following environmental metrics were used: Global Warming Potential, Smog Potential, Acidification Potential, Non-Renewable Energy, Eutrophication Potential, and Ozone Depletion Potential. As inputs to the Impact Estimator, a series of factors related to the building are considered in order to calculate the life-cycle impact of each factor on the above categories. There are five assemblies that consist of information on the project: foundations, floors, columns and beams, roofs, and walls (Figs. 39 and 40).

The figure displays the comparison between different constructional categories that are used in the case study. The report compares the amount of CO_2 emissions that each section can have on the environment (Fig. 41).

A comparison is made between operating and embodied in both primary energy and global warming potentials. Operating accounts for a greater share in both charts.

Figure 42 presents the comparison between different constructional categories that are used in the case study. The report compares the amount of O_3 emissions that each section can have on the environment.

Environmental Dimensions of Climate Change: Endurance and Change in Material... 341

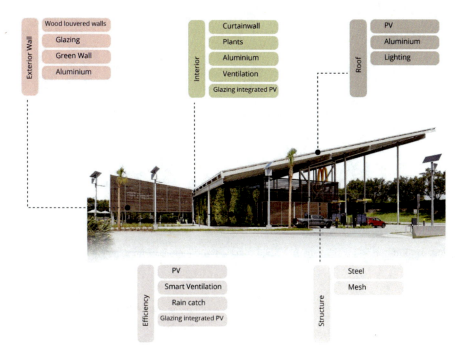

Fig. 39 Envelope materials and integrated sustainable strategies

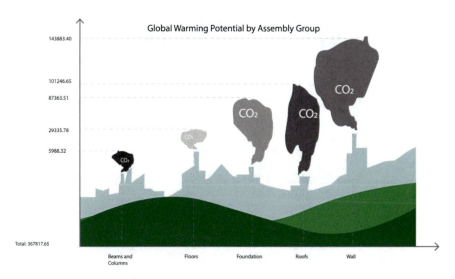

Fig. 40 Global warming potential

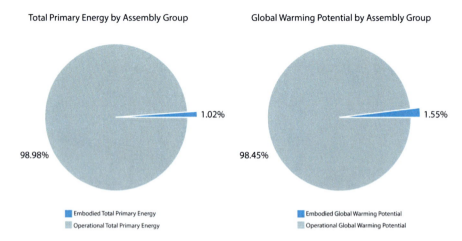

Fig. 41 Left: operational versus embodied energy; right: global warming potential

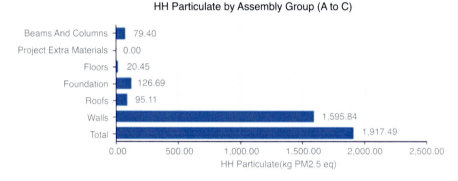

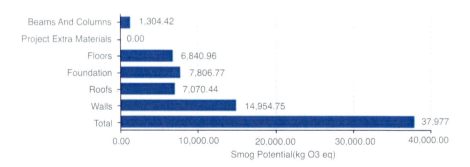

Fig. 42 Human health particulate and smog potential

Conclusion

Since its construction in 2021, the Orlando McDonald's Flagship building has garnered a large amount of press and attention due to the sustainable features included. Both the passive and active strategies employed by Ross Barney Architects are necessary for the LEED Platinum status and regard that the design holds. Strategies like the louvered walls or V-shaped roof that respond to Orlando's humid subtropical climate showcase the effort and effectiveness of these passive strategies in the building's design. These strategies, combined with ingenuitive PV and other energy-generation techniques, all contribute to the realized design of the Orlando McDonald's Flagship building and its place as an innovative and exciting precedent for net-zero fast-service restaurants.

Case Study: McDonald's Flagship, Chicago

Architect: Ross Barney Architects

Introduction

The McDonald's Flagship, located in Chicago, Illinois, not only showcases the company's corporate commitment to sustainability but also responds to Chicago's continental climate and the city's density through incorporation of green spaces. Designed by Ross Barney Architects in 2018, the building incorporates a solar pergola encompassing a pure glass box as the LEED Platinum design strategy. Located on the site of the long-standing, well-known "Rock 'n Roll" McDonald's, the design reuses elements from the 1985 structure while prioritizing pedestrian accessibility. The design brings natural elements, such as light and trees, indoors and utilizes permeable paving to reduce the heat island effect. Serving as the first commercial use of CLT in the city of Chicago, the design has been awarded a LEED Platinum status. This writing aims to investigate the role of material selection in the project with specific attention to the envelope and its relationship to the overall sustainable status of the structure (Fig. 43).

Identified as a continental climate, Chicago's weather is not only affected by the sun and wind but also by Lake Michigan. Chicago is located directly on Lake Michigan and touches the southwestern portion of this piece of the Great Lakes. One way in which Lake Michigan affects the climate of Chicago is by moderating temperature swings due to its thermal mass [54]. Referring to the city's nickname, "the windy city", the lake also allows air to pass over its surface, leading to increased snowfall and high winds. Chicago averages 38 inches of rain per year, 35 inches of snow per year, and slightly below the national average of sunny days per year [55]. The average temperature in Chicago ranges from a high of 84 °F in summer to a low

Fig. 43 McDonald's Flagship, Chicago. (Credit: David Thaddeus)

of 19 °F in winter [55]. During a span of 5 days in 1995, the city of Chicago endured extraordinarily high temperatures that ultimately led to several hundred deaths [56]. Since then, the city of Chicago has made it a priority to identify urban heat areas and to adopt heat-reducing strategies in the construction and design of these areas. With the help from the US Department of Energy, the Passive Solar Industries council has set guidelines for climate-reactive passive strategies employed in the Chicago region [57]. While this information is specifically targeted to residential homes, much of the climate combattant information can also relate to larger structures such as the McDonald's Flagship. These guidelines include techniques like added insulation and sun-tempering to allow for efficient natural heating and cooling of spaces specific to Chicago's climate (Fig. 44).

Sustainable Features

Located in the center of Chicago, a pedestrian-oriented quick-serve restaurant with a rooftop orchard stands out among the traditional McDonald restaurants many associate with the company name. However, the McDonald's Flagship location in Chicago includes not only a sustainable site design but also energy reduction strategies and an efficient material selection. The solar pergola itself, which encompasses the structure below, consists of over 1000 solar panels [58]. The pergola extends well beyond the interior structure beneath it, attempting to connect both of the interior and exterior spaces below. This structure sets a new precedent for car and pedestrian traffic in an attempt to rebalance the two user groups. Taking up an

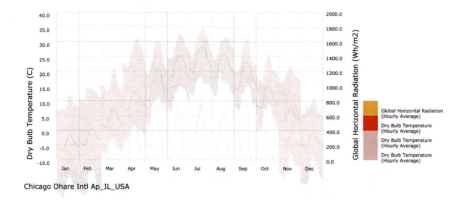

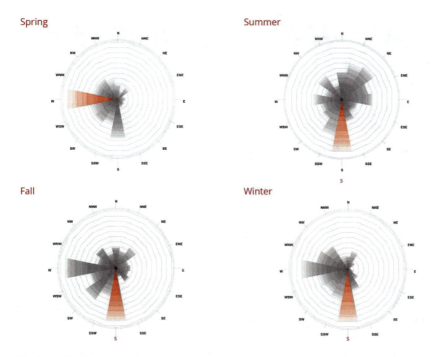

Fig. 44 Climatic data, dry-bulb temperature, wind roses, and psychrometric chart

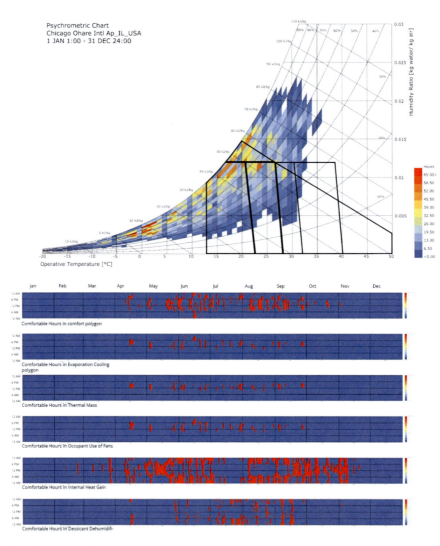

Fig. 44 (continued)

entire city block near the center of Michigan Avenue, the building serves not only as a restaurant but also as a public outdoor space. As there are no public parks within a one-third mile radius of the site, the McDonald's Flagship location provides necessary and usable space for the surrounding community. Considering the walkability and pedestrian-oriented nature of the building, it is no surprise that this design resulted in a 72% increase in pedestrian friendly spaces from the original design [59].

While many buildings that reach a net-zero or LEED Platinum status are designed from the ground-up, the McDonald's Flagship locations strategically utilize kitchen space and existing walls from the prior building into the final design. Ross Barney

Architects claimed that "The most sustainable building is one that is already built" [59]. Remaining in accordance with this quote, both the existing basement and kitchen of the previous building were retained and incorporated into the final building design. Existing walls were re-clad to improve thermal value and contribute to the sustainable status of the building. In addition to the reuse of existing materials, the McDonald's Flagship made history as the first commercial building in the city of Chicago to use timber as the primary structural material. Both CLT and glulam were chosen for their light environmental impact. This primary structural system, together with the additional use of steel, combine to form the 19,000 square feet structure [60]. Both timber and steel elements are visible from the interior of the building's dining area with 27 foot high ceilings, as illustrated in Fig. 45.

Possibly one of the most unique sustainable elements of the McDonald's Flagship is the inclusion of plants and green space in and around the building (Fig. 46). Compared to the prior structure, there is over 400% more green space in the final design [61]. Over 70 trees are placed on-site around the exterior of the building and over 10,500 plants in total [58]. The outdoor plants are situated near permeable outdoor paving that reduces storm-water runoff and minimizes irrigation. Inside the structure, floating glass walls of native ferns and white birch trees can be seen from the dining area. As shown in Fig. 47, a row of harvestable apple trees is even visible from the interior of the dining area. Rooftop trees and a rooftop garden also contribute to the inclusion of plants in the design. All produce grown on-site – including apples, arugula, broccoli, kale, Swiss chard, and carrots – are donated to the Ronald McDonald House [59]. The inclusion of these plants helps improve air quality and create an oasis in the center of Chicago.

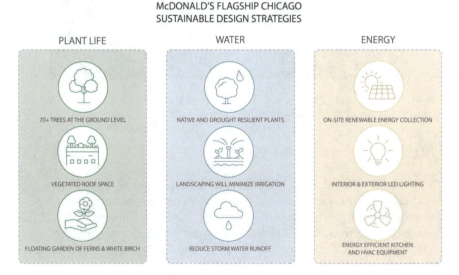

Fig. 45 Sustainable strategies. (Original diagram credit: https://corporate.mcdonalds.com/corpmcd/en-us/our-stories/article/ourstories.new_flagship.html)

Fig. 46 The McDonald's Flagship was the first commercial building in Chicago to use timber framing as a primary structural element. (Credit: David Thaddeus)

Envelope

The envelope of the Chicago McDonald's Flagship could be reasonably broken down into two main categories: the solar pergola and the structure underneath it. The solar pergola, often described as the "big roof," is supported by 12.75-inch diameter hollow structural section (HSS) columns [62]. Acting as both shading and energy-generating elements, the pergola consists of 1062 south-facing solar panels [59]. This large shading device hovers above the interior spaces below as well as the drive-through car line and is outperforming initial estimates modeled for energy generation, as shown in Fig. 48.

Environmental Dimensions of Climate Change: Endurance and Change in Material... 349

Fig. 47 "Floating" white birch and fern trees can be seen from the interior dining room. (Credit: David Thaddeus)

The second "envelope" that can be identified when analyzing the Chicago McDonald's Flagship building is the structure underneath the hovering pergola. A space of 12,720 square feet, more than half of the total area of the building, is a renovated space from the previous "Rock 'n Roll" McDonald's [58]. The materials here were kept in place and covered in a new concrete facade. In contrast to the opaque and heavy concrete used in the existing structure, much of the new restaurant is covered in glazing. This VS-1 vertical facade system curtain wall is unique in that the facade is held to the mullions without visually obtrusive bolts [63]. The slender design produces a sleek and clean appearance to the outdoors and vice versa.

Fig. 48 The solar pergola hovers over the car lane providing shade and generating energy. (Credit: David Thaddeus)

Then **McDonald's Rock 'n Roll. Chicago. (Credit: Caitphoto. Caitlin on Flickr) and** *Now* **McDonald's Flagship, Chicago. (Credit: David Thaddeus)**

As illustrated in Fig. 49, the CLT roof deck is left exposed from the underside and visible to customers. The CLT deck measures 7 inches in thickness at the hanging atrium and 12 inches thick in the dining area [58].

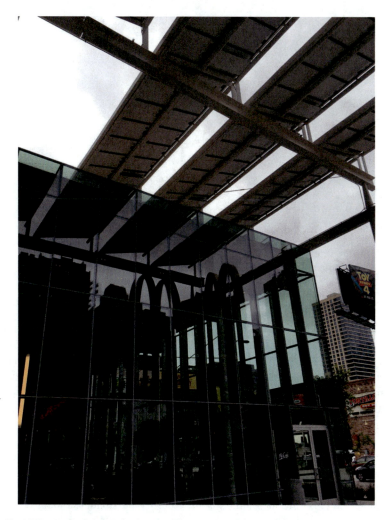

Fig. 49 A VS-1 vertical facade system curtain wall allows for natural light to enter interior spaces and views of the outdoors from the inside. (Credit: David Thaddeus)

Operational Versus Embodied Energy

The Chicago McDonald's Flagship location was awarded a LEED Platinum status due to the sustainable features implemented in the design. While much of the energy generation, specifically the solar pergola, may appear more visually distinctive to the customers at the location, the material selection and embodied energy in the materials used also greatly contribute to this status. As discussed above, the building is encompassed by a large solar pergola made up of over 1000 solar panels. This system generates enough energy to run approximately 60% of the building's use [58]. As illustrated in Fig. 50, the operational energy production involved in the

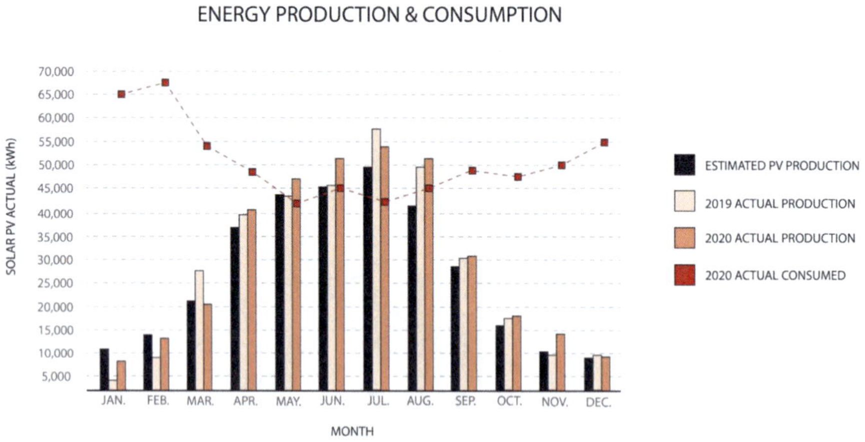

ENERGY PRODUCTION & CONSUMPTION

Fig. 50 Energy production and consumption. (Original diagram credit: https://acrobat.adobe.com/link/track?uri=urn%3Aaaid%3Ascds%3AUS%3A8aa2c8d4-d83e-4e65-969a-997bf2c1cd50#pageNum=1)

design saves McDonald's nearly 50% of the overall energy costs to run the restaurant [58]. Other operational energy systems include interior LED light fixtures with daylight sensors, smart exterior lighting that is designed to reduce light pollution in the city, and electric charging stations for customers. These techniques together help support the operational costs and use of the kitchen and central heating and cooling systems.

Although active techniques are included in the design to reduce operational costs and save energy, passive strategies, such as embodied energy due to the material selection, also play a key role in the sustainable design of the structure. Reuse of the previously existing structure in the new structure ensured that a smaller amount of new materials would be needed for construction. This eliminates the cradle-to-grave process involved in the creation of new materials for buildings. In this way, concrete was able to be reused in the building's structure with a smaller amount needing to be newly constructed. The concrete that was newly created utilized Carbon Cure Concrete, which sequesters recycled CO_2 into fresh concrete mix [58]. In the portion of the structure that was newly constructed, timber frame construction was used as the primary material selection (Fig. 47). It is estimated that the use of CLT in the building when compared to traditional construction methods equates to a saved carbon amount of 34,000 passenger vehicles off the road per year [58]. Not only does a natural and warm wood structure contribute to a more relaxed and inviting atmosphere, but it also contains a low embodied carbon footprint (Fig. 51).

The diagrams below display data obtained by using the ATHENA® Impact Estimator for Buildings. Located in Ontario, Canada, the ATHENA® Impact Estimator for Buildings was developed by the Sustainable Materials Institute. As part of the institute's mission, it leverages the life-cycle assessment in North America to promote sustainability in the built environment [59]. According to the

Fig. 51 Exposed timber roof deck elements can be seen from the interior dining area. (Credit: David Thaddeus)

developers, "robust life cycle inventory databases provide exact scientific cradle-to-grave information about building materials and products, transport, and construction and demolition activities" [63]. The Athena Institute connects designers to the power of life-cycle analysis without requiring them to become LCA experts themselves [63].

Any part of a building has the potential to be modeled using the Impact Estimator when the bill of materials has been provided. Using simple inputs, the Impact Estimator can create a bill of materials for users who do not have one. Examples include [54] foundations, footings, slabs, all below- and above-grade structure and envelope, windows and doors, and building interiors. Based on a 60-year life cycle, the study examines the overall building's life cycle. According to ISO 14040, we can compare up to five design scenarios according to the US Environmental Protection Agency's environmental impact categories [59]. In this study, the following environmental metrics were used: Global Warming Potential, Smog Potential, Acidification Potential, Non-Renewable Energy, Eutrophication Potential, and Ozone Depletion Potential. As inputs to the Impact Estimator, a series of factors related to the building are considered in order to calculate the life-cycle impact of each factor on the above categories. There are five assemblies that consist of information on the project: foundations, floors, columns and beams, roofs, and walls.

Any section of the building must be devoid of various types of materials and constructional elements. The Chicago McDonald's Flagship is mostly made up of CLT and wooden elements, while there are also steel columns and curtain walls. For any other elements that cannot be measured in one of the above assemblies, we can use the "Project extra materials" section. Regarding the operating energy consumption, based on the information provided by the design team, the amount of electricity in kWh per year was entered (Fig. 52).

The diagram shows the comparison between different constructional categories of the case study. In this chart, the highest CO_2 emissions are found in walls and roofs due to the widespread use of glazing and steel components (Fig. 53).

The diagram shows the comparison between different constructional categories of the case study. In this chart, the highest SO_2 emissions are found in walls and roofs due to the widespread use of glazing and steel components (Fig. 54).

The diagram shows the comparison between different constructional categories of the case study. Compared with previous charts, the highest concentrations of nitrogen (N) emissions are found in foundations, whereas in previous charts, it had less warming and acidification impact comparing walls and roofs. In part, this can be attributed to the use of concrete for foundations (Fig. 55).

The diagram shows the comparison between different constructional categories of the case study. It is interesting to note that concrete-made foundations have the

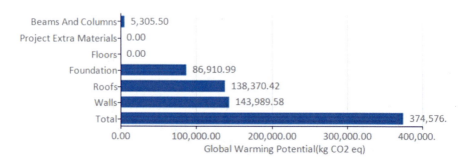

Fig. 52 Global warming potential. (McDonald's Flagship, Chicago)

Environmental Dimensions of Climate Change: Endurance and Change in Material... 355

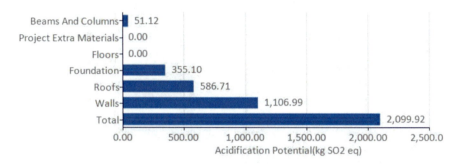

Fig. 53 Acidification potential (McDonald's Flagship, Chicago)

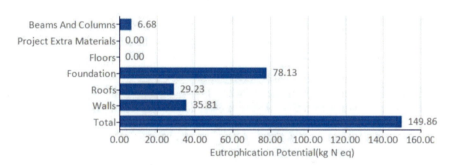

Fig. 54 Eutrophication potential (McDonald's Flagship, Chicago)

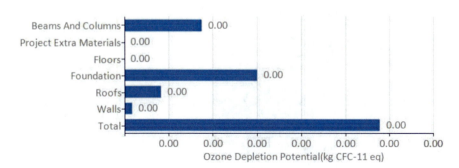

Fig. 55 Ozone depletion potential (McDonald's Flagship, Chicago)

most impact on CFC emissions, which again shows the negative impacts of using concrete even in small quantities (Fig. 56).

The diagram shows the comparison between different constructional categories of the case study (Fig. 57).

The diagram makes a comparison between operating and embodied global warming potentials (Fig. 58).

Figure 59 displays the comparison between different constructional categories that are used in the case study. The report compares the amount of CO_2 emissions that each section can have on the environment (Fig. 60).

The figure displays the comparison between different constructional categories that are used in the case study. The report compares the amount of O_3 emissions that each section can have on the environment (Fig. 61).

A comparison is made between operating and embodied in both primary energy and global warming potentials. Operating accounts for a greater share in both charts.

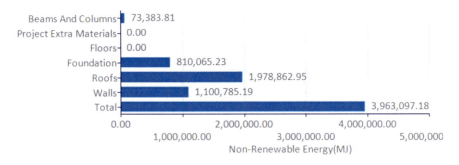

Fig. 56 Non-renewable energy (McDonald's Flagship, Chicago)

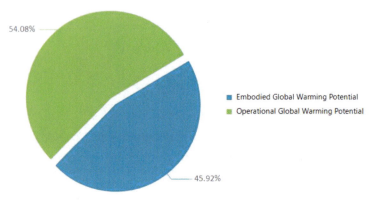

Fig. 57 Non-renewable energy (McDonald's Flagship, Chicago)

Environmental Dimensions of Climate Change: Endurance and Change in Material... 357

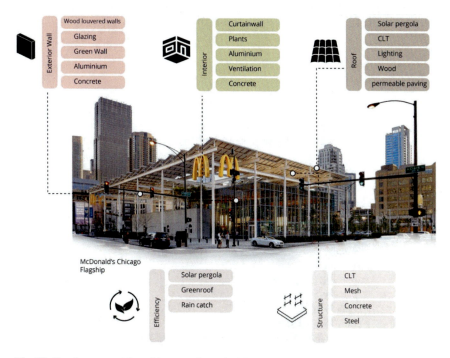

Fig. 58 Envelope materials and integrated sustainable strategies

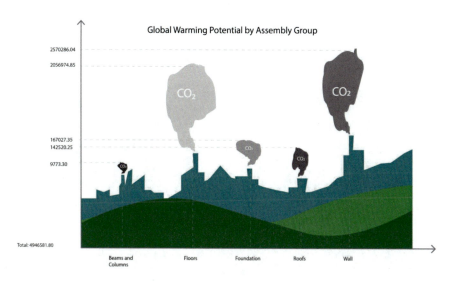

Fig. 59 Global warming potential

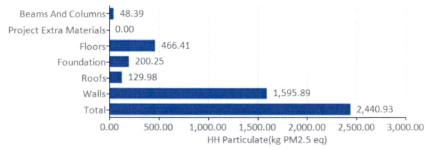
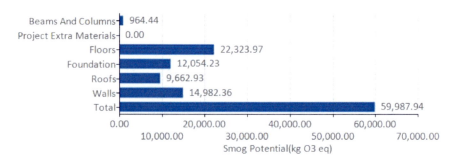

Fig. 60 Human health particulate and smog potential

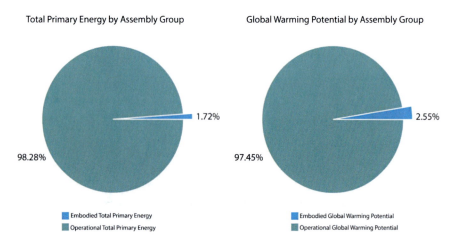

Fig. 61 Left: operational versus embodied energy; right: global warming potential

Conclusion

Since its completion in 2018, the Chicago McDonald's Flagship building has garnered a large amount of press and attention due to the sustainable features included. Both the passive and active strategies employed by Ross Barney Architects are the foundation for the LEED Platinum status and regard that the design holds. Most notably, the structure will forever serve as the first commercial use of CLT in the city of Chicago and has set a precedent for future McDonald Flagship locations. McDonald's calls the transformation of this building the "Experience of the Future" 'with the goal of enhancing customer experience dramatically [60]. At the time of construction, nearly 5000 McDonald's locations had been transformed to meet the "Experience of the Future" guidelines, although it can be argued that none have gained the recognition or utilized the vast amount of sustainable strategies that are in place at the Chicago McDonald's Flagship location.

Port of Portland Headquarters Building

Architect: ZGF

Introduction

Designed by ZGF and constructed in 2010, the Port of Portland Headquarters building is ranked by Forbes as one of the 10 most high-tech sustainable buildings in the world. The building is located in Portland, Oregon, and reacts to the Mediterranean climate of the region. Both a high-performance glazing system and a reflective roof membrane actively minimize heat gain from this climate. The building program includes seven floors of public airport parking and three floors of office space, totaling 205,603 square feet [64]. Awarded the Smart Environments Award by the International Interior Design Association and Metropolis magazine, the design reaches a LEED Platinum status. This writing aims to investigate both operational and embodied aspects of the design with specific attention to the envelope and its relationship to the overall sustainable status of the structure (Fig. 62).

Identified as a Mediterranean continental climate, Portland, Oregon, is characterized by short warm summers and overcast very cold winters. During December, Portland's cloudiest month, the sky is overcast nearly 75% of the time [65]. As a result, December is also the wettest month of the year with 6.8 inches of rainfall [66]. In contrast, Portland's dry season is in summer, and cold season generally lasts from October to March. Portland averages 43 inches of rain per year, 3 inches of snow per year, and well below the national average of sunny days per year [65]. The average temperature in Portland ranges from a high of 81 degrees Fahrenheit in summer to a low of 36 degrees Fahrenheit in winter [65]. With the help from the US

Fig. 62 Port of Portland Headquarters building. (https://inhabitat.com/green-roofed-port-of-portland-headquarters-aims-for-leed-gold/)

Department of Energy, the Passive Solar Industries council has set guidelines for climate-reactive passive strategies employed in the Portland region [67]. While this information is specifically targeted to residential homes, much of the climate combattant information can also relate to larger structures such as the Port of Portland Headquarters building. Some of these guidelines include increasing insulation and adding exposure to the sun from the south. It should be noted that the magnetic north in Portland is 21 degrees of true north and should be corrected when considering light exposure (Fig. 63).

Sustainable Features

While the form, being influenced by the shape of an airplane hull, is striking in itself, it is the sustainable features included in the design of the Port of Portland Headquarters building that has garnered attention from the public, which defines it as a noteworthy case study. The sustainable features of the structure expand beyond the building itself and are present in the site. Constructed wetlands on-site include both tidal and vertical flow wetland cells. Filled with native, naturalized, and flowering plants that avoid attracting birds in close proximity to the airport, these wetlands additionally aid in the wastewater treatment [68]. Coined as a "Living Machine," the

Environmental Dimensions of Climate Change: Endurance and Change in Material... 361

wastewater system is located in the interior lobby to serve a percent for the possibilities of such technology, as illustrated in Fig. 64.

Occupying a total of 700 square feet indoors, the system additionally utilizes the outdoor wetlands to cycle water [70]. Compared to a similar structure of the same size, the Port of Portland Headquarters building decreases water use by 75% from the baseline due to the efficient water features, as shown in Fig. 65 [70]. An

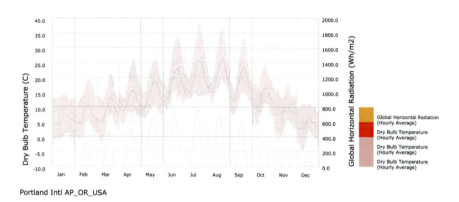

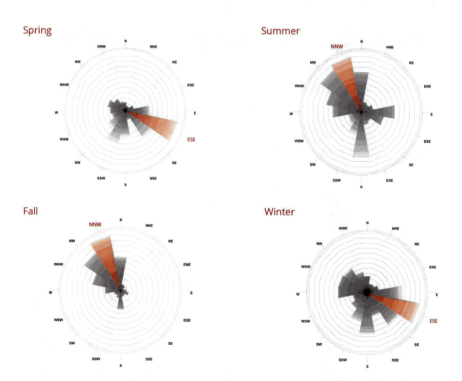

Fig. 63 Climatic data, dry-bulb temperature, wind roses, and psychrometric chart

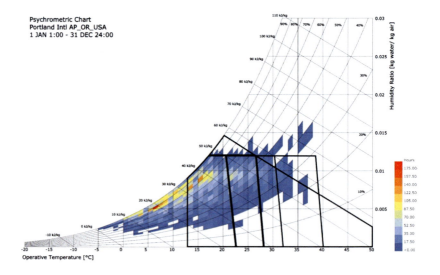

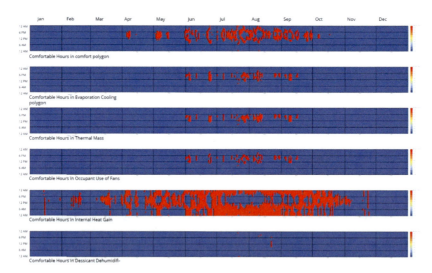

Fig. 44 (continued)

ecological wastewater treatment system, the Living Machine system produces quality fresh water from both gray and black water without odors, chemicals, offensive by-products, or high-energy usage typical of conventional systems [71]. In their tidal flow cells, the Living Machine uses many plants to decontaminate incoming wastewater at levels far exceeding the Oregon Department of Environmental Quality's standards. An additional benefit is that the plants enhance the overall beauty of the site and create a microclimate within and around the building [69].

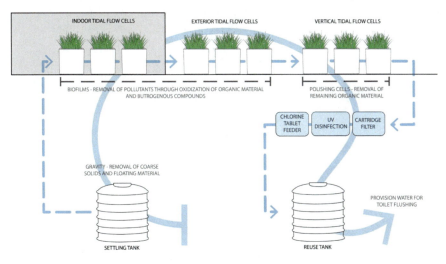

Fig. 64 Living Machine [69]

Fig. 65 The "Living Machine" mimics an interior garden while working to cycle wastewater. (https://www.mayerreed.com/portfolio/port-of-portland-headquarters-parking-garage/)

Above ground and located on the 9th floor of the structure is an extensive green roof amounting to a total of 10,000 square feet, as illustrated in Fig. 66 [70]. Additionally, helping to treat rainwater, this green roof aids in insulating the building and reducing the heat island effect. Other sustainable features included in the design of the Port of Portland Headquarters building include a high-performance glazing system utilized on the exterior of the building.

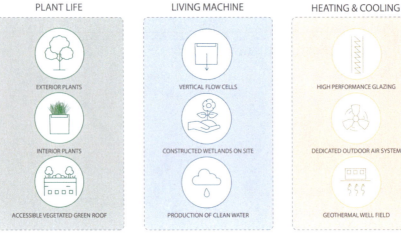

Fig. 66 Sustainable strategies

Envelope

The Port of Portland Headquarters building features a reflective roof membrane and high-performance glazing to minimize heat gain and energy consumption. An extensive eco-roof covers the roof of the 9th floor of the building, which reduces the heat island effect and offers a large surface area for rainwater treatment. Treatment of wastewater is assisted by a Tidal Flow Wetland Living Machine [72]. In addition to the high-performance glazing utilized, sensors are used for maintaining efficient and sustainable occupant comfort levels. By placing sensors and taking advantage of sidelight from windows, the lighting design is optimized. Light and occupancy sensors are included with each fixture as part of the control system in the open office [72]. In order to balance and control daylight, glare, and heat gain, automated exterior shades and light shelves were utilized. Workstations are also equipped with task lighting to reduce the need for overhead lighting [72], as shown in Fig. 67.

Operational Versus Embodied Energy

The Port of Portland Headquarters building utilizes efficient and high-tech solutions to energy generation and utilization. Below ground, 200 geothermal wells aid in managing heating and cooling [73]. The design for air inside the building consists of a dedicated outdoor air system (DOAS) that works in conjunction with the geothermal wells below ground as well as a radiant ceiling system consisting of over

Environmental Dimensions of Climate Change: Endurance and Change in Material... 365

Fig. 67 Green roof. (https://inhabitat.com/green-roofed-port-of-portland-headquarters-aims-for-leed-gold/)

56,000 square feet of metal radiant ceiling panels [72]. The building's plan is primarily an open layout with shared offices divided by half-walls and does utilize an RCP system with underfloor ventilation. A traditional forced-air system, which is used throughout the rest of the building, provides the air conditioning for the smaller, contained break-out and meeting areas [74].

When compared to the average office energy use intensity (EUI) performance from the national CBECS 4 and California CEUS 5 datasets, the Port of Portland Headquarters building shows a drastic reduction of over 40% [75]. The total EUI currently amounts to 46 kBtu/ft^2. When compared to an office building built to the Oregon code 2010, it can be seen that the building uses 30% less energy than a traditional office structure. Although much lower in energy use than traditional office buildings, it should be noted that the Port of Portland Headquarters building uses 15% more energy than the ASHRAE best-practice energy efficiency standard 100 [75]. In addition to energy use, the Port of Portland Headquarters building also shows higher rates of occupant thermal comfort when compared to the baseline. As illustrated in Fig. 68, it should be noted here that a few zones of the Port of Portland Headquarters building have override controls for window blinds and thermostats that residents are given remote access to [74]. Along with the sensor technology, these contribute to efficient occupant comfort levels.

The diagrams below display data obtained by using the ATHENA® Impact Estimator for Buildings. Located in Ontario, Canada, the ATHENA® Impact

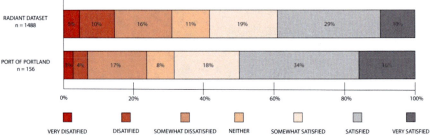

Fig. 68 Occupant comfort levels. (Original diagram credit: Caroline Karmann https://escholarship.org/content/qt3cj9n3n4/qt3cj9n3n4_noSplash_22165b8afc25e451d393ae2571822695.pdf)

Estimator for Buildings was developed by the Sustainable Materials Institute. As part of the institute's mission, it leverages the life-cycle Assessment in North America to promote sustainability in the built environment [68]. According to the developers, "robust life cycle inventory databases provide exact scientific cradle-to-grave information about building materials and products, transport, and construction and demolition activities" [69]. The Athena Institute connects designers to the power of life-cycle analysis without requiring them to become LCA experts themselves [69].

Any part of a building has the potential to be modeled using the Impact Estimator when the bill of materials has been provided. Using simple inputs, the Impact Estimator can create a bill of materials for users who do not have one. Examples include [65] foundations, footings, slabs, all below- and above-grade structure and envelope, windows and doors, and building interiors. Based on a 60-year life cycle, the study examines the overall building's life cycle. According to ISO 14040, we can compare up to five design scenarios according to the US Environmental Protection Agency's environmental impact categories [68]. In this study, the following environmental metrics were used: Global Warming Potential, Smog Potential, Acidification Potential, Non-Renewable Energy, Eutrophication Potential, and Ozone Depletion Potential. As inputs to the Impact Estimator, a series of factors related to the building are considered in order to calculate the life-cycle impact of each factor on the above categories. There are five assemblies that consist of information on the project: foundations, floors, columns and beams, roofs, and walls (Fig. 69).

Figure 70 displays the comparison between different constructional categories that are used in the case study. The report compares the amount of CO_2 emissions that each section can have on the environment.

A comparison is made between operating and embodied in both primary energy and global warming potentials. Operating accounts for a greater share in both charts (Fig. 71).

Figure 72 displays the comparison between different constructional categories that are used in the case study. The report compares the amount of O_3 emissions that each section can have on the environment.

Environmental Dimensions of Climate Change: Endurance and Change in Material... 367

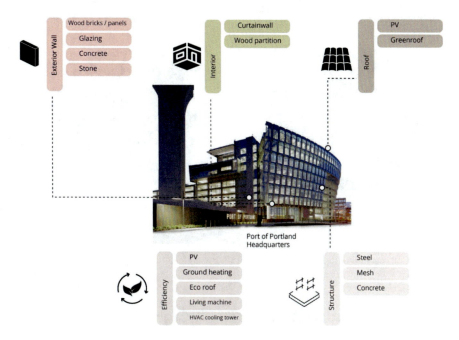

Fig. 69 Envelope materials and integrated sustainable strategies

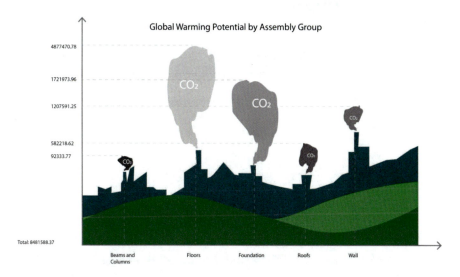

Fig. 70 Global warming potential

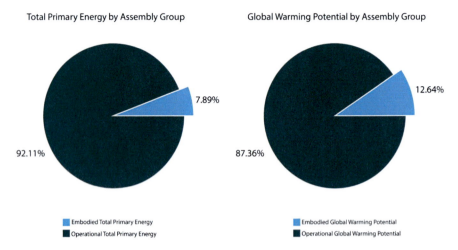

Fig. 71 Left: operational versus embodied energy; right: global warming potential

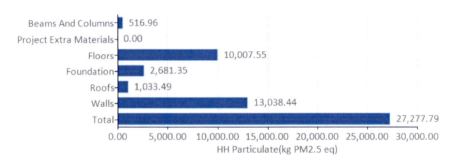

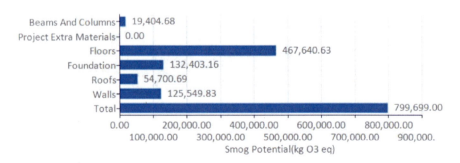

Fig. 72 Human health particulate and smog potential

Conclusion

The Port of Portland Headquarters building reacts to the Mediterranean climate of the region through both passive and active techniques that contribute to the design's label as a LEED Platinum building. Most notably, the sustainable features of the structure expand beyond the building itself and are present in the site as represented through the complex living machine. Due in part to the green roof, PV energy generation, and geothermal wells, the building uses 30% less energy than a traditional comparable office building. These statistics, combined with the ingenuity of the design, mark the Port of Portland Headquarters as a prime example of sustainable design in the USA.

References

Environmental Dimensions of Climate Change: Endurance and Change in Material Culture

1. Carson, R. (2002). *Silent spring*. Houghton Mifflin Harcourt.
2. The Paris Agreement | UNFCCC. Retrieved May 22, 2022.
3. The net-zero transition. What it would cost, what would it bring. McKinsey Global Institute Report. January 2022. The net-zero transition: What it would cost, what it could bring Retrieved May 22, 2022.
4. Ehrlich, P. R., & Holdren, J. P. (1971). Impact of population growth. Retrieved April 19, 2022.
5. McGregor, S. (2021, October 12). *Embodied carbon vs operational carbon: What's the difference, and why does it matter? Renewable Energy World from Embodied carbon vs operational carbon: What's the difference, and why does it matter?* Retrieved April 19, 2022.
6. Cassandra. (2017, July 11). *Could a new facade help make your building carbon neutral? - Locker Group*. Locker Group. from Could a new facade help make your building carbon neutral? - Locker Group. Retrieved April 19, 2022.
7. Getting to net-zero emissions by 2050. National Academy of Sciences. Washington, DC. Retrieved May 26, 2022.
8. Hill, S., Dalzell, A., & Allwood, M. (n.d.). Net-zero Carbon Buildings: Three steps to take now. *Arup*, from https://www.arup.com/perspectives/publications/research/section/net-zero-carbonbuildings-three-steps-to-take-now. Retrieved April 19, 2022.
9. Carroll, C., Alves de Souza, Y., Salter, E., Hunziker, R., De Giovanetti, L., & Contucci, V. (n.d.). Net-zero Buildings: where do we stand? *Arup*. from https://www.arup.com/perspectives/publications/research/section/net-zerobuildings-where-do-we-stand. Retrieved April 19, 2022.
10. Data to the rescue: Embodied carbon in buildings and the urgency of now | McKinsey. Retrieved May 22, 2022.
11. *Operating vs Embodied carbon in the built environment: The difference and why it matters - Sustainable brands*. Retrieved May 27, 2022.
12. NYC's gas ban takes fight against climate change to the kitchen - *The New York Times*. Retrieved May 24, 2022.
13. *Local Law 97 - Sustainable buildings*. Retrieved May 27, 2022.
14. *Architecture 2030*. Retrieved May 27, 2022.
15. *What is SE 2050 overview? – SE2050*. Accessed May 27, 2022.
16. Carbon Smart Materials Palette. CONCRETE. Accessed May 27, 2022.
17. Jackson, N. (n.d.). 1 Triton Square: How can existing buildings combat climate change? *Arup*. Retrieved April 19, 2022, from https://www.arup.com/projects/1-triton-square

Case Study: 888 Boylston Street

18. https://www.climatestotravel.com/climate/united-states/boston
19. https://www.bestplaces.net/climate/city/massachusetts/boston
20. https://www.architects.org/news/passive-house-design
21. https://www.boston.gov/sites/default/files/embed/file/2018-10/resilient_historic_design_guide_updated.pdf
22. http://www.fxcollaborative.com/projects/140/888-boylston-street/
23. https://www.abettercity.org/docs-new/The_Commercial_Net_Zero_Energy_Building_Market_In_Boston.pdf
24. https://www.bxp.com/wp-content/uploads/2021/02/888_Boylston_Case_Study.pdf
25. https://www.usgbc.org/articles/boston-properties-pushes-boundaries-sustainable-design
26. https://builtenvironmentplus.org/wp-content/uploads/2019/10/Green_Building_of_the_Year_Showboard_Package_2019.pdf
27. https://builtenvironmentplus.org/888-boylston-sustainable-design-you-can-experience/
28. https://www.reit.com/news/articles/888-boylston-takes-sustainability-to-new-level-for-boston-properties

Case Study: CLT Passivhaus

29. https://generatetechnologies.com/work-model-c
30. https://passivehouseaccelerator.com/articles/model-clt-multifamily-in-boston
31. https://www.dezeen.com/2020/02/14/model-c-carbon-neutral-clt-apartment-block-boston-generate/
32. https://www.thinkwood.com/projects/clt-passive-house-demonstration-project
33. https://www.placetailor.com/study/201-hampden/

Case Study: Golisano Institute for Sustainability

34. https://www.greenroofs.com/projects/golisano-institute-for-sustainability-gis-green-wall/
35. https://www.weather.gov/buf/ROCclifo
36. https://www.bestplaces.net/climate/city/new_york/rochester#:~:text=The%20annual%20BestPlaces%20Comfort%20Index,other%20places%20in%20New%20York.&text=August%2C%20July%20and%20June%20are,are%20the%20least%20comfortable%20months
37. https://www.nrel.gov/docs/legosti/old/17284.pdf
38. https://static1.squarespace.com/static/5f69141a20665f4000eb34a2/t/60819a03fd898e3c9b433980/1619106308758/CSA_FS1_overview_210421a_low-res.pdf
39. http://www.swbr.com/wp-content/uploads/2014/03/rit-gis-case-study_lo.pdf
40. https://continuingeducation.bnpmedia.com/courses/multi-aia/ultimate-daylighting/4/
41. https://www.rematec.com/news/industry-players-and-markets/new-yorks-golisano-institute-recognized-for-new-building/
42. https://www.stantec.com/en/projects/united-states-projects/r/rit-golisano-institute-sustainability
43. PDF from architects.

Case Study: Orlando McDonald's Flagship

44. https://www.bestplaces.net/climate/city/florida/orlando
45. Climate. https://www.climatestotravel.com/climate/united-states/florida

46. Climate. https://www.carbonae.com/blog/sustainable-residential-design-considerations-for-florida
47. McDonald website. https://corporate.mcdonalds.com/corpmcd/en-us/our-stories/article/our-stories.net-0-energy-disney.html
48. Arch Daily. https://www.archdaily.com/949584/mcdonalds-global-flagship-at-walt-disney-world-resort-ross-barney-architects
49. Green wall: https://sempergreenwall.com/projects/green-wall-mcdonalds-walt-disney-world-orlando/
50. Plant care system. https://sempergreenwall.com/plant-care-system/
51. Medium. https://medium.com/the-belnor-blog/how-disney-worlds-mcdonald-s-flagship-restaurant-became-usa-s-first-zero-emissions-restaurant-8fc8d54d2939
52. https://www.archpaper.com/2020/09/disneys-net-zero-flagship-mcdonalds-is-a-feat-of-climate-responsive-efficiency/
53. Solar Panel. https://www.belnor.com/onyx-solar

Case Study: McDonald's Flagship, Chicago

54. https://www.isws.illinois.edu/statecli/general/chicago-climate-narrative.htm
55. https://www.bestplaces.net/climate/city/illinois/chicago
56. https://www.epa.gov/arc-x/chicago-il-uses-green-infrastructure-reduce-extreme-heat
57. https://www.nrel.gov/docs/legosti/old/17188.pdf
58. https://www.world-architects.com/en/architecture-news/reviews/mcdonald-s-chicago-flagship
59. https://www.r-barc.com/work/mcdonalds-global-flagship-chicago
60. https://corporate.mcdonalds.com/corpmcd/en-us/our-stories/article/ourstories.new_flagship.html
61. https://www.archdaily.com/902882/mcdonalds-chicago-flagship-ross-barney-architects?ad_medium=gallery
62. https://www.castconnex.com/projects/mcdonalds-chicago-flagship-restaurant
63. https://www.christopherglasschicago.com/mcdonalds-chicago-flagship-restaur

Case Study: Port of Portland Headquarters Building

64. https://www.zgf.com/work/1153-port-of-portland-headquarters-amp-long-term-parking-garage
65. https://weatherspark.com/y/757/Average-Weather-in-Portland-Oregon-United-States-Year-Round
66. https://www.bestplaces.net/climate/city/oregon/portland
67. https://www.nrel.gov/docs/legosti/old/17292.pdf
68. http://sustainablewater.com/wp-content/uploads/2013/07/POP-Case-Study-070213.pdf
69. https://urbanecologycmu.wordpress.com/2016/10/14/port-of-portland-headquarters/
70. https://www.portofportland.com/PDFPOP/HQ_Green_Fact_Sheet.pdf
71. https://awwa.onlinelibrary.wiley.com/doi/full/10.1002/j.1551-8833.2010.tb10038.x?saml_referrer
72. https://newbuildings.org/wp-content/uploads/2017/09/Radiant_PortofPortland.pdf
73. https://inhabitat.com/green-roofed-port-of-portland-headquarters-aims-for-leed-gold/
74. https://web.p.ebscohost.com/ehost/pdfviewer/pdfviewer?vid=1&sid=db7a2ab1-d2e8-44ad-8af9-0c02d40681e1%40redis
75. https://escholarship.org/content/qt3cj9n3n4/qt3cj9n3n4_noSplash_22165b8afc25e451d393ae2571822695.pdf

Towards Climate Neutrality: Global Perspective and Actions for Net-Zero Buildings to Achieve Climate Change Mitigation and the SDGs

Mohsen Aboulnaga and Maryam Elsharkawy

> "Start carbon neutrality now! Each country, city, financial institution and company should adopt plans for transitioning to net-zero emissions by 2050 and take decisive action now".
> Antonio Guterres, Secretary General of the United Nations [1]

> **Box 1 Low-Carbon City**
> Low-carbon City (LCC) means a sustainable urbanisation approach that focuses on curtailing the anthropogenesic carbon footprint of cities by virtue of minimizing or eliminating the utilisation of energy sourced from fossil fuel. It combines the futures of Low-carbon society and low-carbon economy while supporting partnerships among governments, private sectors and civil societies. Sustainable Cities and Communities, Springer [2]

Introduction

It all begins with a question! Are buildings efficient enough to meet their energy demand and rely on renewable sources to generate clean energy? The term low-carbon city including low- carbon buildings (Box 1) has recently been used in response to the climate crisis. By and large, the answer simply is not, and buildings nowadays consume a large amounts of energy and emit a colossal amount of greenhouse gases (GHG), mainly carbon dioxide – CO_2 (Fig. 1). In the old days, buildings were built from locally sourced materials and responded effectively to climate

---

M. Aboulnaga (✉)
Department of Architecture, Faculty of Engineering, Cairo University, Giza, Egypt
e-mail: maboulnaga@eng.cu.edu.eg

M. Elsharkawy
Architecture Engineering &Technology Program (AET), Faculty of Engineering, Cairo University, Giza, Egypt

conditions, hence, consuming less energy and emitting less carbon as shown in Figs. 2, 3, and 4.

The evolution of building solutions in the construction industry does not stop, and with the increasing consciousness about the depletion of resources and environmental awareness, the sustainability dimensions of any building is becoming mandatory in climate mitigation [3]. The concept of climate neutrality has been expanding in all building types to mitigate CO_2 emissions to a minimum level and offset the remaining emissions as well [4]. The avoidance of all building-related GHG emissions means that all buildings are considered "climate-neutral." Global initiatives worldwide aimed at enhancing climate neutrality and achieving net-zero energy levels are considered one of the most important steps in pushing for creating climate action plans in cities. Action plans provide more specific guidelines and solutions required in order to mitigate climate change in each city [5]. The use of renewable energy resources is the first step utilized to reach net-zero energy models meaning that the production of the energy in the system will approximately cover the total system energy operational needs [6]. Figure 5 presents a vivid exemplary model of a net-zero complex, the Terra – The Sustainability Pavilion at Expo 2020 Dubai. Another significant step is the application of sustainable and green building solutions to act against pollution and emissions production, waste management, and energy saving by reducing total energy consumption [7]. Nevertheless, green building doesn't mean net-zero. Box 2 presents the definition of net-zero.

Fig. 1 Buildings nowadays consume large amount of energy from nonrenewable sources and emit the same of carbon emissions. (Image credit and source: Mohsen Aboulnaga)

Fig. 2 Buildings were built in Brussels, Belgium, in the seventeenth century to respond to climatic conditions and use less amount of energy due to locally sourced materials. (Image credit and source: Mohsen Aboulnaga)

Energy consumption in buildings has been rapidly increasing in the past years more than in other main sectors such as the transportation and industrial sectors [8]. Residential and commercial buildings, according to Ortiz et al., have steadily increased their contribution to global energy consumption, with estimates ranging from 20% to 40% in developed countries [9]. It is highly expected that energy consumption will increase in the next few years due to the continuous demand for energy to achieve higher comfort levels inside buildings, as well as the increasing demand for building services due to the massive population growth all around the world. The energy demand and the growing energy use caused and still causing a lot of adverse effects on the environment, during the period from 1984 to 2004, carbon dioxide (CO_2) emissions have increased by 43%, so the energy usage in buildings could be considered as one of the main reasons of global warming and ozone depletion [10]. According to the International Energy Agency (IEA), global energy-related CO_2 emissions will increase by 6% to 36.3 billion tonnes in 2021, the highest level ever [11]. Hence, the urge to build low-carbon buildings (environment-friendly buildings) or net-zero buildings that use all energy needed from renewable and clean sources while reducing carbon emissions is exclaimed as net-zero building [12].

Box 2 Net-Zero-Energy Building
Net-zero energy building means a building that combines energy efficiency and renewable energy generation to consume only as much energy as can be produced onsite through renewable resources over a specified time period. Department of Energy – Office of Energy Efficiency & Renewable Energy, USA [13]

Fig. 3 Buildings in the medieval city of, in Belgium respond to climatic conditions and use less amount of energy due to locally sourced materials. (Image credit and source: Mohsen Aboulnaga)

Fig. 4 Medieval stones and slates of buildings in Ghent, Belgium, to respond to the climatic conditions and use less energy due to locally sourced materials. (Image credit and source: Mohsen Aboulnaga)

Why Net-Zero?

The energy crisis, in terms of skyrocketing prices increase and current shortage of supply in the third quarter of 2021, coupled with the impact of the COVID-19 crisis and rapid population growth, will force city leaders and local governments to rethink and develop cities to meet climate neutrality. In 2022, this crisis becomes catastrophic due to the Russian-Ukraine war, which not only forces the oil price to reach almost 120.00 US$ since June 6, 2022, but also severely distributes and/or halts the energy and food supply worldwide [14]. According to Khassan et al., a zero-energy building can be identified as a building that has zero-net energy consumption, or in other words, it is the building that depends on renewable energy only. This means that the overall consumption of a building is being used from renewable energy sources such as the solar panels, wind turbines, and any other clean sources [15]. According to Kaewunruen et al., the use of solar energy has become critical for obtaining a zero-energy building since it provides the building with a good amount of energy, especially at places where the sun is always available [16]. Referring to the European Union (EU), from the beginning of 2020, the new buildings should be near-zero-energy buildings. According to El Sayary and Omar, the first step towards achieving a net-energy balance of zero would be to reduce the amount of electricity

Fig. 5 A net-zero model (Solar Complex) Terra – The Sustainability Pavilion at Expo 2020 Dubai – by Grimshaw Global. (Image credit and source: Mohsen Aboulnaga & Amina El-Haggan)

committed and then meet the remaining demand with onsite renewable energy sources [17]. Hence, a net-zero-energy building (NZEB) cannot rely only on the amount of energy generated from clean energy, but it also depends on other techniques and materials used to reduce the need for excessive energy consumption. As part of a research project to create a detailed strategy to achieve net-zero-energy performance level in Egyptian office buildings, Suzuki and Sumiyoshi conducted building energy simulation towards developing a guideline for NZEBs in Egypt [18]. According to Pye et al., an existing residential building was modeled and redesigned to achieve net-zero-energy consumption; this residential building was mocked up in three different versions to compare enhancements in energy performance, such as boosting the thermal efficiency of the building envelope, increasing wall thickness, and inserting smart windows (switchable windows). When compared to the original townhouse, these three solutions can save energy and cost by 8.16%, 10.16%, and 14.65% respectively [19]. Fankhauser et al. also defined the net-zero meaning and how to get it right [20].

What Is a Net-Zero City?

The Climate Adaptation Platform (CAP) defined net-zero city as follows: "It is a city that promotes energy efficiency and renewable energy in all its sectors and activities, extensively promotes green solutions, applies land compactness with mixed land use and social mix practices in its planning systems, and anchors its local development in the principles of green growth and equity. The concept of net-zero carbon emissions operates through social, political, and financial considerations" [21]. However, another net-zero city definition is portrayed in Box 3 [22]. Figure 6 shows examples of renewable energy and energy efficiency manifested in Fujisawa smart, low-carbon, and sustainable city in Yokohama, Japan, and Masdar net-zero city in Abu Dhabi, UAE.

> **Box 3 Definition of Net-Zero City**
> It is a city that produces sustainable, carbon free energy in amounts equal to, or exceeding, the amount which it consumes, which may be accomplished through a combination of the following (as well other) strategies: (i) integrating renewable and distributed energy generation sources (i.e., supply side; (ii) Smart grid technologies to increase energy efficiency, increase the usage of clean power, ad reduce overall consumption, particularly at peak demand periods; and (iii) energy storage technologies. Council Climate Oversight Group (CCOG) [22]

What Is Climate Neutrality?

"Climate neutrality" is identical to the net phaseout of all GHG emanations, while "GHG neutrality" has the same sense, in spite of the fact that it is more particular than climate neutrality. However, "carbon neutrality" encounters the same perception but only includes CO_2 emissions. Climate neutrality in general indicates zero GHG emissions over the building's entire life cycle. The term "GHG neutrality" or

(a) Integrated solar PV panels with green roofs   (b) Solar PV panels installed on buildings' rooftops

Fig. 6 Elements of net-zero in Fujisawa smart, low-carbon, and sustainable city in Yokohama, Grater Tokyo, Japan, and Masdar's net-zero city in Abu Dhabi, UAE (Source: Mohsen Aboulnaga). (**a**) Integrated solar PV panels with green roofs. (**b**) Solar PV panels installed on buildings' rooftops

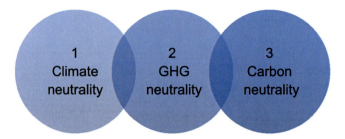

Fig. 7 Various types of building neutrality to mitigate climate change. (Source: developed by authors after [3])

"climate neutrality" refers to the total amount of different emission classifications, which include emissions from embodied energy, operational energy, industrial processes, and non-CO2 GHG emissions [3]. Figure 7 lists the various types of building neutrality to mitigate climate change, while Fig. 8 presents the definition of these types.

The Importance of Net-Zero Cities

The concept of net-zero carbon emissions operates through social, political, and financial considerations. In this context, seven characteristics of net-zero, which are basic to create a viable framework for climate action, were recognized [23]. The properties emphasize the necessity for social and characteristic judgment. This implies that carbon dioxide (CO_2) removals have to be utilized cautiously and the utilization of carbon offsets effectively.

Despite the fact that net-zero target being set, but a query arises on how current net-zero target align [23]. Net-zero must be balanced with broader viable progression objectives, which recommend an impartial net-zero move, socio-ecological supportability, and the interest of wide money-related support. Therefore, limiting temperature rise requires a change in the rate of CO_2 release into the atmosphere and removal into sinks. This point of view offers an arrangement of translations of what net-zero implies and how it ought to be accomplished. Nonetheless, these clarifications guarantee consistency with worldwide temperature objectives, while inserting net-zero into sociopolitical and lawful settings. In debate, it is conceivable to adjust net-zero with maintainable improvement targets, permit for distinctive stages of improvement, and secure zero-carbon success [24], and in any case, there are a few clear limitations.

Net-zero commitments are not an elective to critical and comprehensive outflows cuts. The "net" in net-zero is fundamental, but the requirements for social and natural keenness force firm imperatives on the scope, timing, and administration of both carbon dioxide expulsion and carbon offsets [24]. To understand the various actions to achieve net-zero, there are about six principles that assist in attaining such actions,

Fig. 8 Definitions of neutrality types to mitigate climate change. (Source: Developed by authors after [3]

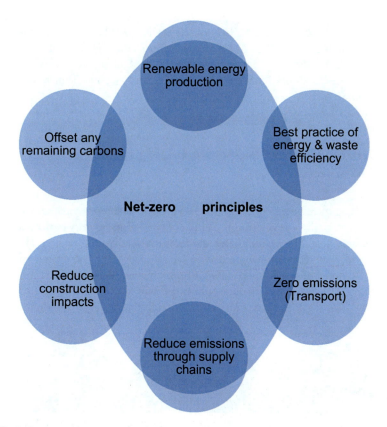

Fig. 9 Actions to achieve net-zero principles. (Source: Developed by authors)

including (a) renewable energy production, (b) best practice of energy and waste efficiency, (c) zero emissions (transport), (d) reduction of emissions through supply chains, (e) reduction of construction impacts, and (g) offset any remaining carbons (Fig. 9).

There are many examples of net-zero in various sectors in cities, but these are not enough to reach the COP26 and GCP 2021 aiming at mitigating 45% of CO_2 by 2030 (almost 8 years today) and achieve net-zero by 2050 [24]. Therefore, it is imperative to highlight that many examples that are considered low-carbon or near-zero buildings, net-zero energy, or net-zero heating whether in buildings or in transport worldwide could guide the architects and local authorities to follow if not enhance future models.

Recent Exemplary Model of Net-Zero Buildings, 2021 to 2022

Between the years 2021 and 2022, there are some vivid examples that have adopted the concept of low-carbon and NZEBs and have showcased clean energy manifestation. These are the iconic BEEAH Headquarters in Sharjah, UAE, designed by Zaha Hadid Architects; the UAE Pavilion, designed by Santiago Calatrava; and The Terra – The Sustainability Pavilion – designed by Sir Nicholas Grimshaw at EXPO 2020 Dubai (Fig. 10).

BEEAH Headquarters in Sharjah UAE by Zaha Hadid Architects

The smart and sustainable building – the last project, designed by the late renowned architect Zaha Hadid – reveals highly efficient operational energy performance and smart technologies. The building, the first fully artificially intelligent (AI) complex in the world, was recently inaugurated on March 31, 2022. The design concept is inspired from nature – the desert – and it mimics sand dunes while creating curvilinear roofs. The design is centered on a two-pillar strategy, which addresses sustainability and digitalization [25, 26]. In addition, a large number of solar

Fig. 10 Examples of net-zero and near-zero buildings. (Source: Developed by authors)

(a) Climatic building shape of interlocking dunes   (b) Flowing efficient envelope of the smart building

Fig. 11 The iconic net-zero, smart, and artificial intelligent building – BEEAH Headquarters in Sharjah, UAE. (Images' credit and source: Fearandloathingindubai https://commons.wikimedia.org/wiki/File: BEEAH_headquarters.jpg). (**a**) Climatic building shape of interlocking dunes. (**b**) Flowing efficient envelope of the smart building

photovoltaic (PV) arrays are installed onsite to enhance the sustainable energy production, hence contributing to mitigating the CO_2 emissions of the building [27]. Its design concept incorporated overall shapes of interlocking dunes that respond to the desert climate with specific orientation as presented in Figs. 11 and 12. The LEED-certified building also exploited local materials and smart technologies, such as onsite water treatment, energy storage via Tesla batteries, and glass-reinforced skin that regulates solar heat gain [28].

By examining the BEEAH artificial intelligent and net-zero building, it is clear that such a building, which spreads over a floor area of 7000 m², incorporates the following smart and green technologies integrated within this iconic headquarters [27, 28]:

- Installed solar power plant to generate 20,000 kW (20 MW) to achieve net-zero emissions.
- Lightweight glass fiber reinforced concrete (GFRC) – approximately 4000 panels to reduce solar radiation and maximize emittance.
- AI integrated ERP systems' streamlines all business functions to ensure speed and efficiency in the workspace (smart meeting rooms).
- AI systems developed by Microsoft, Johnson Controls, and EVOTEQ.
- High-tech infrastructure allows the entire environment to be monitored in real-time.
- Analyze volumes of data to optimize building performance.
- A zero-energy strategy with optimum energy efficiency.
- Smart parking navigation and spot preservation.
- Smart security and access control using facial recognition.
- Office access and building navigation tools.
- Companion app for user guidance before entering the building.
- Productive visitor interactivity through AI concierge.
- Peak working conditions and comfort as well as light and temperature control.

Fig. 12 General view of the envelope and roof of BEEAH Headquarters' net-zero and smart building at night. (Image credit and source: Hufton Crow, https://www.zahahadidarchitect.com)

- Above 90% of materials used for constructing the building are locally sourced and recycled.
- Onsite solar PV farm with Tesla pack battery.

The Terra – The Sustainability Pavilion EXPO 2020 Dubai

The complex encompasses 19 energy trees (E-trees); each of the 18 E-tree canopies – 130 meters wide – emulates the sun flower as shown in Fig. 13. All E-trees rotate 180 degrees to increase the efficiency of the cells that generate 4 GWh of renewable energy as well as produce water from recycled rainwater [29]. There are 4912 ultraefficient monocrystalline PV panels, covering an area of more than 6000 sq.m, which are installed on the 18 trees (each ranging from 15–18 m in diameter) and implanted in glass panels [29, 30]. The Terra – Sustainability Pavilion – generates 28% of the energy needed to power the building [29 27] as presented in Fig. 13. The combination of the cell and the glass casing permits the building to harness solar energy, while granting shade and daylight for visitors. Each E-tree has a canopy serving as a large storm water harvesting and dew that replenishes the building's water systems [30] and harnesses energy (Fig.13). In addition, the Pavilion after Expo will be used as a museum of sustainability to reflect such legacy. Figure 14 illustrates the elements used in the Terra Pavilion.

Fig. 13 The Terra – The Sustainability Pavilion generating 4 GWh annually from renewable energy. (Images' credit and source: Mohsen Aboulnaga and Aya Ghobashy)

The Sustainability Pavilion is designed by the famous British firm Grimshaw Global. According to Grimshaw Global, key sustainability issues are centered on six main factors, including (a) project site, (b) transect zone/climate zone, (c) ecoregion, (d) operation energy/carbon, (e) embodied energy, and (f) water [31]. In alignment with the SDGs, the Pavilion achieved five SDGs, including goal 6 (clean water and sanitation), goal 7 (clean and renewable energy), goal 9 (innovation and infrastructure), goal 12 (responsible consumption and production), and goal 13 (climate action), in addition to goal 4 (quality education in the Legacy mode). A summary of the key sustainability features are listed below [31]:

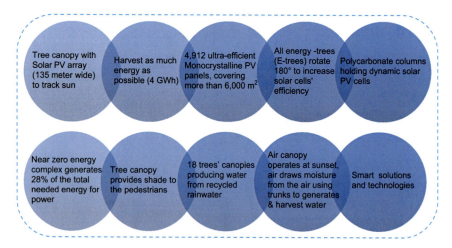

Fig. 14 Elements exploited in the Terra – The Sustainability Pavilion at EXPO 2020 Dubai. (Image source: Developed by authors after Grimshaw Global)

- Energy trees – Eighteen E-trees that support a smaller axially rotating PV array dish on a long stem, furnishing 4000 m^2 of solar PV panels generating about 2.6 GWh power yearly.
- Canopy – The 135-meter-wide multifunction canopy encompasses 1055 solar PV panels covering an area of 8000 m^2.
- The Pavilion energy demand – Produces its own power of 4 GWh which is made available by energy-saving techniques through the building passive design.
- Daylight – Captured where appropriate and a range of light pipe and fiber-optic systems are incorporated to furnish daylight to deep spaces.
- Canopy shading – The shading provided by the canopy reduces energy use of the internal exhibition spaces by decreasing the solar radiation impinging on them.
- Buried accommodation – The Pavilion sits partially below ground, providing a thermal effect which is generally cooler than the ambient temperature in summer.
- Night cooling – Low landscape to southeast of the site maximizes inflow of cool night air.
- Canopy materials – It was built from steel with 97% recycled content and manufactured at a site 15 minutes away from the Pavilion.
- Cement and embodied carbon reduction – Strategies to mitigate the use of cement included constructing about 10,900 m^2 of upper floors with bubble decks that uses around 25% less concrete and resulting in less steel compared to solid concrete slabs.
- Comfort – Earth below ground is fairly cooler due to its high thermal mass; this provides comfort during used periods.
- Water – Rainwater is harvested and percolated into landscape to recharge the groundwater, which is then being extracted and treated for use as potable water within the building, corresponding to a net-zero water ambition.

- Ecology – Integrated into this landscape of native and adapted species are new crops that produce zero-km food and biofuels.
- Education – The design of each exhibit space promotes learning by school groups and other arranged uses, specifically after Expo and in Legacy.

UAE Pavilion EXPO 2020 Dubai

The second recent near-zero-energy building is the UAE Pavilion, which was inspired from the desert nature biodiversity (Falcon). The iconic pavilion features many green technologies and smart solutions (Fig. 15). The pavilion, which was designed by the famous architect, Santiago Calatrava, is spread over 15,000 square meters as illustrated in Fig. 16 [32]. The building has 28 floating movable wings that emulate the falcon's wings and cover the building's roof. Each wing has solar PV arrays to generate clean energy and is covered by dynamic rips to protect these panels from rain and dust during sand storms [33–35] as presented in Figs. 16 and 17 .

Global Examples of Net-Zero-Energy Buildings

NZEBs around the world incorporate different feasible vitality arrangements that consolidate and progress the building CO_2 outflows, i.e., cooling framework coefficient of execution COP26 and producing clean energy. All these arrangements

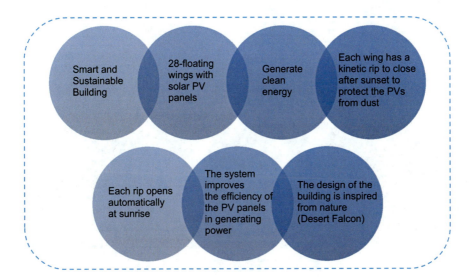

Fig. 15 Features integrated in the UAE Pavilion at EXPO 2020 Dubai. (Image source: Developed by authors)

Fig. 16 The iconic UAE Pavilion with 28 floating wings integrating solar PV panels, EXPO 2020 Dubai. (Image credit and source: Mohsen Aboulnaga and Amina El-Haggan)

Fig. 17 The floating wings generate clean energy from renewable sources for the building energy annual consumption. (Image credit and source: Mohsen Aboulnaga and Amina El Haggan)

have a more noteworthy viewpoint towards climate moderation and offered to worldwide directions and developments [36]. Global examples of net-zero buildings have exempted huge transitions in terms of innovation and clean energy production to ensure the lowest carbon emissions by buildings and adoption of GHG offset measures. There are multiple techniques and strategies that can be applied that serve the needs of each case to adapt to the surrounding conditions. Figure 18 portrays the examples selected for case studies that address net-zero energy targets.

Dalian Convention Centre in Dalian, China

The new conference center and opera house of Dalian are living proof of the Chinese sustainable construction goals. It is designed by Austrian architect Wolf D. Prix, where energy consumption levels are kept at minimum through the use of thermal energy, passive ventilation, and solar energy [37]. Allowing the building to produce energy using seawater helps in cooling the building in summer and its heating in winter as presented in Figs. 19. As illustrated in Fig. 19a, the building is covered by a huge number of integrated PV cells on the curved roof to generate clean energy. Figure 19b portrays the magnificent flowing building at night. The façade is created to protect the building from solar radiation, while maintaining an adequate amount of sunlight and natural light inside the center's spaces (Fig. 19c, d). In the smart and green buildings, the following features are adopted and integrated:

- Net-zero building
- Use of renewable energy (solar PV) covering the rooftop to generate clean energy
- Integrate high emissivity and reflectance metal material on the building's smart envelope
- Achieve energy efficient, smart, and green building to mitigate CO_2 emissions
- Use seawater for cooling and heating by virtue of heat pumps
- Enhance comfort by controlling daylight
- Provide good sunlight inside the spaces while minimizing heat and glare

National University of Singapore, Singapore

Within Asia, another net-zero building has been developed and operated in Singapore. This net-zero building is the Singapore Design and Environmental School where solar PV arrays are integrated to generate clean energy. At the National University of Singapore, the University Hall, which is a remarkable net-zero building, is designed by Serie Architects and Singapore studio, Multiply Architects (Fig. 20). As shown in Fig. 20a, b, the building is intended to operate as a NZEB and generate its own energy needs from clean sources [38]. Figure 21 presents a side view of the University Hall, while Fig. 22 shows the shaded glass by the roof

1. Dalian Convention Centre in Dalian, China
2. University Hall, National University of Singapore
3. CIC zero-carbon Building, Hong Kong
4. The NIER–CCRC Building, Incheon, SK

5. Incheon-Seoul Int'l Airport, Incheon, S. Korea
6. Incheon National University, S. Korea
7. UN Headquarters, G-Tower in Incheon, SK
8. Lumen building, Wageningen University, The Netherlands

9. The Edge Office building in Amsterdam
10. GSK's Lab. building, Nottingham University, UK
11. DeepStone House in Scotland, UK
12. Solar Settlement & Sun ship, Germany

13. Plus-energy house, Germany
14. Kunsthaus Bregenz Ar Museum, Austria
15. A Holiday Home, Spain
16. Unisphere Building, USA
17. NREL campus, Colorado, USA
18. La Jolla Commons, CA

Fig. 18 Selected global buildings' examples integrating net-zero-energy strategies. (**1**) Dalian Convention Centre in Dalian, China. (**2**) University Hall, National University of Singapore . (**3**) CIC zero-carbon Building, Hong Kong. (**4**) The NIER–CCRC Building, Incheon, SK . (**5**) Incheon-Seoul Int'l Airport, Incheon, S. Korea . (**6**) Incheon National University, S. Korea. (**7**) UN Headquarters, G-Tower in Incheon, SK. (**8**) Lumen building, Wageningen University, The Netherlands. (**9**) The Edge Office building in Amsterdam. (**10**) GSK's Lab. building, Nottingham University, UK. (**11**) DeepStone House in Scotland, UK. (**12**) Solar Settlement & Sun ship, Germany. (**13**) Plus-energy house, Germany. (**14**) Kunsthaus Bregenz Art Museum, Austria. (**15**) A Holiday Home, Spain. (**16**) Unisphere Building, USA. (**17**) NREL campus, Colorado, USA. (**18**) La Jolla Commons, CA

(a) Dalian Conference Center's external façade acts as a screen light penetration

(b) General view of the vivid Dalian Conference Center in Dalian city at night.

(c) The building system controls facades and maintains ventilation and provides sunlight

(d) The interior consists of green areas to purify the atmosphere and skylights permit daylight.

Fig. 19 Dalian Convention Center in Dalian China, a near-zero-energy building. (Image credit and source (**a–d**): Forgemind ArchiMedia, https://commons.m.wikimedia.org/wiki/File:Dalian_ International_Conference_Center.jpg). (**a**) Dalian Conference Center's external façade acts as a screen light penetration. (**b**) General view of the vivid Dalian Conference Center in Dalian city at night. (**c**) The building system controls facades and maintains ventilation and provides sunlight. (**d**) The interior consists of green areas to purify the atmosphere and skylights permit daylight

canopy and inclined columns of Yong Siew Toh Conservatory at the National University of Singapore in Singapore.

The CIC Zero Carbon Building in Hong Kong, China

The third example of net-zero-energy building is the CIC Zero Carbon Complex in Hong Kong, China. It is a pioneering project that exhibits the state of the art in zero-carbon building (ZCB) technologies and raise community awareness of sustainable living in Hong Kong, China. The CIC Zero Carbon Building achieves significant energy savings through daylighting, cross-ventilation, a high-performance façade, high-volume-low-speed fans, radiant cooling, and desiccant dehumanization (Figs. 23 and 24). The entrance of the Zero Carbon Building is covered by a large area of solar PV arrays which is illustrated in Fig. 25. Besides, local materials, such

(a) The entrance of the University Hall building, National University of Singapore

(b) Close up of the University Hall building, the National University of Singapore

Fig. 20 The net-zero University Hall Building, National University of Singapore, Singapore (Images credit and source: (**a**) Sengkang, https://commons.wikimedia.org/wiki/File:NUS,_University_Hall,_Nov_06.JPG); (**b**) Alex.ch, https://commons.wikimedia.org/wiki/File:University_Hall,_National_University_of_Singapore_-_20070125.jpg). (**a**) The entrance of the University Hall Building, National University of Singapore. (**b**) Close-up of the University Hall Building, the National University of Singapore

Fig. 21 Side view of the net-zero University Hall, National University of Singapore, Singapore. (Image credit and source: Joshua Rommel Hayag Vargas https://commons.wikimedia.org/wiki/File:University_Hall,_National_University_of_Singapore,_February_2020.jpg)

as glass and timber, were used to further lower the carbon footprint of the building and onsite PVs as well as a large-scale biodiesel generator that produces more energy than the building consumes [39].

Fig. 22 Glazed façade of the Yong Siew Toh Conservatory at the National University of Singapore, Singapore. (Image credit and source: Joshua Rommel Hayag Vargas, https://commons.wikimedia.org/wiki/ File:YST_ Conservatory, _National_University_of_Singapore,_February_2020.jpg)

National Institute of Environmental Research (NIER) Building in Incheon, South Korea

The fourth iconic example that demonstrates solutions and technologies for zero-carbon building is the National Institute of Environmental Research (NIER) in Incheon, South Korea. The NIER and Climate Change Research Building in Incheon is South Korea's first zero-carbon building; it was completed in 2010 and ranked the world's first carbon-neutral and net-zero-energy commercial-scale office building as well as the world's fourth largest NZEB as of 2012. The building includes the

Fig. 23 The CIC Zero Carbon Building layout. (Image credit and source: Wpcpey, https://commons.m.wikimedia.org/wiki/File:CIC_Zero_Carbon_Building_Overview_201708.jpg)

Fig. 24 Back view of the CIC Zero Carbon Building showing the solar PV panels on top of the roof. (Image credit and source: Ceeseven, https://commons.m.wikimedia.org/wiki/File:CIC_Zero_Carbon_Building.jpg)

office of Climate Change Research Center (CCRC). It is considered as the world's 4th zero-carbon building as of 2012. The building's facades are cladded by solar PV arrays as shown in Fig. 26. The NIER building exhibits the state of the art in zero-carbon building technologies [40]. Figure 27 also illustrates different types of solar

Fig. 25 The entrance of the CIC Zero Carbon Building showing the solar PV arrays. (Image credit and source: Ceeseven, https://commons.m.wikimedia.org/wiki/File:CIC_Zero_Carbon_Building.jpg)

PV panels installed on the façade of the building to generate the whole energy demand needed to power and operate the building over the year.

Incheon International Airport Transportation Center in Incheon, South Korea

Incheon-Seoul International Airport in South Korea is portraying not only a low-carbon building but also a smart and efficient transportation center. Incheon Aerotropolis is a global leader in aviation-linked commercial development and smart technology applications. The building is integrating panoramic curved glass walls allowing generous natural light to nurture extensive interior gardens as shown in Figs. 28, 29, and 30. The center's sustainable techniques include water and waste management, low-carbon management, maximization of the use of natural illumination and ventilation, and reduction of carbon footprint [41].

Most conventional solar cells use visible and infrared light to generate electricity. In contrast, the innovative new solar cell also uses ultraviolet radiation. It replaces conventional window glass, or placed over the glass, and installation surface area could be large, leading to potential uses that take advantage of the combined

Towards Climate Neutrality: Global Perspective and Actions for Net-Zero Buildings... 397

Fig. 26 The NIER and CCRC zero-carbon building in Incheon, South Korea. (Image credit and source: Mohsen Aboulnaga)

Fig. 27 Various types of solar PV panels installed on the façade of NEIR/CCRC building and solar collector arrays on rooftop. (Image credit and source: Truth Leem, Reuters https://www.reuters.com/articles/us-korea-green-/south-korea-unveils-completely-eco-friendly-building-idIN TRE74J1FE20110520)

Fig. 28 Incheon-Seoul Airport Terminal envelope engulfs cells that contain titanium oxide coats with a photoelectric dye for generating electrical power for lighting and temperature control. (Image credit and source: Mohsen Aboulnaga, 2015)

functions of power generation, lighting, and temperature control. Another type of transparent PVs is "translucent photovoltaics" (transmit half of the light that impinges on them). Similar to inorganic PVs, organic PVs are capable of being translucent. Incheon-Seoul airport solar panels' curved structure uses a tin oxide coating on the inner surface of the glass panes to conduct current out of the cell as illustrated in Figs. 28, 29, and 30. The cell contains titanium oxide that is coated with a photoelectric dye [42, 43].

Incheon National University Campus in Songdo – Incheon, South Korea

The new campus of Incheon National University (INU) in the smart city of Songdo in Incheon, South Korea, is another vivid example of a low-carbon campus, with nearly 20% of the electricity energy consumption in the campus and its buildings generated from clean sources, as shown in Fig. 31. As illustrated in Fig. 31a, a solar

(a) The terminal's façade fixed with smart glass by Saint Gobain to protect it from sun radiation.

(b) Incheon International Airport Station with low carbon transportation.

Fig. 29 Incheon-Seoul Airport Terminal, South Korea, a near-zero-energy building (Image credit and source: (**a**) Minseong Kim, https://commons.m.wikimedia.org/wiki/File:IncheonTerminal1. jpg#mw-jump-to-license (**b**) Siqbal, https://commons.m.wikimedia.org/wiki/File:Incheon_Hall. jpg). (**a**) The terminal's façade fixed with smart glass by Saint-Gobain to protect it from sun radiation. (**b**) Incheon International Airport Station with low-carbon transportation

Fig. 30 Interior of Incheon-Seoul Airport Terminal, South Korea, a near-zero-energy building. (Image credit and source: Eliazar Parra Cardenas, https://commons.m.wikimedia.org/wiki/File:Incheon_airport.jpg)

plant comprising 80 solar PV panels, each produces 6 kWp. Hence, the total power of this array amounts to 480 kWp. Figure 31b also shows another solar plant with a 90 PV array that generates 540 kWp output. Figure 32 depicts the installation of two additional solar PV plants with 90 panels each (360 panel array) within the landscape of the INU Campus. According to a campus visit conducted by the lead author from October 18 to 21, 2015, the INU campus manifests near-zero-energy buildings.

United Nations' Headquarters in G-Tower, Incheon, South Korea

The United Nations Economic and Social Commission for Asia and Pacific (ESCAP) Headquarters in Incheon is an excellent example of a smart and net-zero-energy building, where smart solutions and technologies are incorporated in each floor and the building as a whole to achieve a near-zero-carbon exemplary facility as shown in Figs. 33, 34, and 35. The 33-storey high building is also equipped with smart light movement and occupancy sensors to reduce energy usage (Fig. 33). Th building is also a zero-waste system, where all waste is collected, burned in a special container in the basement floor, and then converted into energy at zero waste. This was observed during the visit by the lead author with A.NERGY delegates to the UN building in Incheon in November 2016 (Figs. 34 and 35).

Many buildings located in Asia, and Europe, particularly in the Netherlands, the United Kingdom, Germany, and Austria, demonstrate the concept of low-carbon, zero-carbon, and net-zero building through the use of smart and green solutions and innovative technologies. These examples include Lumen Building at Wageningen University in Wageningen and the Edge Office Building in The Netherlands; GlaxoSmithKline net-zero Laboratory at Nottingham University Jubilee Campus in England and DeepStone House in Scotland, UK; and The Solar Settlement and Sun Ship and the Plus-Energy House in Frankfurt, Germany, as well as Kunsthaus Bregenz Art Museum in Bregenz, Vorarlberg, Austria.

The Lumen Building at Wageningen University, Wageningen, The Netherlands

The Lumen, known as the Lumen Greenhouse Building, Wageningen University in The Netherlands, portrays a model when it comes to low-energy and near-zero-carbon concept. The interior has a green space under a glass controlled roof, in a high-quality environmental condition at the campus [44]. Figure 36 shows the interior of the building spaces, which fully receive natural light through a glazed wide atrium. The green areas reduce the cooling loads, hence the energy needed (Fig. 36a, b). In terms of a zero-carbon transport on campus, students, staff, and visitors at the

Towards Climate Neutrality: Global Perspective and Actions for Net-Zero Buildings... 401

(a) Font view of a 180-solar PV array installed at Incheon National University green campus

(b) View of a 180-solar PV plant installed at Incheon National University (INU) campus

Fig. 31 Solar PV plants to generate clean energy at INU low-carbon campus in Songdo, Incheon, South Korea. (Image credit and source: Mohsen Aboulnaga). (**a**) Front view of a 180 solar PV array installed at INU green campus. (**b**) View of a 180 solar PV plant installed at INU campus

campus are traveling using a clean means of transportation electrical bus called "self-moving WEpod shuttle buses" at the Wageningen University Campus [45]. The driverless six-seater shuttle bus drives at 40 km/h [45] as illustrated in Fig. 37.

(a) Another Solar PV plant with 90 panels capacity installed at Incheon National University

(b) A close up view of two arrays of 180-soalr panels, each 90 PV panels installed at INU

Fig. 32 Solar PV plants (540 cells) to generate clean energy at INU low-carbon campus in Songdo, Incheon, South Korea. (Image credit and source: Mohsen Aboulnaga). (**a**) Another solar PV plant with a 90-panel capacity installed at INU. (**b**) A close-up view of two arrays of 180 solar panels, each 90 PV panels installed at INU

Towards Climate Neutrality: Global Perspective and Actions for Net-Zero Buildings... 403

Fig. 33 View of the smart and net-zero G-Tower, the United Nations Headquarters in Songdo, Incheon, South Korea. (Image credit and source: Piotrus, https://commons.m.wikimedia.org/wiki/Category:G-Tower,_Incheon#/media/File%3ASongdo_International_Business_District_11.JPG)

Fig. 34 The interior of the UNDP Asia Headquarters, fully equipped with a smart and near-zero-carbon technologies, Songdo, Incheon, South Korea. (Image credit and source: Mohsen Aboulnaga)

Fig. 35 General view of the smart and net-zero G-Tower, the headquarters of United Nations offices in Songdo, Incheon, South Korea. (Image credit and source: http://www.undog.org/page/sub3_1_view.asp?sn=37&page=4&search=&SearchString=&BoardID=0004)

The Edge Smart and Net-Zero Office Building in Amsterdam – The Netherlands

The Edge net-zero office building of Deloitte Headquarters in Amsterdam, The Netherlands, is the greenest and smartest building globally based on the British Rating Agency BREEAM at a score of 98.4%. The iconic building design strategy is centered on resource efficiency, information technology, and Internet of Things (IoT) as well as rentable energy through solar energy, where the building generates more electricity than it consumes [46]. The 15-storey north-facing glass atrium smart building is equipped with about 28,000 sensors to control motion, light, temperature and humidity, and infrared to provide excellent indoor environmental quality as shown in Fig. 38. The atrium is the building's key center of solar system, and its air volume has mesh panels between each floor to let musty air fall into open spaces, where it rises and is exhausted through the roof creating natural ventilation and assuring excellent air quality as depicted in Fig. 40 [47].

On the southern façade of the Edge building, there is a checkerboard of solar panels and windows as shown in Figs. 39a, b. The thick load-bearing concrete assists in regulating heat, and deeply recessed windows reduce the need for shades, though they are directly exposed to the sun [46]. The roof incorporates solar PV (OVG) panels to enable the office building to generate more electricity than it uses. The Edge building also consumes 70% less electrical energy than any typical office

Towards Climate Neutrality: Global Perspective and Actions for Net-Zero Buildings... 405

(a) Interior of Lumen greenhouse building illuminated naturally by daylight through glazed atrium.

(b) Large green space covering most the interior under glass roof to cool and save energy.

Fig. 36 Lumen Greenhouse Building at Wageningen University, a low-energy concept and near-zero-carbon building, Wageningen, The Netherlands. (Image credit and source: (**a**) Vincent, https://upload.wikimedia.org/wikipedia/commons/9/99/Lumen_Builiding_Greenhouse.jpg (**b**) Vincent, https://upload.wikimedia.org/wikipedia/commons/f/f0/Wageningen_Uinveristy_-_Building_Lumn.JPG). (**a**) Interior of Lumen Greenhouse Building illuminated naturally by daylight through glazed atrium. (**b**) Large green space covering most of the interior under glass roof to cool and save energy

building. In addition, the building pumps warm water (more than 37 m^2 deep) into the aquifer beneath the office building in the summer months, where it sits, insulated, until winter, when it's exploited for heating. This is considered the most efficient thermal storage worldwide [47]. Moreover, the Edge building is an open smart space with no office desks and is naturally illuminated (Fig. 40a, b). It uses superefficient LED panels, powered by the same cables that carry Internet data. The building is wired with a vast network of two types of long blue tubes: one binds data (Ethernet cables) and the other holds water for radiant heating and cooling (Fig. 40c).

Fig. 37 The self-moving WEpod shuttle bus, a clean means of transportation passing by the Lumen Greenhouse Building at Wageningen University campus, Wageningen, The Netherlands. (Image credit and source: ArjanH, https://commons.m.wikimedia.org/wiki/File:Wepods_WUR.jpg#)

GlaxoSmithKline Carbon-Neutral Building, Nottingham University, Nottingham, UK

The GlaxoSmithKline (GSK)'s carbon-neutral building (CNB) at Nottingham University Jubilee Campus in Nottingham, England, United Kingdom, is the first laboratory building with neutral carbon life cycle. The GSK building presents an excellent example of a low-energy building or a zero-carbon building. The CNB is constructed according to the highest energy efficient standards and is cladded by timber (a high-performance insulation material) to protect the building from heat losses in winter and heat gain in summer (Fig. 41) [48]. The remaining energy requirements are also met by onsite sustainable biomass. The excess energy generated onsite will pay back the embodied carbon associated with its construction within 25 years [48]. Figure 42 shows the façades with efficient glass and roof solar chimneys exploited for natural ventilation and cooling in summer. Figure 42a, b present the entrance, façade, and timber cladding. In addition, the CNB has green roofs on 45% of its surface [49].

The building occupies 4500 square meters over two floors in addition to laboratory space for around 100 researchers. This building, which was opened on February 27, 2017, is part of GSK's target of reaching a carbon-neutral value chain by 2050. The GSK's building was awarded both the BREEAM Outstanding and LEED Platinum certifications [50]. Such a facility has energy-intensive cooling systems

Fig. 38 The Edge, the greenest office building in Amsterdam. (Image credit and source: MrAronymous, https://upload.wikimedia.org/wikipedia/commons/c/c1/Zuidas_20210512_%286%29_uitsnede.jpg)

that are necessary to stop temperatures reaching levels where solvents will evaporate. In the meantime, recovering access heat from processes can be challenging due to the risk of chemical and fume corrosion on the ventilation systems. In terms of power generation, the GSK building has an onsite system that produces 125 kWe of biofuels combined heat and power (CHP) furnishing the majority of heat required for the buildings. In addition to mitigating carbon emissions, the CHP system exports the excess heat to nearby buildings on the Nottingham University Jubilee Campus [50].

Moreover, the GSK's building exhibits the state-of the-art technologies where about 231 kWp solar PV array is mounted on the main building's roof covering 45% of its area. For energy efficiency, the building is fitted throughout with LED lighting at an average of 5.4 watts per square meter [50]. In general, the GSK building saves more than 60% on energy and uses only 15% of the heat required by a conventional building design. Excess energy generated by this building (near 40 MWh) provides enough carbon credits over 25 years to offset the construction phase as well as being utilized to heat the nearby office development onsite [50].

(a) A checkerboard of solar panels and windows to regulate heat and respond to weather conditions to minimise energy use

(b) The façade with smart glazing to control heat losses in winter and heat gain in summer months

Fig. 39 The Edge's smart façade in Amsterdam. (Image credit and source: (**a**) MrAronymous, https://upload.wikimedia.org/wikipedia/commons/9/9e/Zuidas_20210512_155017_uitsnede.jpg (**b**) MrAronymous, https://upload.wikimedia.org/wikipedia/commons/9/9e/Zuidas_20210512_ 155017_uitsnede.jpg). (**a**) A checkerboard of solar panels and windows to regulate heat and respond to weather conditions to minimize energy use. (**b**) The façade with smart glazing to control heat losses in winter and heat gain in summer months

Towards Climate Neutrality: Global Perspective and Actions for Net-Zero Buildings... 409

(a) Open office spaces fitted with IoT and mobiles' App to provide livable and social prolific space   **(b)** Close up of the open and smart spaces inside the Edge net-zero office building   **(c)** Long blue tubes, one binds data (Ethernet cables); and second holds water

Fig. 40 The Edge's smart interior utilizing IoT in operation, Amsterdam. (Images' credit and source: (**a**, **c**) Raymond Wouda; (**b**) Tom Randall https://www.bloomberg.com/features/2015-the-edge-the-worlds-greenest-building/). (**a**) Open office spaces fitted with IoT and mobiles' app to provide livable and social prolific space. (**b**) Close-up of the open and smart spaces inside the Edge net-zero office building. (**c**) Long blue tubes, one binds data (Ethernet cables), and the other holds water

DeepStone House in Scotland, United Kingdom

The DeepStone House in Scotland, United Kingdom, is an example of a near-zero-energy small building, where it generates its electrical energy demand from renewable sources. The house – designed by Simon Winstanley Architects – is overlooking the Solway Firth in southwest Scotland, UK. The low-energy house has inclined roof fitted with an array of solar PV panels to track the sunlight and produces electrical energy it nearly consumes. Figure 43 presents the near-zero-energy building with its tilted solar PV array on the rooftop as well as the high-insulation exterior walls and triple-glazed thermal windows. It is clear from Fig. 43 that the envelope is cladded with timber to conserve energy in winter and reduce heat gain in summer [51].

The Solar Settlement and Sun Ship in Schlierberg, Germany

The Solar Settlement and Sun Ship complex in Schlierberg, Germany, is considered one of the first housing communities in the world. All houses are net-zero and generate a positive energy balance and are emission-free and CO_2 neutral [52]. As of 2022, it is the largest residential roof-integrated PV system. The south-facing roofs

Fig. 41 GlaxoSmithKline Carbon-Neutral Building at Nottingham University Jubilee Campus, England, UK. (Image credit and source: Michael Thomas, https://upload.wikimedia.org/wikipedia/commons/e/e1/GlaxoSmithKline_Carbon_Neutral_Building._Nottingham_University_Jubilee_Campus.jpg)

are covered with solar PV arrays. Figure 44 shows a bird's-eye view of the solar settlement, while Figs. 45, 46, and 47 present the houses with solar PV arrays on top of the tilted roofs.

The concept is centered on achieving the structure's extreme energy efficiency, so that it holds a positive energy balance and produces more energy than it uses. Each house is built from a timber skeleton and integrated with eco-friendly materials with 91 solar PV arrays (Figs. 46 and 47). Also, the Sun Ship's building consists of retail, commercial, and residential functions. The premises remain vehicle-free, thanks to the parking garage underneath the Sun Ship as shown in Fig. 48 [52].

Prototype of a Plus-Energy House, Frankfurt, Germany

Frankfurt City in Germany showcases many prototypes of plus-energy houses. The basic concept of the energy system utilized in the project is mainly to reduce the energy demand and produce it sensibly and efficiently. This has been achieved by an optimal interplay of various passive (low-tech) and active (high-tech) elements [53]. In this concept, the integral and sensible combination of the individual subsystems

(a) Entrance to GSK Laboratory and its façades with efficient glass and cladding to save energy.

(b) Close up of the roof line with chimneys on the rooftop to promote natural ventilation and clean air

Fig. 42 The façade and roof of the first laboratory building with neutral carbon life cycle at Nottingham University, UK. (Image credit and source: (**a**) Michael Thomas, https://upload.wikimedia.org/wikipedia/commons/c/c7/GlaxoSmithKline_Carbon_Neutral_Building_Entrance._Nottingham_University_Jubilee_Campus.jpg (**b**) Michael Thomas, https://upload.wikimedia.org/wikipedia/commons/e/e8/GlaxoSmithKline_Carbon_Neutral_Building_for_Sustainable_Chemistry._-_40557421815. jpg). (**a**) Entrance to GSK laboratory and its façades with efficient glass and cladding to save energy. (**b**) Close-up of the roof line with chimneys on the rooftop to promote natural ventilation and clean air

Fig. 43 Side view of DeepStone net-zero House in Scotland, UK. (Image credit and source: Aswinstanley, https://upload.wikimedia.org/wikipedia/commons/c/c5/Side-view1-photoshop-filtered.jpg)

Fig. 44 The Solar Settlement in Schlierberg, Germany. (Image credit and source: Andrewglaser https://commons.m.wikimedia.org/wiki/File:SoSie%2BSoSchiff_Ansicht.jpg#mw-jump-to-license)

Fig. 45 Close-up of the Solar Settlements houses with PVs covering rooftops. (Image credit and source: Andrewglaser, https://commons.m.wikimedia.org/wiki/File:LuftSS.jpg#mw-jump-to-license)

is very important for an optimized and innovative overall energy system that combines building components and technologies and exploits synergies. In the interests of an integral view of the building, already in the design process, construction materials and systems were taken into consideration in order to provide a comfortable

Towards Climate Neutrality: Global Perspective and Actions for Net-Zero Buildings... 413

Fig. 46 A Plus-energy house in the Solar Settlement in Germany. (Image credit and source: Andrewglaser https://commons.m.wikimedia.org/wiki/File:SSHaus.jpg#mw-jump-to-license)

Fig. 47 View of the 91 solar PV array covering the tilted roof of the net-zero houses. (Image credit and source: Andrewglaser, https://commons.m.wikimedia.org/wiki/File:SoSie%2BSoSchiff_Ansicht.jpg#mw-jump-to-license)

Fig. 48 The Sun Ship's retail, commercial, and residential integrated building. (Image credit and source: Andrewglaser, https://commons.m.wikimedia.org/wiki/File:SoSchiff_Ansicht.jpg#mw-jump-to-license)

Fig. 49 Plus-Energy House in Frankfurt, Germany. (Image credit and source: DontWorry https://commons.wikimedia.org/wiki/File:Plus-energie-haus-ffm-033.jpg#mw-jump-to-license)

Fig. 50 Kunsthaus Bregenz Art Museum in Bregenz – Vorarlberg, Austria. (Image credit and source: Raymond, https://commons.wikimedia.org/wiki/File:Bregenz_kunsthaus_zumthor_2002_01.jpg)

and an energy-efficient indoor climate without technical aids as depicted in Fig. 49 [53].

Kunsthaus Bregenz Art Museum in Bregenz – Vorarlberg, Austria

Another example which is not far from Germany is the Kunsthaus Bregenz Art Museum in Bregenz – Vorarlberg, Austria. The art museum, which is made of glass and steel and a cast concrete stone mass, stands in the light of Lake Constance. The vivid glass building endows the interior of the building with texture and spatial composition as presented in Fig. 50. The exterior of the museum absorbs the changing light of the sky and the haze of the lake and then reflects light and color to furnish an intimation of its inner life according to the angle of vision, the daylight, and the weather as illustrated in Fig. 51 [54].

Fig. 51 Exterior and facades of the Art Museum. (Image credit and source: DVD RW, https://commons.wikimedia.org/wiki/ File:Bregenz_kunsthaus_zumthor_2002_02.jpg)

Fig. 52 The holiday housing in Natural Park of Valles Pasiegos, Northern Spain. (Image credit and source: Lizzie Crook, https://www.dezeen.com/2019/04/19/villa-slow-laura-alvarez-architecture/)

A Holiday Home in Rural Spain

Spain also showcases a small-scale zero-energy building model. The holiday home in the Natural Park of Valles Pasiegos located in northern Spain is designed by Architect Laura Alvarez (Fig. 52). This home is living proof of a NZEB, and it produces energy more than it consumes. Also, a heat pump system is connected to the city electric grid, and it is able to generate five times the energy it uses from the grid. In addition, the building uses local ancient stones to maximize heat insulation and minimize the heat gain into indoor spaces as shown in Fig. 52. Because of the locally sourced materials, the NZEB not only provides a good indoor environment but is also considered a sustainable building [55].

From Europe to USA, the concept of NZB has been adopted in many buildings, such as the Unisphere Building in Maryland, the National Renewable Energy Laboratory (NREL) in Colorado, and the La Jolla Commons net-zero-energy office building in San Diego, California, USA.

The Unisphere, Maryland, USA

The Unisphere office building in Maryland incorporates a wide range of zero-energy solutions including a geothermal system that relies on soil heating-cooling context in addition to an electromagnetic building envelope that provides a high insulation level by changing the tint level based on weather conditions. The 3000 panels of PV roof array provide 1175 MWh of clean energy every year, in excess than the building consumes, enough to power 100 homes, and the excess power is sent back to the utility grid [56]. The façade's operable windows and panels of the building are naturally ventilated and harvest daylighting to minimize artificial lighting, thus reducing energy demand as shown in Fig. 53. Additionally, the Unisphere building has 52 closed-loop, dual circuited geo-exchange wells 46.5 m^2 beneath it to provide energy storage [56].

National Renewable Energy Laboratory, Colorado, USA

The National Renewable Energy Laboratory (ENRL) is among the first buildings that were developed based on a NZEB's concept. The ENRL was built in 1947 and is considered a state-of-the-art example of sustainable features and technologies for energy production including solar energy, wind energy, and aiming at mitigating emissions and setting example for others.

The building integrates a wide range of energy-efficient strategies and solutions that are continuously developed to achieve climate mitigation as shown in Figs. 54a, b. These developments include renewable energy solutions such as biochemical

Fig. 53 The Unisphere net-zero office building and its efficient glazed envelope, Maryland, USA. (Image credit and source: Rdellbillings, https://www.wikiwand.com/en/United_Therapeutics)

(a) The NREL campus in Colorado fully covered by Solar PV panels to generate clean energy

(b) The NREL buildings integrated with energy systems for a zero-energy target

Fig. 54 The National Renewable Energy Laboratory Campus in Colorado, USA. (Images credit and source: (**a**) NREL, https://www.nrel.gov/workingwithus/partnering-facilities.html (**b**) NREL, https://www.nrel.gov/workingwithus/partnering-facilities.html). (**a**) The NREL campus in Colorado fully covered by solar PV panels to generate clean energy. (**b**) The NREL buildings integrated with energy systems for a zero-energy target

(a) Façades with low-emissive coating reflects IR heat   (b) Fuel cells to generate 5.4 MWh at the back

(c) Close up of the buildings' fuel cells generating 5.4 MWh

Fig. 55 The La Jolla Commons net-zero building in San Diego, CA, USA. (Source: (**a–c**) https://www.researchgate.net/publication/341264745_Net-Zero_Energy_Buildings_Principles_and_Applications). (**a**) Façades with low-emissive coating reflects IR heat. (**b**) Fuel cells to generate 5.4 MWh at the back. (**c**) Close-up of the buildings' fuel cells generating 5.4 MWh

conversion, thermochemical conversion, micro-algal biofuels, and biomass processing [57].

La Jolla Commons, San Diego, California

Another example of a large-scaled zero-energy building is La Jolla Commons in San Diego, California, USA. The building, which is a 13-storey office space and is one of the largest NZEBs in the United States, was designed by architect Paul Danna (AECOM) between 2006 and 2008. It is one of the first examples built in the USA with a net-zero-energy strategy. Figure 55 presents the La Jolla Commons zero-energy building's façades and fuel cells [58]. The envelope incorporates an insulated double glazing (Figs. 55a) in addition to an efficient systems to assist in reducing energy use by controlling heat losses and gains [58]. Fuel cells are also used at a capacity of 5.4 MWh, while the historical consumption of the La Jolla Commons building consumption is about 4.5 MW as presented in Figs. 55b, c [58].

Net-Zero Heating Building's Concept

The concept of net-zero-heating building (nZHB) or near-zero-heating building is a strategy in which such a building has essentially zero-heating energy demand described to be less than 3.0 kWh/m² annually. It is intended for use in heating-dominated areas, and it is used to supersede NZEBs as a way to bring building-related GHG emissions to zero [59]. Many examples will be presented and discussed in the following section, such as (a) the American Geophysical Union (AGU) headquarters in Florida, USA; (b) the zero-heating Samling Library in Nord Odal, Norway; (c) the net-zero-heating office building in Rakvere, Estonia; and (d) the net-zero-heating office building in the Netherlands.

The American Geophysical Union Headquarters in Florida, USA

The American Geophysical Union (AGU) headquarters in Florida, USA, has been upgraded into a model of energy efficiency to achieve net-zero goals. The changes encompass strategy in shading and envelope better insulation, daylighting, and new window glass leading to energy efficient walls, and it has a 4.88-meter-tall rooftop solar PV array installed on a projecting canopy as presented in Fig. 56. Also, it uses a sewer heat exchange, through which AGU would capture the energy flowing through a large Florida Avenue sewer main and run it back into the building [60]. As shown in Fig. 56, the retrofitted envelope and solar PV array on rooftop generate clean energy and mitigate carbon emissions.

The Zero-Heating Samling Library in Nord Odal, Norway

The Samling Library, located in the village of Sand at Nord Odal in Norway, is another example of net-zero-heating building (nZHB). The façades consist of a timber batten cladding of vertical wooden slats around the entire building in horizontal bands. The building has an efficient fabric with Air® 6-pane glazing by Reflex, Slovenia, and glass U-value of 0.26 W/m²K [61]. The timber cladded skin and the large glass surfaces of the public spaces provide a depth effect, yet create a transparent skin between the interior and exterior, enriching the visual connection (Fig. 57). The interior materials and environment are also constructed from timber. The nZHB inspires a more sustainable approach by utilizing a significant amount of local wood, representing the cultural tradition of wooden buildings and the local wood industry. The distinctive timber ceiling hides the integrated technical fixtures, while serving as bookshelves and sunshades [61].

Towards Climate Neutrality: Global Perspective and Actions for Net-Zero Buildings... 421

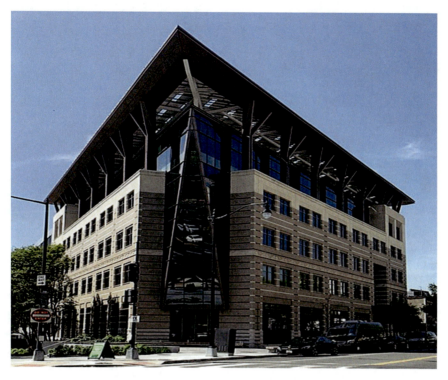

Fig. 56 The American Geophysical Union Headquarters. (Image credit and source: APK, https://commons.wikimedia.org/wiki/File:American_Geophysical_Union_Headquarters.jpg)

Fig. 57 The exterior of the Samling net-zero heating library in Nord Odal, Norway. (Image credit and source: Alek14, https://commons.wikimedia.org/wiki/File:Six-pane_application_in_Nord_Odal,_Norway.jpg)

Fig. 58 The main façade of the near-zero-heating office building in Rakvere, Estonia. (Image credit and source: Alek14, https://www.wikiwand.com/en/Zero_heating_building)

The Near-Zero-Heating Office Building in Rakvere, Estonia

The third example of near-zero-heating building (nZHB) is the office building in Rakvere, Estonia (Fig. 58). The façades are composed of a metal and timber batten cladding. There is a large canopy standing over a 3-storey floor with staggered black columns to provide shading on building facades. The fabric of the nZHB is also created from Air® 6-pane glazing by Reflex, Slovenia, and glass U-value of 0.26 W/m²K. Also, horizontal breakers are installed on the façade to reduce sunlight impinging on the building facades as presented in Fig. 58. In addition, the building has a double skin façade with glazed windows to provide the interior spaces with daylight and control glare, enhancing the visual connection with the outside view (Fig. 58) [59, 62–64].

The Net-Zero-Heating Office Building in The Netherlands

The fourth example of near-zero-heating buildings is an office building in The Netherlands, where the building adopted the same strategy for the envelope efficiency to achieve the goal, i.e., exploiting Air® 6-pane glazing by Reflex, Slovenia, and glass U-value of 0.26 W/m²K. The building was built in 2017, and its façades are composed of metal reflective materials on the façade vertical long windows to provide enough daylight inside the indoor space and reduce energy demands. In this building, ultralow U-value glazing is used; thus the window U-values approach 0.3 W/m²K and the heating demands diminish as shown in Fig. 59. In this context, the nZHB would not require a winter power reserve, and obviously it would not need any seasonal energy storage [59].

How Net-Zero Buildings Contribute to the Net-Zero Target and Climate Neutrality?

Net-zero-energy building (NZEB) is a term, subject to uncertainty, that might be utilized to portray a building with characteristics such as breakeven with energy utilization, altogether reducing energy consumption and energy costs equaling zero or net-zero (GHG) outflows. In spite of missing a definitive description of NZEBs, this moderately unused developing concept in Australia gives critical options to diminish GHG emissions, energy utilization, and operational energy costs for building properties. This chapter points to investigate the existing NZEB models, survey the movement of NZEB technologies, recognize key arrangements empowering NZEB improvement, and perceive potential ranges of global policies [65]. Since buildings consume about 40% of global energy and emit 33% of GHG [66], in Europe buildings are responsible for 40% of EU energy consumption and 36% of the energy-related GHG emissions [67]. In the MENA region, the share of building sector is 31% and 25% of the total primary energy supply and total final energy consumption, respectively [68]. Therefore, developing net-zero buildings would reduce almost the same percent of energy and mitigate related carbon emissions.

Current Policies, Actions, and Initiatives Worldwide

Polices, laws, action plans, and initiatives globally play a vital role in promoting NZEBs. Wang et al. (2020) provided a comprehensive sustainable management agenda which covers the sustainable principles of planning, transformation, environmental awareness, and climate mitigation [69]. The agenda summarizes low-carbon transformation policies followed in international cities, to prepare a new management matrix for enhancing the sustainable development of cities. More global policies and regulations are applied to leakage of pollutants such as GHG emissions and similar gases.

With special consideration to imposing rewards to low-pollution production firms [70], it is critical to notice that the climate mitigation has multidimensional aspects including political, social, and economic facets, whereas the social dimension confirms the gap between the future-oriented society and the present-oriented

Fig. 59 The exterior of the net-zero-heating office building in The Netherlands. (Image credit and source: Alek14, https://sl.m.wikipedia.org/wiki/Slika:Six-pane_application_in_Netherlands.jpg)

society in accepting high costs in return for long-term benefits. The most enduring firms and entities are those who are willing to adopt policies and regulations for climate change mitigation [71].

Current Policies, Activities, and Initiatives in Egypt Towards Net-Zero-Energy Buildings

Policies, laws, and initiatives relating to low-carbon buildings and cities in Egypt have recently grown and manifested in order to achieve green and sustainable cities. The following studies revealed successful applications, strategies, and attempts for the local private and public authorities. The local practice of climate change mitigation in Cairo has been highlighted by Dabaieh et al. in 2021, in which they listed the main factors and current adaptation measures including low-carbon buildings, cities, and transportation activities [72]. In addition, sustainable cities have become a priority over individual sustainable buildings and in the forefront of Egypt's current urban development policies and strategic plans. All new cities, including Madinaty, New Al-Ameen City on Egypt's northwest coast, and the New Administrative Capital, east of Cairo, are designed not only to be low carbon but also to mitigate climate change by focusing on carbon and air pollution reduction and utilizing renewable energy to generate clean energy [73].

The capital sustainable cities in the world are studied and examined by Armanuos et al. (2021), to provide learned lessons for the creation of sustainable cities. The study focuses on providing sustainable guidelines for the creation of Egypt's new capital city by 2050 [74]. The research analyzes existing cases and provides alternative scenarios for the new urban development to be considered as low-carbon city [74]. Another study reviewed and analyzed Egyptian and non-Egyptian new cities with a focus on achieving both high thermal performance in urban spaces and high energy efficiency inside building spaces. A new evaluation tool has been provided to allow for enhancing urban cluster forms coupled with efficient ground green cover and vegetation [75]. This study also focused on housing projects as a new sustainable model for low-carbon cities. In 2018, Dabaieh and Johansson examined a high-energy-efficient building which is located in Bahira – Delta region, Egypt, where the building is fitted with solar PV panels to generate 3.5 kWp. The building is analyzed, and sustainability measures are recommended and applied; it is provided as one of the main elements in the sustainable urban development strategies to mitigate climate change and rely on renewable energy resources [76].

Low-Carbon Cities and Zero-Carbon Emission Transportation in Egypt

In the context of low-carbon cities, the government strategy is mainly centered on a revolutionary approach to reach zero-emission transport. The new monorail was built to link the New Administrative Capital (NAC) to New Cairo and Adly Mansour main interconnected station (Fig. 60). The New Administrative capital has manifested many projects that address the concept of low-carbon city such as Cairo Monorail with a length reaching 56.5 km and passes through 21 stations [77] as depicted in Fig. 61. The aim is to achieve an eco-friendly transportation plan in greater Cairo, specifically the NAC. The green transportation plan reveals the significance of providing a variety of low-carbon to zero-emission transportation modes, and green alternatives to current vehicles, mainly electrical buses. In the same context, 6th of October has manifested a bold project that addresses the concept of low-carbon city (LCC) by virtue of the new monorail, which has been also constructed at a length of 42 km connecting Mohandiseen district to 6th of October City, and it passes by 12 stations [77] as shown in Fig. 61. In 2020, an analysis of new sustainable solutions to existing spaces in Alexandria, Egypt, was carried out [78]. The study revealed an example for carbon-neutral spaces via applying

Fig. 60 The Cairo Monorail, a zero-carbon emission transportation means to the NAC, Egypt. (Image credit and source: Ahram, https://english.ahram.org.eg/Media/News/2021/10/6/41_2021-637691497718843259-884.jpg)

Fig. 61 The 6th of October City new monorail, zero-carbon emission transportation in Giza, Egypt. (Image credit and source: Ahram, https://English.ahram.org.eg/NewsContent/50/1202/393529/ AlAhram_Weekly/ Economy/Egypts-first-monorail-Building-the-high-ride.aspx

environmental analysis and sustainable urban development concepts such as solar PV parking covers, shaded pedestrian paths, smart infrastructure, solar panels installation on a roof top example, wind energy, waste allocation, solar heating, gray water utilization, envelope installation, and green transportation alternatives [78].

Role of Zero-Carbon and Managing "Transition" in Cities and Regions

Numerous nations have transferred long-term objectives with respect to climate change by coupling these goals with innovation thinking for the future. The Paris agreement, signed at COP21 by 196 nations to address low-carbon challenges and achieve a circular economy, was the first step towards such global goal [79]. Further to COP21, it is clear that COP22, COP23, COP24, COP25, and recently COP26 manifested global determination towards mitigating GHG emissions and asked governments to work on developing Climate Adaptation Plans (CAP).

Can Cities Meet COP26 Outcomes and the Glasgow Climate Pact 2021?

In context with COP21 meeting in Paris in December 2015, the EU committed itself to limit GHG emissions' outcomes required to remain below 2 °C rises in normal worldwide temperature. The EU has agreed on energy policies which was proposed by the European Commission (EC) in November 2016, entitled "Clean energy for all Europeans." The adopted climate mitigation targets and new energy levels are coupled with GHG emission reduction (40% to 45%) less than 1990 levels, enhanced energy efficiency (32.5% lower than estimated in 2007), and renewable energy production (32% as a share of gross energy consumption) within the year 2030. The EU 2030 aims at reducing CO_2 emissions by 80% in 2050. On the other hand, the Paris agreement stated that the optimum efforts could only limit the temperature to rise above 1.5 °C, phasing out GHG emissions by 2050. However, carbon neutrality is still considered [80, 81].

Therefore, more approaches to reach climate neutrality, including energy utilization, have been added to the current policies to consider the possibility of a decarbonized long-term economy by 2030 climate and energy approach. What if climate neutrality by 2050 cannot be achieved by conventional fuel utilization? What if carbon neutrality is not affordable? What elements should be added to promote the current agenda? The deployment of current neutrality approaches can only be utilized with cost subsidiaries and support to overcome the additional cost barriers.

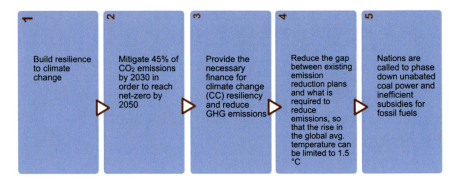

Fig. 62 Glasgow Climate Pact's decisions – COP26. (Image credit and source: Developed by authors after COP26)

Glasgow Climate Pact 2021 and Net-Zero?

The GCP, part of COP26 outcomes, revealed five agreed global decisions as shown in Fig. 62. The most imperative goal that requires vast mobilization is decision two, which states "Mitigate 45 percent of CO_2 emissions by 2030 in order to reach net-zero by 2050." Such mobilization is urgently required in view of the recent IEA analysis in 2021, which reveals that the global energy-related CO_2 emissions rose by 6 percent to 36.3 billion tones, which is considered the highest level recorded.

However, to achieve the net-zero target, it is imperative to digest the definition of low-carbon city and net-zero city, which is highlighted in Box 4, and consequently it can assist in achieving the COP26 goals [82].

> **Box 4 Definition of Carbon Neutrality or Low-Carbon Cities**
> Carbon neutrality is a state of net-zero carbon dioxide (CO_2) emissions. This can be achieved by balancing emissions of CO_2 with its removal or by eliminating emissions from society by vertical green walls. It is used in the context of CO_2-releasing processes associated with transportation, energy production, agriculture, and industry [82]

Conclusion

From the above review of global examples of low-carbon and net-zero buildings, it is still not enough to meet the COP26 goal. It is clear from these examples that low-carbon or net-zero buildings are gaining momentum, but not enough to reach COP26 goals and GCP of mitigating carbon emissions by 45% in 2030 and reach net-zero by 2050.

Low-carbon city or net-zero city including buildings can be achieved to mitigate 45% CO_2 emissions by 2030 and to reach net-zero goal by 2050 – a goal set by

COP26 and GCP in November 2021 – if mobilization and finance of US$ 100billion per year, agreed by COP26 for developing countries, are provided.

As indicated by research, low-carbon measures could cut emissions from urban areas by almost 90% by 2050. This would be achieved through four main sectors by 58%, 21%, 16%, and 5% from buildings, transport, materials efficiency, and waste, respectively [83]. Furthermore, investing in 16 low-carbon measures in cities could reduce global urban emissions by 90% by 2050 and has a net present value of nearly $24 trillion, which is nearly one-third of global GDP in 2018 [83]. Moreover, investments required to reduce urban emissions by 2050 are estimated to be US$ 1.83 trillion – about 2% of global GDP per year [83].

As of 2021, all new buildings in the EU member state countries must be NZEBs, and since 2019, all new public buildings in the EU should be NZEBs [67]. Finally, carbon-neutral hydrocarbons are to be considered in the future zero-energy emission models. Low-carbon cities including net-zero carbon buildings have recently been used in response to the climate crisis, but the current world perspective is looming.

References

1. Rowlatt, J. (2020). Humans warning 'suicidal war' on nature – UN Chief Antonio Guterres. *BBC News*. Retrieved 12 June 2022, from https://www.bbc.com/news/sciene-environment-55147647
2. Vidya, H., & Chatterji, T. (2020). SDG 11 sustainable cities and communities. In *Actioning the global goals for local impact* (pp. 173–185). Springer.
3. Liu, Z., Deng, Z., He, G., Wang, H., Zhang, X., Lin, J., & Liang, X. (2022). Challenges and opportunities for carbon neutrality in China. *Nature Reviews Earth & Environment, 3*(2), 141–155.
4. Höhne, N., Warnecke, C., Day, T., & Röser, F. (2015). *Carbon market mechanisms in future international cooperation on climate change*. New Climate Institute.
5. Grafakos, S., Viero, G., Reckien, D., Trigg, K., Viguie, V., Sudmant, A., & Dawson, R. (2020). Integration of mitigation and adaptation in urban climate change action plans in Europe: A systematic assessment. *Renewable and Sustainable Energy Reviews, 121*, 109623.
6. Deng, S., Wang, R. A., & Dai, Y. J. (2014). How to evaluate performance of net zero energy building? – A literature research. *Energy, 71*, 1–16.
7. State-of-the-art sustainable approaches for deeper decarbonization in Europe–An endowment to climate neutral vision. *Renewable and Sustainable Energy Reviews, 159*, 112204.
8. Chen, X., Shuai, C., Wu, Y., & Zhang, Y. (2020). Analysis on the carbon emission peaks of China's industrial, building, transport, and agricultural sectors. *Science of the Total Environment, 709*, 135768.
9. Ortiz, L., González, J. E., & Lin, W. (2018). Climate change impacts on peak building cooling energy demand in a coastal megacity. *Environmental Research Letters, 13*(9), 094008.
10. Neale, R. E., Barnes, P. W., Robson, T. M., Neale, P. J., Williamson, C. E., Zepp, R. G., & Zhu, M. (2021). Environmental effects of stratospheric ozone depletion, UV radiation, and interactions with climate change: UNEP Environmental Effects Assessment Panel, Update 2020. *Photochemical and Photobiological Sciences, 20*(1), 1–67.
11. *Global CO2 emissions rebounded to their highest level in history in 2021 – news – IEA*. (2022). IEA. Retrieved 18 June 2022, from https://www.iea.org/news/global-co2-emissions-rebounded-to-their-highest-level-in-history-in-2021

12. Ur Rehman, H., Haider, S. A., Naqvi, S. R., Naeem, M., Kwak, K. S., & Islam, S. R. (2022). Environment friendly energy cooperation in neighboring buildings: A transformed linearization approach. *Energies, 15*(3), 1160.
13. *Zero energy buildings*. (2022). Energy.gov. Retrieved 22 June 2022, from https://www.energy.gov/eere/buildings/zero-energy-buildings
14. *Oil market report – June 2022 – Analysis – IEA*. (2022). IEA. Retrieved 23 June 2022, from https://www.iea.org/reports/oil-market-report-june-2022
15. Khassan, A., Donenko, V. I., & Ischenko, O. L. (2021). The use of BIM to achieve zero energy building. *Metal Science and Heat Treatment of Metals, 1*(92), 59–65.
16. Kaewunruen, S., Rungskunroch, P., & Welsh, J. (2018). A digital-twin evaluation of net zero energy building for existing buildings. *Sustainability, 11*(1), 159.
17. El Sayary, S., & Omar, O. (2021). Designing a BIM energy-consumption template to calculate and achieve a net-zero-energy house. *Solar Energy, 216*(4), 315–320.
18. Suzuki, S. & Sumiyoshi, D. (2018). *Building energy simulation towards developing a guideline for NZEBs in Egypt* (master thesis), Graduate School of Human-environment Studies, Kyushu University, Japan, https://doi.org/10.13140/RG.2.2.20261.14569.
19. Pye, S., Broad, O., Bataille, C., Brockway, P. E., Daly, H. E., Freeman, R., Gambhir, A., Geden, O., Rogan, F., Sanghvi, S., Tomei, J., Vorushylo, I., & Watson, J. (2021). Modelling net-zero emissions energy systems requires a change in approach. *Climate Policy, 21*(2), 222–231.
20. Fankhauser, S., Smith, S. M., Allen, M., Axelsson, K., Hale, T., Hepburn, C., Kendall, J. M., Khosla, R., Lezaun, J., Michell-Larson, E., Obersteiner, M., Rajamani, L., Rickaby, R., Seddon, N., & Wetzer, T. (2022). The meaning of net zero and how to get it right. *Nature Climate Change, 12*, 15–21.
21. *Climate change and the crucial role of low-carbon infrastructure – climate adaptation platform*. (2020). Climate Adaptation Platform. Retrieved 15 June 2022, from https://climateadaptationplatform.com/climate-change-and-crucial-role-of-low-carbon-infrastructure
22. *Net zero city definition | law insider*. (2022). Retrieved 10 July 2022, from https://www.lawinsider.com/dictionary/net-zero-city
23. *What is a net zero? – Net zero climate*. (2022). Net Zero Climate. University of Oxford, Retrieved 10 July 2022, from https://netzeroclimate.org/what-is-net-zero/
24. UNFCCC. (2021). *Glasgow climate pact – Key outcomes from COP26*. Unfccc.Int. Retrieved 23 June 2022, from https://unfccc.int/process-and-meetings/the-paris-agreement/the-glasgow-climate-pact-key-outcomes-from-cop26
25. *Zaha Hadid Architects completes Bee'ah HQ in Sharjah with dune-shaped volumes blended into desert*. (2022). World Architecture Community. Retrieved 23 June 2022, from https://worldarchitecture.org/article-links/emmcp/zaha-hadid-architects-completes-bee-ah-hq-in-sharjah-with-dune-shaped-volumes-blended-into-desert.html
26. *BEEAH group officially opens ground-breaking new headquarters, unveiling the 'Office of the Future' – Zaha Hadid Architects*. (2022). Zaha-hadid.com. Retrieved 15 June 2022, from https://www.zaha-hadid.com/2022/03/31/beeah-group-officially-opens-ground-breaking-new-headquarters-unveiling-the-office-of-the-future/
27. Vara, V. (2019). *Bee'ah headquarters, Sharjah, UAE*. Design Build Network. Retrieved 15 June 2022, from https://www.designbuild-network.com/projects/beeah-headquarters-sharjah-uae/
28. *Beeah Solar PV Park, United Arab Emirates*. (2022). Power Technology. Retrieved 20 June 2022, from https://www.power-technology.com/marketdata/beeah-solar-pv-park-united-arab-emirates/
29. Sahar Ejaz, S. (2021). *50% of expo 2020 Dubai's energy to come from renewable sources*. Gulfnews.com. Retrieved 11 Oct 2021, from https://gulfnews.com/expo-2020/news/50-of-expo-2020-dubais-energy-to-come-from-renewable-sources-1.1627207021737
30. Grimshaw Architects. (2021). *Terra –The sustainability pavilion expo 2020 Dubai/ GRIMSHAW*. Grimshaw.global. Retrieved 11 October 2021, from https://grimshaw.global/projects/dubai.expo-2020-sustainability-pavilion
31. Grimshaw Architects. (2022). *Terra — The sustainability pavilion expo 2020 Dubai: Case study/GRIMSHAW*. Grimshaw.global. Retrieved 25 Feb 2022, from https://grimshaw.global/sustainability/expo-2020-sustainability-pavilion-case-study/

32. Patterson, Z. (2021). *Everything you need to know about the UAE Pavilion at Expo 2020 Dubai*. TimeOut. Retrieved 2 Oct 2021, from https://www.timeoutdubai.com/news/473575-uae-pavilion-at-expo-2020-dubai
33. Calatrava, S. (2021). *Santiago Calatrava tops UAE Pavilion at Dubai Expo with 28 opening wings – Santiago Calatrava – Architects & Engineers*. Calatrava.com. Retrieved 10 Oct 2021, from https://www.calatrava.com/news/reader/Santiago-calatrava-tops-uae-pavilion-at-dubai-expo- with-28-opening-wings.html
34. Cull, N. J. (2022). The greatest show on earth? Considering expo 2020, Dubai. *Place Branding and Public Diplomacy, 18*, 49–51.
35. Liptow, J. (2021). *Alter-expo: Expo 2020 and the future of international expositions* (doctoral dissertation, Kent State University).
36. Hermelink, A., Schimaschar, S., Boermans, T., Pagliano, L., Zangheri, P., Armani, R., Voss, K., & Musall, E. (2013). *Towards nearly zero energy buildings: Definition of common principles under the EPBD* (pp. 1–22). ECOYS 2012 by order of: European Commission. Retrieved from https://ec.europa.eu/energy/sites/ener/files/documents/nzeb executive_ summary.pdf
37. *Coop-himmelbI(au)*. (2022). World-architects. Retrieved 10 July 2022, from https://www.world-architects.com/en/coop-himmelb-l-au-vienna/project/dalian-international-conference-center
38. Hien, W. N., Tan, E., Seng, A. K., Mok, S., & Goh, A. (2012). Performance of greenery systems in zero energy building of Singapore. In *ICSDC 2011: integrating sustainability practices in the construction industry* (pp. 74–80).
39. *Hong Kong's first zero carbon building*. (2012). Arup.com. Retrieved 9 June 2022, from https://www.arup.com/projects/cic-zcb
40. *South Korea unveils a completely eco-friendly building*. (2011). Reuters. Retrieved 21 June 2022 from, https://www.reuters.com/articls/us-south-korea-unveils-completely-eco-friendly-building-idINTRE74J1FE20110520
41. Incheon Airport. (2017). *Incheon International Airport corporation – Green Report 2017* [Ebook]. Retrieved 25 June 2022, from https://www.airport.kr/co_cnt/en/cyberpr/publicat/pbooks/pbooks.do
42. Pavlovic, T., Tsankov Ts., P., Cekić Dj., N., & Radonnjić Mitić, I. S. (2019). *The sun and photovoltaic technologies* (pp. 45–193). Springer. https://link.springer.com/chapter/10.1007/978-030-22403-5_2
43. Majd, B. K., Maddahi, S. M., & Soflaei, F. (2021). Presenting an energy-efficient model for the envelope of high-rise office buildings case study: Cold and dry climate. *International Journal of Sustainable Energy and Environmental Research, 10*(2), 85–100.
44. *Sustainability*. Sites.wageningenur.nl. (2022). Retrieved 12 July 2022, from https://www.wageningencampus.nl/en/campus/about/Sustainability.htm
45. Woollaston, V. (2016). *Driverless shuttle bus to take to Dutch public roads in world first*. Mail Online. Retrieved 9 July 2022, from https://www.dailymail.co.uk/sciencetech/article-3420837/Driverless-shuttle-bus-Dutch-public-roads-world-first.html
46. Randall, T. (2016). *The smartest building in the world*. Bloomberg.com. Retrieved 10 July 2022, from https://www.bloomberg.com/features/2015-the-edge-the-worlds-greenest-building/
47. Hutt, R. (2017). *Future of the environment: Is this the world's greenest, smartest office building?* World Economic Forum. Retrieved 10 July 2022, from https://www.weforum.org/agenda/2017/03/smart-building-amsterdam-the-edge-sustainability
48. *Case studies | carbon neutral lab Nottingham*. World Green Building Council. (2022). World Green Building Council. Retrieved 10 July 2022, from https://www.worldgbc.org/advancing-net-zero/case-studies-0
49. *The carbon neural laboratory – The University of Nottingham*. (2022). Nottingham.ac.uk. Retrieved 10 July 2022, from https://www.nottingham.ac.uk/chemistry/research/centre-for-sustainabile-chemistry/the-carbon-neural-laboratory.aspx
50. *CASE STUDY: GSK Centre for sustainability Chemistry – UKGBC – UK Green Building Council*. (2022). UKGBC – UK Green Building Council. Retrieved 10 July 2022, from https://www.ukgbc.org/solutions/case-study-gsk-centre-for-sustainability-chemistry/

51. Winstanley, S. (2022). *Low energy house designs – Stylish sustainability in Scotland*. Trendir.com. Retrieved 12 July 2022, from, https://www.trendir.com/low-energy-house-designs-stylish-sustainability-in-scotland/
52. MJoaoEnes. (2022). *Solar settlement and sun ship, Freiburg, Germany, Sustainablearchitecturenews.blogspot.com.* Retrieved 3 July 2022, from http://sustainablearchitecturenews.blogspot.com/2015/08/solar-settlement-and-sun-ship-germany.html
53. Stokes, R. (2022). *Zero energy house surplushome. By Manfred Hegger*. Networking und automation intelligent buildings. Quo vadis, sustainability? Docplayer.net. Retrieved 3 July 2022, from http://docplayer.net/52608619-Zero-energy-house-surplushome-by-manfred-hegger-networking-und-automation-intelligent-buildings-quo-vadis-sustainability.html
54. Kroll, A. (2022). AD Classics: AD Classics: Kunsthaus Bregenz/Peter Zumthor. *ArchDaily*. Retrieved 3 July 2022, from https://www.archdaily.com/107500/ad-classics-kunsthaus-bregenz-peter-zumthor
55. Crook, L. (2019). *Laura Álvarez Architecture transforms stone ruin into zero-energy Villa Slow*. Dezeen. Retrieved 19 June 2022, from https://www.dezeen.com/2019/04/19/villa-slow-laura-alvarez-architecture/
56. *Sustainability – United Therapeutics*. (2018). Utunisphere.com. Retrieved 25 June 2022, from https://www.utunisphere.com/home/sustainability/
57. *National Renewable Energy Laboratory/SmithGroupJJR*. (2013). *ArchDaily*. Retrieved 20 June 2022, from https://www.archdaily.com/443969/national-renewable-energy-laboratory-smithgroupjjr
58. Shehadi, M. (2020). Net-zero energy buildings: Principles and applications. *Intechopen*. https://doi.org/10.5772/intechopen.92285
59. *Zero heating building*. (2019). En.wikipedia.org. Retrieved 13 June 2022, from https://en.wikipedia.org/wiki/Zero_heating_building
60. Neibauer, M. (2016). *World's largest association of scientists plans D>C> HQ upgrade with a net-zero goal*. Bizjournals. Retrieved 4 June 2022, from https://www.bizjournals.com/washington/breaking_ground/2016/03/world-s-largest-association-of-scientists-plans-d.html
61. Pintos, P. (2020). *Samling Library/Helen & Hard*. ArchDaily. Retrieved 16 June 2022, from https://www.archdaily.com/945294/samling-library-helen-and-hard
62. Thalfeldt, M., Kurnitski, J., & Mikola, A. (2013). Nearly zero energy office building without conventional heating. *Estonia Journal of Engineering, 19*(4), 309–328. https://doi.org/10.3176/eng.2013.4.06
63. Thermia. (2022). *Estonian smart house competence center powered by Thermia geothermal solutions* (pp. 1–2) Retrieved from https://thermia.com/wp-content/uploads/2017/05/case_Conference-Centre-powered-by-heat-pumps.pdf
64. Erhorn, H., & Erhorn-Kluttig, H. (2014). *Selected examples of nearly zero-energy buildings: Detailed report* (pp. 22–24) Concerted Action: Energy Performance of Buildings. Retrieved from https://www.epbd-ca.eu/wp-content/uploads/2011/05/CT5_Report_Selected_examples_of_NZEBs-final.pdf
65. Wells, L., Rismanchi, B., & Aye, L. (2018). A review of net zero energy buildings with reflections on the Australian context. *Elsevier, 158*, 616–628. Retrieved 7 June 2022, from https://www.sciencedirect.com/science/article/abs/pii/S037877881733445X
66. Tricoire, J. P. (2021). *Buildings are the foundation of our energy-efficient future*. The World Economic Forum. Retrieved 11 July 2022, from https://www.weforum.org/agenda/2021/02/why-the-buildings-of-the-future-are-key-to-an-efficient-energy-ecosystem
67. *Energy performance of buildings directive*. (2018). Retrieved 11 July 2022, from https://energy.ec.europa.eu/topics/energy-efficiency/energy-efficient-buildings/energy-performance-buildings-directive_en
68. UN-ESCWA. (2018). *Addressing energy sustainability issues of the buildings sector in the Arab region*. Beirut, Lebanon. Retrieved from https://archive.unescwa.org/sites/www.unescwa.org/ files/publications/ files/ addressing-energy-sustainability-issues-buildings-sectori-arab-

region-english.pdf&sa=U&ved= 2ahUKEwi31umHhvL4AhVwSkEAHROtBkoQFnoECAY QAg&usg=AOvVaw2pprc1KRrSQeIRa1o2cfhV
69. Wang, C., Zhan, J., & Xin, Z. (2020). Comparative analysis of urban ecological management models incorporating low-carbon transformation. *Technological Forecasting and Social Change, 159*, 120190.
70. Bushnell, J., Peterman, C., & Wolfram, C. (2020). Local solutions to global problems: Climate change policies and regulatory jurisdiction. *Review of Environmental Economics and Policy*.
71. Cai, M., Murtazashvili, I., Murtazashvili, J. B., & Salahodjaev, R. (2020). Patience and climate change mitigation: Global evidence. *Environmental Research, 186*, 109552.
72. Dabaieh, M., Maguid, D., Abodeeb, R., et al. (2021). The practice and politics of urban climate change mitigation and adaptation efforts: The case of Cairo. *Urban Forum*. https://doi.org/10.1007/s12132-021-09444-6
73. MPMAR – Ministry of Planning, Monitoring and Administrative Reform, 2018. *Egypt's voluntary national review 2018, Cairo: Ministry of Planning and Economic Development*. https://www.arabdevelopmentportal.com/sites/default/fles/publication/vnr-egypt-2018.pdf. Accessed Sept 2021.
74. Armanuos, R., & Rashed, A. (2021). Study of scenarios of the world capital sustainable cities as input to Egyptian capital sustainable city 2050. *Mansoura Engineering Journal, 37*(3), 1–13. https://doi.org/10.21608/bfemu.2021.156754
75. Fahmy, M., Mahmoud, S., Elwy, I., & Mahmoud, H. (2020). A review and insights for eleven years of urban microclimate research towards a new Egyptian era of low carbon, comfortable and energy-efficient housing typologies. *Atmosphere, 11*(3), 236.
76. Dabaieh, M., & Johansson, E. (2018). Building performance and post occupancy evaluation for an off-grid low carbon and solar PV plus-energy powered building: A case from the Western Desert in Egypt. *Journal of Building Engineering, 18*, 418–428. https://doi.org/10.1016/j.jobe.2018.04.01
77. El-sayed, J. (2022). *Explainer: What you need to know about Egypt's first monorail – Urban & Transport – Egypt*. Ahram Online. Retrieved 11 July 2022, from https://english.ahram.org.eg/News/470435.aspx
78. Mohamed, A. A. F. (2020). Carbon neutral urban spaces under climate change case study: Renovation of Sidi Gaber neighborhood in Alexandria, Egypt. *Architecture and Planning Journal, 25*(1), 3.
79. Gills, B., & Morgan, J. (2020). Global climate emergency: After COP24, climate science, urgency, and the threat to humanity. *Globalizations, 17*(6), 885–902.
80. *Climate change negotiations outcomes at the COP26 Glasgow – climate adaptation platform*. (2022). Climate Adaptation Platform. Retrieved 3 July 2022, from https://climateadaptation-platform.com/climate-change-negotiations-outcomes-at-the-cop26-glasgow/
81. *CNCA*. Carboneutralcities.org. (2022). Retrieved 1 July 2022, from https://carbonneutralcities.org
82. Brilhante, O., & Klaas, J. (2018). Green City concept and a method to measure green city performance over time applied to fifty cities globally: Influence of GDP, population size and energy efficiency. *Sustainability, 10*(6), 2031. https://doi.org/10.3390/su10062031
83. Lazer, L., Haddaoui, C., & Wellman, J. (2022). *Low-carbon cities are a $24 trillion opportunity*. World Resources Institute. Retrieved 3 July 2022, from https://www.wri.org/insights/low-carbon-cities-are-24-trillion-opportunity

Index

A
Academic, 72, 221, 283
Adaptive thermal comfort, 116, 129
Average buildings, 235–248

B
Battery energy storage, 254
Bioclimatic building design, 255–256
Bioelectricity, 45–47, 53, 58–60
Building facade, 12, 149, 150, 184, 288, 301, 422
Building operation, 82, 283, 297
Building typologies, 117, 131, 236–238, 247, 248
Built environment, 63, 72, 73, 76, 82, 87, 152, 154, 165, 166, 181, 283, 286, 296, 366

C
Carbon dioxide (CO_2) emission, v, 101, 135, 195, 197, 206, 215, 237, 238, 283, 294, 298, 299, 311, 320, 329, 337, 349, 352, 361, 368, 369, 373, 377, 383, 421, 422
Carbon emissions, 195, 196, 199, 200, 210, 215, 216, 219, 221, 223, 235, 294–301, 374, 375, 379, 380, 389, 407, 420, 423, 428
Carbon neutrality, 195, 295, 373, 379, 427
2050 Challenge, 196
Climate change, 23, 48, 49, 136, 137, 153–158, 195–197, 199–201, 203–206, 251, 284, 285, 289, 294, 295, 373–429
Climate goals, 63
Climate neutrality, 373–429
Coating texture, 26
Community, 44, 51, 53, 54, 58–61, 155, 157, 158, 173, 175–179, 182, 196, 207, 208, 210, 224–226, 284, 289–291, 391, 409
Cooperative learning, 230
Cradle to gate, 297
Creative thinking, 230

D
Decarbonization, 92, 296
Deduction discovery method (DDM), 222
Design, 1, 2, 4, 6, 12, 13, 18, 23, 25, 38, 45, 47, 48, 51–54, 58, 61, 63–75, 79, 84–86, 102, 109, 140, 165–168, 171–173, 178–181, 183–193, 200, 201, 210, 213–230, 237, 251, 252, 255, 256, 283–289, 295, 297, 298, 300, 301, 366, 382, 383, 386, 387, 389, 404, 407, 412
Design Builder, 156, 188, 189
Design for Disassembly (DfD), 296
Dubai, 214, 224–226, 374, 378, 382, 384–388
Dynamic energy performance simulations, 137, 151, 154, 155

E

Embodied carbon, 286, 296, 297, 299, 301, 386, 406
Energy, 6, 12, 23, 24, 43–46, 49, 50, 52–55, 60, 63, 64, 72–83, 85, 86, 91–99, 101, 104, 105, 109, 110, 112, 113, 131, 135–158, 165–168, 171–173, 177, 179–185, 188–191, 193, 195–197, 199–201, 203, 207, 208, 210, 213–221, 228–229, 235–238, 240–248, 251–278, 283–290, 294–298, 300, 301, 366, 373–389, 391, 393, 396, 400–402, 404–412, 417–420, 423, 425, 427
Energy certification, 131
Energy consumption, 2, 14, 23, 24, 49, 63, 75, 80–83, 86, 91, 92, 135, 141, 142, 155, 180, 184, 188–190, 195, 199–203, 214, 218, 223, 226, 235, 238, 251, 252, 256, 260, 262, 263, 265, 267, 271, 275, 276, 278, 284, 294, 300, 354, 374, 375, 377, 378, 389, 398, 423, 427
Energy efficiency, 11, 23, 24, 43–61, 63, 64, 75–77, 81, 92, 109, 171, 181, 195, 196, 200, 201, 210, 214, 215, 224, 238, 251, 297, 301, 379, 383, 407, 410, 420, 425, 427
Energy-efficient refurbishment, 240, 247
Energy management system (EMS), 136, 252, 254, 265–278
Energy positive buildings and communities, 196, 206–208, 210
Energy profile, 236, 247
Environmental protection, 294, 310

F

Future developments, 110

G

Green architecture, 165–182
Green building, 215, 223, 295, 374, 389
Greenhouse gas emissions, 63, 197, 198, 215, 218, 238, 319, 334
Green urbanism, 166, 171–173, 181

H

Heat pump, 180, 216, 218, 223, 254, 257–260, 271, 275, 278, 298, 389, 417
Highland zone, 3, 6, 11, 12, 16
HVAC, 63, 64, 78, 79, 85, 86, 111, 112, 141, 142, 217–218, 221, 223, 224, 229, 254, 255

I

Indicators, 69, 83, 104, 113, 196, 235–248
Induction discovery method (IDM), 220, 222
Input weather data, 111
Integral design, 65, 67, 68, 71
Integrated design studio (IDS), 213, 218–221, 223, 227, 230
Integration, 37, 58, 70, 102, 104, 131, 173, 199, 203, 213, 219–224, 251, 257–265, 289, 291
International Energy Agency (IEA), 75, 76, 136, 200, 214, 218, 375
Iran, 104, 168, 184

L

Life cycle analysis (LCA), 296
Living architecture, 53–56, 58, 61
Living technologies, 54
Louver, 183, 184, 186, 188–191, 193
Low-carbon cities (LCCs), 373, 425–429

M

Metabolism, 43–45, 47, 49–53, 55, 56, 61
Microbes, 44, 46, 50, 53–60
Microgrid, 252–255, 257–268, 270, 272–274, 277
Monitored data, 110, 111, 113, 115, 116, 119, 120, 123, 125, 128, 129
Morphological chart, 66–70, 72, 74, 85
Multi-objective optimization, 104, 105

N

Net-zero, 63–87, 91–105, 195–210, 283–290, 294–296, 298, 301, 373–429
Net-zero concept, 104
Net zero emission (NZE), 101, 105, 229, 284, 285, 289, 290
Net zero energy building (NZEB), 101, 105, 209, 210, 251–278, 283–290, 378, 417, 423
Net zero perception, 196, 203–206

O

Obstruction angle, 2, 3, 6, 7, 9, 10, 12, 18
Operational carbon, 182, 297–300
Orientation, 3, 6–8, 10–12, 14, 15, 18, 30, 32–35, 68, 70, 104, 171, 183, 216, 228, 237, 238, 255, 288, 300, 383
Outdoor thermal comfort, 3, 5–8, 10–12, 14, 16, 18

Index

P
Palm particles, 28–30, 32, 33, 36, 37
Parametric solar envelope (PSE), 3, 11–14, 16, 18
Parametric tool, 3
Paris Agreement, 195, 284, 294, 296, 427
Performance analyses, 148, 156, 221
Performance gap, 75, 81, 82, 109–111, 113, 115–117, 121, 123, 128, 131
Performance modeling, 284
Performative Design, 230
Policies, 140, 148, 201, 205, 209, 214, 220, 242, 247, 423–427
Profession, 289, 290
Python semi-Realtime Energy DYnamics and Climate Evaluation (PREDYCE), 110–117, 119–131

R
Regenerative architecture, 182
Regional architecture, 167
Renewable energy sources (RES), 101, 181, 196, 199, 200, 207, 288, 377, 378
Residential building, 92, 118, 143, 145, 147, 150–152, 155, 179, 184, 188, 235–248, 378
Resource limits, 50
Royal Institute of British Architects (RIBA), 64

S
Shading device, 183–185, 190, 193, 217, 300
Shadow, 26, 28, 31, 35–37
Simulink, 257, 258
Social commitment, 196, 203–206
Solar radiation, 1, 2, 8, 15, 20, 26, 32, 95–97, 100, 104, 136, 138–140, 143, 144, 149, 156, 179, 183, 383, 386, 389
Stakeholders, 84, 220, 284, 298
Surface temperature, 26, 28–37, 195, 197
Sustainability, 54, 83, 166, 168–173, 180, 181, 208, 215, 225, 289, 295, 374, 378, 382, 384–387, 425
Sustainable architecture, 226, 286

T
Temperate humid climate, 185
Thermoeconomics, 45, 46
Traditional cities, 157, 172
Typical weather data (TWD), 137, 138, 153

U
Urban microclimate, 145, 146, 148, 150–153, 201
Urban rules, 2, 3, 5, 6, 9, 12
Urban street profile, 1–20

Z
Zero-energy houses, 235–248
Zero net energy (ZNE), 213, 214, 216, 217, 223, 226, 229

Printed in the United States
by Baker & Taylor Publisher Services